E SOURCE Technology Atlas Series

LIGHTING

E SOURCE, INC.
Boulder, Colorado USA

LIGHTING
TECHNOLOGY ATLAS

Lindsay Audin,
David Houghton PE, Michael Shepard, and Wendy Hawthorne

OVERVIEW

CONTEXT *describes where and how this volume's material fits into the world. It defines the boundaries of the lighting systems and explains how and why a systems approach is critical to capturing lighting savings. It discusses the interaction of human behavior, productivity, and visual comfort with lighting systems. It shows trends in the manufacturing and sales of lighting equipment, as well as lighting's energy and demand impacts.*

TECHNOLOGY *characterizes the existing stock of lighting technology and the more efficient alternatives available today and in the future. It ranges from theory to practice, addressing such questions as: What is the technology? How does it work? How long does it last? What does it cost? How much energy does it save? How does it interact with other end uses in the building?*

RESOURCES *contains a set of tools useful for lighting designers, building managers, and specifiers, including a review of lighting software, a summary of the lighting provisions of the ASHRAE 90.1 standard, and a sample lighting specification. A reference list of lighting-related organizations, glossary, and index to this Atlas are also included.*

LIGHTING
TECHNOLOGY ATLAS

Managing Editor
Jeanne Travisono

Illustrations and Layout
John Heiland
Patricia Tait
Max W Chadwick

Electronic Publishing
John Heiland
Max W Chadwick

ISBN 1-58167-002-8 (Volume I)
ISBN 1-58167-000-1 (Five-volume set)
ISBN 1-58167-007-9 (Five-volume set plus CD)

Printed in the USA with soy-based inks on recycled paper.

E SOURCE, INC.
1033 Walnut Street
Boulder, Colorado 80302-5114 USA
TEL 303.440.8500 FAX 303.440.8502
E-MAIL esource@esource.com
WEB www.esource.com

Context

TECHNOLOGY

5 DAYLIGHTING

6 INCANDESCENT LIGHTING

7 FULL-SIZED FLUORESCENT LIGHTING

8 Compact fluorescent lamp technologies

9 High-intensity discharge lighting

RESOURCES

Acknowledgments

Several people made important contributions to this *Technology Atlas*. Amory Lovins, of Rocky Mountain Institute, the principal author of Competitek's lighting atlas, provided spiritual guidance for the project and a substantial review of the manuscript. Rose Burgett, of Astra Lighting Design, put together the information on lighting software (Resources section), and Lori Garcia, of the British Columbia Buildings Corporation, provided the research for lamp and ballast recycling (**3.3**), particularly the Canadian perspective. Debbie Davia provided data gathering and fact checking, and Brad Davids supplied information on window products (**5.3**). Helen Kessler of Sieben Energy Associates provided a helpful review of the Technology section. The 1997 update of this Atlas was managed by David Houghton, with substantial efforts by Lindsay Audin, Robert Sardinsky, Chris Calwell, My Ton, Joan Gregerson, and John Fetters. Linda Yellin of E SOURCE assembled the new lighting organization directory (Appendix E). The authors also wish to thank the numerous colleagues who provided particularly valuable information, insights, and perspectives, including:

Warren Anderson	Osram Sylvania
David Apfel	David Apfel Lighting Design
Jim Benya	Pacific Lightworks
Sam Berman	Lawrence Berkeley National Laboratory
John Betchkal	GE Lighting
George Biegel	Stocker & Yale
Nancy Clanton	Clanton Engineering
Gary Conkright	Profile Systems
John Corbin	Lighting Sciences
Dave Evans	Hewlett-Packard
Ira Faberman	Iftron Labs
Jerry Felper	Tek-Tron Enterprises
Steve Goldmacher	Philips Lighting
Bill Gregory	GE Lighting
Jim Hagar	MagneTek
Brian Halliwell	Lights of America
Nick Hutchins	Hutchins International
Jeff Kessel	University of California
Tim Lamberth	Osram Sylvania
Don MacLennan	Fusion Lighting
Joey Maniquis	Toshiba
Paul Manley	Flexlight
Evan Mills	Lawrence Berkeley National Laboratory
Michael Ostaffe	Advance Transformer
Danny Parker	Florida Solar Energy Center
Michael Peschel	Peschel Energy
Steve Purdy	Advance Transformer
Cyrus Rafii	GE Lighting Controls
Paul Royalty	Electronic Ballast Technology
Francis Rubenstein	Lawrence Berkeley National Laboratory
Michael Siminovitch	Lawrence Berkeley National Laboratory
Elmer Smolnisky	Augustana College
Jack Wolpert	E-Cube
Kevin Youngquist	ProLight

Context

1 BASIC CONCEPTS

Lighting profoundly affects the way people feel, work, and interact. It uses large amounts of energy—about one-fifth of U.S. electricity—but offers potentially vast savings that can cut the capital and operating costs of other building systems. In many respects, lighting is the key to efficient building design. To realize the full potential of energy efficient lighting, however, requires careful attention to many interacting variables, embodied in five core concepts outlined here and expanded throughout this Atlas.

1 Mine the Mother Load first

As every good prospector knows, it pays to mine the richest seams of ore first. Lighting is just that, the Mother Load of energy efficiency. Moreover, like many rich seams, lighting efficiency is linked to other veins, some large and some small. By providing access to these other opportunities, lighting offers powerful leverage over other building systems if measures are pursued in the proper sequence. Not only can lighting upgrades yield quick savings on energy bills, but more efficient lights release less heat, which means air-conditioning systems can be smaller and cheaper to buy and to operate.

2 Start with the task

People don't want light for its own sake; they want what it provides, like vision and beauty. Quality is far more important than quantity. But the quality, amount, and type of light that people need depend entirely on the task at hand. Efforts to promote efficient lighting that ignore the users' needs are likely to fail. While this seems obvious, it is an error that many programs make by promoting particular lighting measures in cookie-cutter fashion, with little regard for how or where they will be applied. For great results—whether measured in kilowatt-hours, dollars, clear vision, or esthetic delight—improve task quality first, then lighting quality, then lighting quantity, and only then lighting equipment.

3 Go beyond efficiency

Consider the broader impacts of lighting choices. While the capital and energy costs of lighting systems are significant, they are dwarfed by worker salaries, so think about the influences of lighting quality—positive or negative—on the health and productivity of occupants. Be aware of the time and money needed for maintenance and replacement equipment. To a building maintenance manager or an elderly householder, the fact that compact fluorescents last ten times as long as incandescents may be more valuable than the energy savings. Utilities need to consider the persistence of efficiency measures, which has implications for the types of technologies they promote and the ways they promote them.

4 Think systems

Lighting is a complex but comprehensible system. Truly optimal lighting solutions can only be reached by considering interactions between daylight, lamps, power supplies, fixtures, controls, building configuration, materials, and furnishings. Retrofits that change lamps and ballasts but ignore daylight, reflectors, delamping, task lighting, and other options skim the cream of energy savings but leave behind some of the richest opportunities. The all too frequent specter of brand new buildings having their lights redone brings to mind the old saw: "If you can't afford to do it right the first time, how can you afford to do it twice?"

5 Stay up to date

Lighting is one of the fastest moving areas of energy efficient technology. One U.S. lighting manufacturer alone introduced over 300 new lighting products in 1993.[1] New technologies are emerging constantly, from microwave sulfur lighting, to fluorescent dimming systems, to photovoltaic-powered direct-current lighting systems, to daylight-sensitive glazings. New technologies could change the future of lighting, or might just be the latest snake oil. Staying informed on what's new, what works, and what doesn't can spell the difference between innovative progress and costly mistakes.

1.1 Mine the Mother Load

About 17 percent of U.S. electricity is used directly by lighting, and another four percent goes to remove the waste heat from those lights. Lighting electricity use is a big target, and one well worth pursuing. With good design, lighting energy use in most buildings can be cut at least in half, and in many cases by more than 80 percent, while maintaining or improving lighting quality. Such designs typically pay for themselves in energy savings alone within a few years, but offer even larger benefits in terms of the potential for smaller and less costly cooling systems and more productive work environments. Two brief examples illustrate how such large energy savings can be achieved.

Converting downlights. An estimated 500 million incandescent downlights—that's right, half a billion—operate in U.S. buildings. Most of these downlights use 75- to 150-watt incandescents. That means they consume as much energy each year as is produced by about twenty 1,000-megawatt power plants.[2] We estimate that two-thirds of these fixtures can be converted to compact fluorescent sources that use 50 to 75 percent less energy and pay for themselves in energy savings within a few years. Properly executed, such retrofits maintain or improve lighting quality. Moreover, the CFLs last about ten times as long as incandescents, so lamp replacement and labor costs are dramatically

Figure 1-1

Conversion of incandescent downlights to CFLs yields huge savings

Switching two-thirds of the half-billion incandescent downlights in the U.S. to compact fluorescent technology would save customers $3 billion per year in electricity bills, slash maintenance costs, and free up 7,000 megawatts of peak capacity.[3]

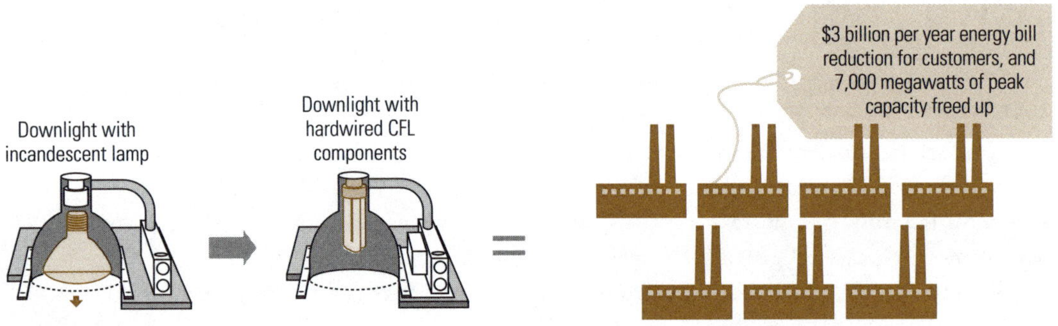

reduced as well. How much energy and money would be saved if two-thirds of the downlights in the U.S. were retrofitted to compact fluorescents? About 40 terawatt-hours, saving customers about $3 billion per year and freeing up 7,000 megawatts of peak capacity representing a utility investment of about $7 billion (**Figure 1-1**).[3] That's a Mother Load!

Improving fluorescent lamps and fixtures. The typical pre-1980s U.S. office space is illuminated with 2-foot by 4-foot lensed troffers, each containing four standard cool-white 40-watt fluorescent lamps and two standard magnetic core-coil ballasts each dissipating 16 watts, for a total rated power draw of 192 watts (nominal 176 watts adjusted for ballast factor and temperature). For about $85 in equipment costs and $40 or less in labor and project management,[4] many such fixtures can be retrofitted with a specular reflector, one electronic ballast, two T8 lamps, and a new lens or diffuser. The new system will draw about 60 watts, for a two-thirds energy savings. (Even deeper savings can be achieved with dimming ballasts linked to dimming and occupancy controls.) Although the new system will not *generate* as many lumens as the old four-lamp troffer, in many cases it will *deliver* as much usable light to the space because the old fixture trapped much of its output inside the fixture through poor geometry, low reflectance, and yellowed, dirty lenses.

Even if a delamping retrofit reduces ambient lighting levels, it may contribute to a higher-quality lighting strategy. Many spaces are overlit by glary overhead lighting, and can benefit both functionally and esthetically from lower ambient lighting levels coupled with high-efficiency task lighting. **Figure 1-2** illustrates the evolution of the lighting scheme in the offices of National Life of Montpelier, Vermont. The firm had retrofitted its lighting in the mid-1980s to fewer overhead fixtures coupled with compact fluorescent task lighting. Nevertheless, it was cost-effective to do another upgrade in 1994, which dropped ambient light levels even further—to 25 footcandles—reduced glare, and installed additional task lighting. Result: the office looked better and *people could see better.* There are hundreds of thousands of offices like National Life's, with open-plan spaces, modular furniture, and a computer screen on most desks. If National Life can benefit from this kind of energy saving improvement in lighting quality, so can they.

Figure 1-2

Evolution of office lighting

National Life's office circa 1965, in 1985 after a task/ambient lighting upgrade, and after the 1994 lighting upgrade.

1965

1985

1994

Source: EPA Green Lights

1.2 START WITH THE TASK

People want light for lots of reasons, and any successful lighting designer must keep those reasons in mind. Although this seems obvious, many installations have been designed to provide footcandles or to meet outdated codes or specifications rather than to meet the needs of people. When the task is inadequately factored into the design or selection of lighting technology, the result may be energy efficient, but it will probably fail in its broader mission.

Here is a sampling of reasons that people want light, and some of the implications they hold for lighting design.

Efficiency upgrades often improve lighting quality, in many cases with less light, not more light.

- *To get things done.* Different tasks require different amounts and types of light. The surgeon needs lots of light with low glare and excellent color rendering. The office worker needs modest ambient light levels, task lighting on desktop reading material, and minimal glare from paper and computer screens.

- *To create a mood or impression.* Restaurants and their diners often want low light levels, warm tones, and a feeling of intimacy. Corporate boardrooms call for lighting that reinforces a feeling of importance and success, while adapting to audio-visual presentations.

- *To make products look nice.* Retailers want to make their merchandise sparkle so it draws shoppers and encourages them to buy.

- *To provide safety.* Whether it's to keep vehicles and people from colliding or to deter vandals, security lighting helps us identify things that go bump in the night. If our eyes are adapted to the dark and we just need to make out the shapes of objects—a dog in the road or a robber in the alley—absolute light levels can be generally quite low while still providing the needed service.

Think quality before quantity. Lighting quality is far more important than lighting quantity. Efficiency upgrades often *improve* lighting quality, in many cases with less light, not more light. Lighting spaces in which people work with video display terminals (VDTs) provides a good example. Many offices that were designed to handle typing and similar horizontal office tasks are now filled with desktop computers and workstations, often having reflective surfaces, that require consideration of both horizontal and vertical illumination. Veiling reflections on VDT screens, such as reflected fixture images that are as bright as screen text, are made worse by brighter fixtures that lack tight cutoff angles to reduce screen glare. Existing ceiling fixtures can be fitted with low-glare distributing media, such as parabolic louvers or specially designed prismatic lenses, or can be replaced with indirect lighting systems that bounce light off the ceiling and walls. VDT tasks usually require lower illumination levels to avoid creating bright areas, such as a directly lit sheet of white paper, that make the relatively dim screen look too dark by comparison. Proper lighting design incorporates such needs, eliminating visual problems and energy waste at the same time.

To do this right, though, often requires the involvement of a skilled lighting designer. They don't have to be involved in every job, however. A well-thought-out design for a given kind of application—be it a supermarket or open-plan office—can often be replicated successfully in other spaces or facilities without the designer's detailed involvement, by providing the proper lamp and fixture specifications for a facility.

How much light do we need for different tasks? Defining proper lighting levels for specific activities is a difficult and largely subjective process. Illuminating Engineering Society (IES) or Commission Internationale de l'Eclairage (CIE) chapters in various nations publish recommended lighting levels. Some are extremely detailed. The North American IES specifies recommended levels for about 250 different interior activities in nonresidential buildings and about 300 specific industrial applications.[5] Such specificity might lead one to think that the factors used to determine lighting recommendations are rigorously quantified and unchanging. This is not the case. These values are arrived at through a consensus process, and inevitably involve compromise between lighting professionals who have differing views of the appropriate levels for given tasks. Because this process operates independently in different countries, a wide range of recommendations exists throughout the world.

Lawrence Berkeley Laboratory researcher Evan Mills and Swedish researcher Nils Borg compared recommended lighting levels from 14 countries in North America, Western Europe, and the former Soviet Union, and found variations as great as a factor of eight for a given activity[6] (**Figure 1-3**). France and the Netherlands have some of the highest levels, the former Soviet Union, Sweden, and Germany some of the lowest, and the North American values are average in most cases, though lowest for video display terminal tasks. Moreover, recommended levels have varied widely over time. As Mills and Borg summarize the situation:

> The general pattern is an increase of up to a factor of ten between the 1930s and the early 1970s, followed by stabilization or decline. . . .
>
> Among the more dramatic cases, the [United Kingdom's] retail lighting recommendations increased from ~100 lux in 1936 to ~500 lux in 1972. In the former Soviet Union, general office lighting was

Recommended lighting levels for given activities are quite subjective—one study found variations up to a factor of eight among different countries.

Figure 1-3

Range of recommended lighting levels in 14 industrialized countries

There is no consensus among countries as to the "right" level of light for a specific task and building type.

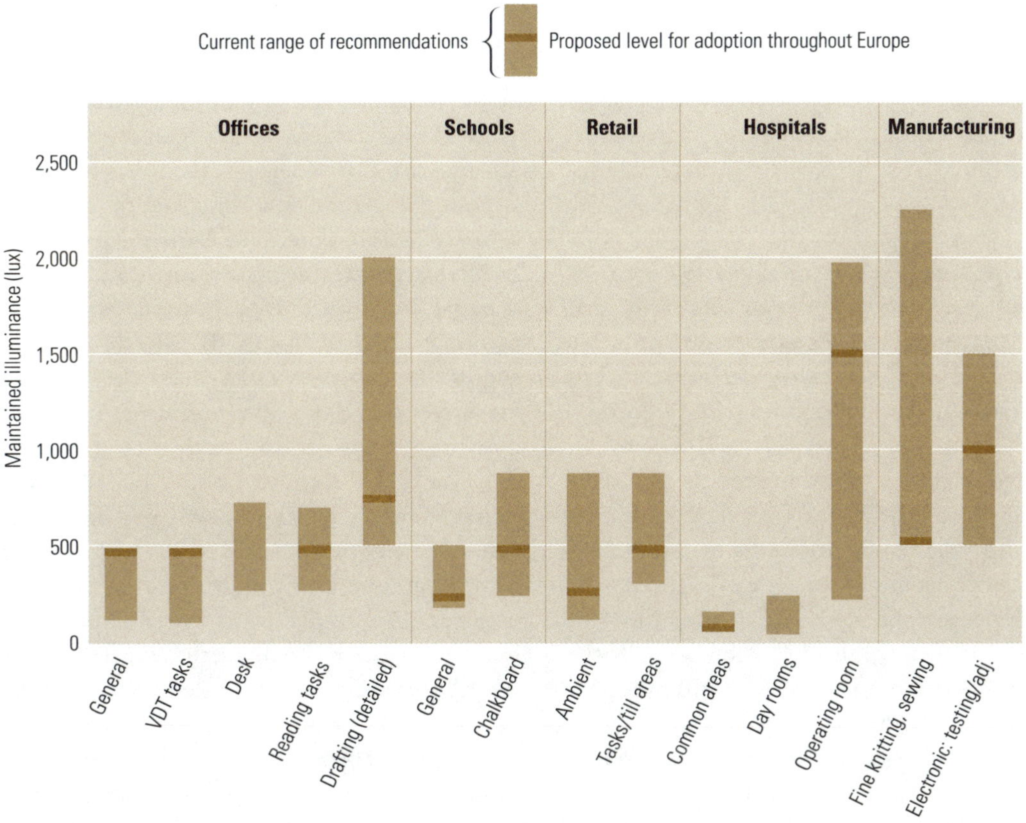

Source: Mills and Borg [5]

~25 lux in 1930, rising to 300 lux in 1979. In North America, recommendations for chalkboard lighting rose from ~150 lux in 1938 to 1,400 lux in 1972.

After the early 1970s, however, the trend either leveled out or changed direction. General office lighting in Finland fell from 450 lux in 1974 to 225 lux in 1985. Dutch recommendations for reading fell from 750 lux in 1970 to 400 lux in 1991. Even very demanding tasks, such as detailed drafting work and reading the chalkboard, show reductions of 50 percent or more.

How can there be such wide variation in recommended lighting levels? Several factors influence standard practice in a given place at a given time. Subjective perceptions of the IES committee members are key. Economics play a major role as well. As Mills and Borg note, "The commercialization of the fluorescent lamp in the 1930s made it possible to dramatically increase light levels without paying a corresponding penalty in energy costs or excessive heat." In the past two decades, increased energy costs, new perspectives on lighting design, and greater use of task lighting have helped lower many recommended levels.

The wide variation in current national guidelines suggests that more sophisticated quantitative methods and visualization tools are needed to identify lighting design solutions that deliver maximum quality while minimizing cost and energy use. Given that international consensus on these matters is not likely to emerge soon, the most sensible course of action for today's practitioner is to use the IES or CIE recommendations as *guidelines,* but to be willing to experiment in view of the subjectivity of such guidelines. Moreover, in the field, users often confuse IES guidelines for given *tasks* as applying to *ambient* lighting. A fine-detail task may require 100 footcandles, but that doesn't mean the entire space needs to be at 100 footcandles. It could be achieved by ambient lighting of about 30 to 35 footcandles throughout the space and task lighting that provides the added light on the work surface. Every situation is unique, and the true measure of success is the satisfaction of the users, not adherence to a subjective and evolving industry standard.

What options does this leave for a utility that is trying to foster more efficient lighting practice while maintaining or improving lighting quality? Few utilities actually play a role in the design of most lighting installations for which they provide financial incentives. This distance from the design process means that the most practical course for utilities is to require that any lighting installations eligible for utility incentives meet applicable IES standards for minimum lighting levels. To preserve flexibility, however, DSM programs should include an option for building owners and designers to diverge from IES guidelines with appropriate justification. Those utilities that do offer lighting design assistance and particularly those with access to lighting design laboratory facilities can simulate advanced designs to confirm their performance.

Pacific Gas & Electric and other utilities have deployed an interesting approach that combines IES guidelines and best efficient lighting practice into a set of lighting power density guidelines for new construction or major renovation in specific kinds of spaces. Lighting rebates are provided on a sliding scale by how much they beat the watts-per-square-foot guidelines (**Table 1-1**). This avoids the problem created by prescriptive incentives that pay a set amount per fixture or measure but provide no reason to minimize the number of fixtures or the total lighting power density in the space. For example, many utilities provide a fixed-dollar rebate for each incandescent downlight that is converted to a compact fluorescent source. But if the space is overlit to begin with, and half of the downlights really aren't needed, incentives are being paid for upgrading fixtures that shouldn't be there at all. The PG&E approach avoids this problem by ensuring that IES recommended lighting levels are maintained, but only pays a rebate if the design exceeds an ambitious lighting power density target.

Sometimes customer expectations of lighting levels are more important than IES guidelines. As discussed in **8.4**, New England Electric System learned that a number of residential customers who had received compact fluorescent lamps through the utility's DSM programs had removed the lamps because they didn't put out as much light as the incandescents to which they were accustomed. While people aren't necessarily concerned

Table 1-1

PG&E lighting power density (LPD) thresholds for new construction program (1994)

Rebates are paid on a sliding scale for beating the thresholds shown. This avoids the problem created by prescriptive incentives that pay a set amount per fixture but provide no reason to minimize the number of fixtures or the total lighting power density in the space.

	Watts per square foot		
	Prescriptive allowed LPD (1992 Title 24)	Actual LPD incentive level 1 (max)	Actual LPD incentive level 2 (max)
Complete building method—lighting power density values (Title 24 Table No. 1-M)			
Building type			
General commercial or industrial work buildings	1.2	1.0	0.8
Grocery stores	1.8	1.5	1.2
Industrial or commercial storage buildings	0.8	0.6	0.5
Medical buildings & clinics	1.5	1.2	1.0
Office buildings	1.5	1.2	1.0
Religious worship, auditorium & convention centers	2.0	1.6	1.3
Restaurants	1.5	1.2	1.0
Retail & wholesale stores	2.0	1.6	1.3
Schools	1.8	1.5	1.2
Theaters	1.5	1.2	1.0
All others	0.8	0.6	0.5
Area category method—lighting power density values (Title 24 Table No. 1-N)			
Area type			
Auditorium	2.0	1.6	1.3
Bank/financial institutions	1.8	1.5	1.2
Classrooms	2.0	1.6	1.3
Convention, conference or meeting centers	1.6	1.3	1.0
Corridors, restrooms, or support areas	0.8	0.6	0.5
Dining	1.2	1.0	0.8
Exhibit	2.3	1.9	1.5
General commercial & industrial work	1.3	1.0	0.8
Grocery	2.0	1.6	1.3
Hotel function	2.3	1.9	1.5
Industrial & commercial storage	0.6	0.5	0.4
Kitchen	2.2	1.8	1.4
Lobbies: Hotel lobby	2.3	1.9	1.5
Main entry lobby	1.6	1.3	1.0
Malls, arcades & atria	1.2	1.0	0.8
Medical & clinical care	1.8	1.5	1.2
Office	1.6	1.3	1.0
Precision commercial and/or industrial work	2.0	1.6	1.3
Religious worship	2.2	1.8	1.4
Retail sales, wholesale showrooms	2.2	1.8	1.4
Theaters: Motion picture	1.0	0.8	0.7
Performance	1.5	1.2	1.0

Source: PG&E

The reduced maintenance requirement of efficient lighting systems can be worth more than their energy savings.

about the brightness of some lights in their home, such as in hallways or laundry rooms, they are very particular about the light output of reading lamps and other heavily used fixtures. By accepting the sometimes exaggerated claims about CFLs' light output equivalence relative to incandescents, and failing to distinguish those applications where absolute light levels are critical, utilities may inadvertently encourage the use of a CFL with lower light output than is needed to serve the task. This may sour the customer's opinion of the very technology that the program is trying to encourage.

1.3 Go Beyond Efficiency

While energy efficiency is an attractive goal for many reasons, successful lighting designs must consider a host of other factors, including the health and productivity of occupants, the mood, beauty, and amenity of the space, and—of particular concern to utilities—the persistence of the efficiency measures.

Productivity. In a typical office, it costs about 300 times as much to pay the workers as it does to run the lights (**Figure 1-4**).[7] That means increasing the workers' productivity by 0.2 percent (four hours per *year*) can be worth about the same as cutting the lighting bill in half. A one percent productivity gain, if sustained, can be worth more to the collective bottom line of the building owner and operator than *eliminating* all of the energy bills.

The links between lighting and productivity are difficult to measure, but evidence is mounting that better quality lighting with reduced glare, better color rendering, no flicker, and quieter operation is likely to have a positive effect on worker performance, job satisfaction, and morale. In many cases, such gains can be achieved while reducing lighting energy use and improving efficiency. These issues are discussed in greater detail in **3.5**.

Health. Many claims have been made over the years about the link between lighting and health. One of the subjects receiving the most rigorous scientific scrutiny is the relationship between light levels and depression. Therapy with high levels of light has been shown to mitigate depression for many individuals who get "the blahs" during the dark winter months. But what about the effects of light levels on people who don't suffer severe depression? Is there a link between mood, productivity, and light that has subtle but pervasive effects throughout the economy? More research is needed; the current state of knowledge is summarized in **3.4**.

Maintenance. Another element of non-energy-related performance is the maintenance requirements of lighting technologies. Compact fluorescent lamps last about ten times as long as incandescents. That means the CFL saves more than its cost in reduced relamping labor alone. There's also a significant safety benefit given that many lamp replacements require someone climbing a ladder or scaffold, often to precarious heights. Changing incandescent lamps in an Orlando hotel, for example, was once observed to

Figure 1-4

Worker salaries dwarf the cost of lighting their work space

Even tiny improvements in worker productivity can quickly pay for the incremental cost—if any—of high quality lighting.

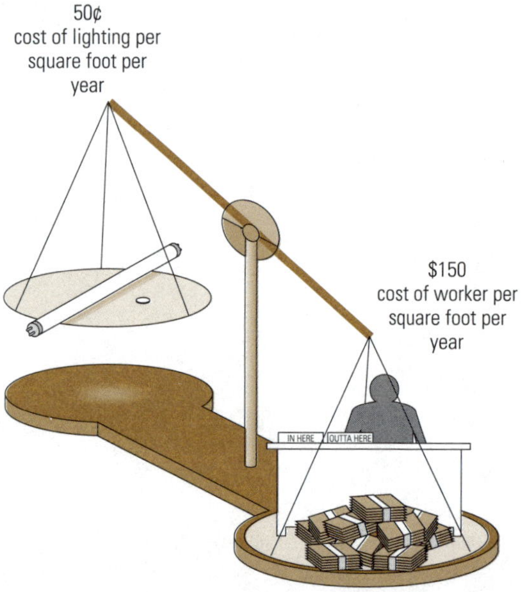

require a maintenance man to be suspended upside down on a line more than fifty feet above the crowded lobby.[8] On a more mundane level, an elderly person living alone climbing a ladder only a tenth as often can greatly reduce the risk of a broken hip, which is worth far more than the energy savings.

Persistence. As utilities build their "conservation power plants," they need to be sure that lighting efficiency measures remain in service for their full rated lives and are replaced with efficient equipment upon failure. Screw-based compact fluorescents can be replaced, with the twist of a wrist, by a far less efficient incandescent lamp, while hardwired CFLs will accept only compact fluorescent replacement lamps for the life of the fixture. Antitheft sockets are another technique to ensure that CFLs stay put until the maintenance staff need to replace them.

1.4 THINK SYSTEMS

The full potential for energy efficient lighting comes only through intelligent integration of many system variables. These system issues range from the most minute details of lamp design, through the blending of lamps, ballasts, reflectors, lenses, and other components, to the macro level of whole buildings. At the micro level, for instance, the quality and quantity of light emitted by a fluorescent lamp is a function of many variables, including the mixture of gases inside the tube and the phosphors coating the tube. Not all fluorescent lamps are created equal. T8 lamps, for example, are coated with tristimulus phosphors, which offer higher efficacy and much better color rendering than conventional phosphors, because they emit light matched to the most sensitive wavelengths of the eye's red, green, and blue retinal receptors (see **2.2.6**).

It's not enough to select good lamps and other components, however. One must also understand how these components behave in the field. In the lab, fluorescent lamps are typically rated in open-air fixtures at 77°F ambient temperature with a reference ballast that drives the lamp to its full rated output. In the field, however, many ballasts underdrive or overdrive lamps. This so-called "ballast factor" and the temperature of the lamps in field conditions can vary light output by 20 percent or more. Lamp position—whether it is installed base up or base down—can also have a 10 to 20 percent effect on light output from certain sources, particularly compact fluorescents. CFLs are rated in the base-up position, but they produce less than their rated light in base-down applications, such as table lamps. This may be why some homeowners don't like to use CFLs in their base-down reading lamps. Utilities may find that a higher wattage CFL that *really* puts out as much as the incandescent it replaced will better satisfy their customers, while still saving energy.

In most applications, lamps are part of a fixture that contains several light delivery components, including reflectors and lenses or louvers. The materials, geometries, and maintenance of these delivery elements can strongly affect the success of an installation (see case study, next page).

Beyond the fixture, lighting systems and their energy use are profoundly influenced by controls. For example, one of the first rules of good lighting design is to make use of natural daylight (see **CHAPTER 5**). Daylight keeps people in touch with the natural world and is widely regarded as a boon to productivity—if done well. But making optimal use of daylight also requires smart controls that sense the levels of natural light in a space and adjust electric light levels accordingly.

Controls can also cause problems if they are deployed improperly. Lamp life is closely related to the frequency of on-off cycling and to the maintenance of adequate current levels in the lamp electrodes. Use of occupancy sensors in spaces that are alternately occupied and vacated many times per day, such as restrooms, may lead to dramatically reduced lamp life from frequent cycling. Moreover, occupancy sensors and electronic ballasts can

CASE STUDY **REFLECTOR REPLACEMENT FOILS, THEN FIXES, LIGHTING RETROFIT**

In one commercial lighting retrofit,[9] hundreds of fixtures with T12 lamps were retrofitted with T8 lamps and electronic ballasts. The retrofit reduced lighting energy use and provided light of higher color rendering, but office workers complained bitterly about the new system.

The lighting designer inspected the space and determined that the T8 lamps had been installed at the same height in the fixtures as the original T12 lamps. Because the T8s are narrower and brighter than T12s, they were causing intense glare. By raising the T8 lamps *one-quarter inch higher* in the fixture, the designer was able to eliminate the glare problem and turn a disaster into a success. This example illustrates the sensitivity of lighting installations to system level interactions that are all too often ignored in the field.

Lighting technology is evolving very rapidly and it is important to stay abreast of the latest advances.

damage or disable each other if they are not specified and installed properly.

Lighting systems also interact with other building systems, often in unanticipated ways. Electromagnetic radiation from certain high-frequency ballasts and lamps can sometimes disrupt other electronic equipment, such as hearing aids, powerline carrier signals in energy management systems, and the remote controls on televisions and video players (see **8.7**).

The waste heat emitted from lights also affects heating and cooling systems. Most commercial buildings need more cooling than heating. In those buildings, more efficient lights may yield a "cooling bonus" that can boost the direct lighting energy savings by as much as 40 percent (see **3.2**). Capitalizing on this effect, however, may require readjustment of various HVAC components, such as zoned electric reheat coils.

As this discussion illustrates, high-quality, energy efficient lighting is not just a matter of screwing in a better light bulb. It is a systems challenge, requiring complete solutions.

1.5 STAY UP TO DATE

Lighting technology has a rich history of evolution, and the pace of change and innovation has accelerated in recent years. Just when it seems we've exhausted all the possibilities, along comes a new technology that lasts longer or performs better or costs less than what we currently think is state of the art. The potential payoff is so great that it makes sense to keep abreast of the latest developments. To gain some perspective on how far we've come, consider the following description, from a 1980 lecture,[10] of the astonishing progress in the efficiency of lighting technology since the ordinary candle: a source which, though a great advance over the open fire, consumed about 80 watts of chemical energy to emit only 12 lumens of light. The quotient of these two values, the candle's "efficacy," was only 0.15 lumens per watt. In a mere seven hours' burning, too, the candle went up in smoke.

The candle was succeeded by various kerosene, oil, and gas lights, and ultimately by the early electric light—the carbon-filament incandescent lamp. This used one-quarter less energy at the lamp (60 watts of electricity), emitted 15 times as much light (180 lumens), hence had an efficacy 20 times higher (three lumens per watt), and lasted about ten times as long (50 to 100 hours) as the candle. Next came another quadrupling of efficacy, via the modern tungsten-filament incandescent lamp, with 60-W input, 730-lm output, 12-lm/W efficacy, and 1,000-h nominal life.

The tungsten-filament lamp *reduced the cost of lighting services by nearly 600-fold* relative to the candle, while markedly improving reliability, convenience, and lighting quality. One tungsten-filament lamp could match the lifetime lumen-hour output of 8,100-odd candles, to say nothing of the cost of matches and replacement labor; yet that lamp and its electricity cost only about as much as 14 candles. The lamp provided 80 times as much light per unit of energy as a candle did—*a combined gain of more than 10,000-fold* in lumen-hours delivered per watt and per lighting-source lifetime.

Technology didn't stop there, however. The modern mix of incandescent, fluorescent, and high-intensity discharge lamps in a country like the United States averages about 44 lm/W—nearly four times the efficacy of tungsten-filament incandescent lamps—and has an average lifetime probably in the vicinity of 12,000 hours. Thus the diversification of lighting sources during the past few decades has increased the product of efficacy times lifetime by a further factor of about 45 compared to the tungsten-filament lamp, or roughly *half-a-million-fold* altogether since the candle (**Figure 1-5**).

The story continues as ever more versatile and effective lighting sources and practices emerge and take hold in the market. In 1988, there were a few dozen models of screw-in compact fluorescent lamps on the U.S. market. Today there are over 500. In 1988 fewer than nine million CFLs were sold in the U.S. In 1995 the number was close to 40 million.

Some new technologies attract a lot of media attention as revolutionary developments that may or may not be warranted. The E-Lamp (see **10.5**) hit the press with a big splash in the summer of 1992, promising to make the compact fluorescent "the eight track tape of lighting."[11] One major utility—American Electric Power—invested in one version of the technology, which promised to match the CFL in efficacy but last much longer and offer other benefits, including dimming and a more compact size. As of late 1997, the product is still not on the North American market (although Philips and General Electric have introduced similar models in Europe). If it meets the claims made by its proponents, it could be a major advance in lighting, so it is worth tracking. So far, however, it is more a marketing phenomenon than a competitive product.

Figure 1-5

A brief history of lighting

The compact fluorescent has improved the product of efficiency times lifetime by 45-fold compared to the tungsten-filament lamp and by half-a-million-fold compared to the candle.

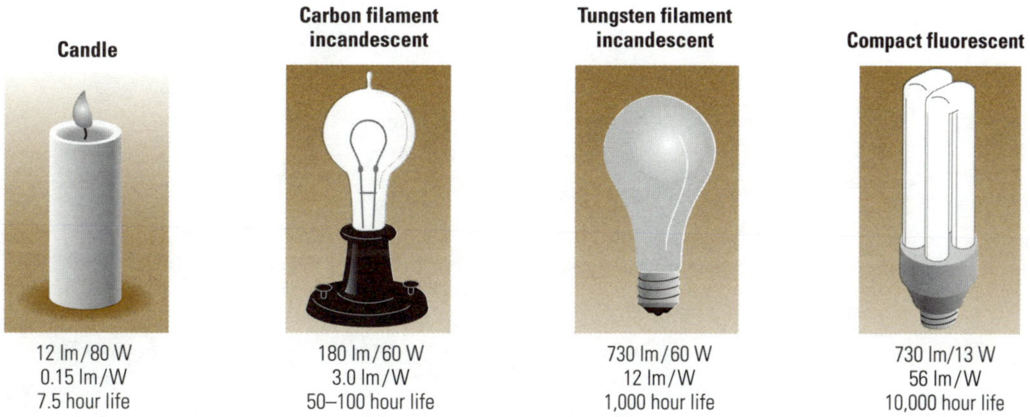

On the other hand, some innovations that clearly can have a profound impact can remain obscure and unrecognized for years. Research at Lawrence Berkeley Laboratory, for example, has documented how lamp orientation and temperature affect the performance of fluorescent lamps, particularly CFLs. LBL has developed prototype fixtures and cooling fins that keep CFLs cooler and thereby boost their light output by up to 20 percent in many settings. Some fixture manufacturers have picked up on this discovery by developing vented fixtures, but none have yet commercialized LBL's innovative and inexpensive concept of a small metal fin to "wick" heat away from the lamp surface (see **8.4.5**).

The bottom line is that new technologies are coming along all the time. Lighting users should use the best of what's available today, but keep an eye on the horizon for the latest developments—be it microwave cavity lighting, fiber optics, photovoltaic-powered direct-current lighting, or the inexpensive, reliable dimming ballast.

1 Kempton Smith, personal communication (June 1994), General Electric, Nela Park, Building 336, 1975 Noble Road, Cleveland, OH 44112, tel 216-266-8300, fax 216-266-2987.

2 Assuming an average of 100 watts per downlight and 2,000 hours per year of operation (which is high for residential but lower than in many commercial settings—see Table 4-1), these downlights use on the order of 100 TWh per year. A 1,000-MW generating station operating at 60 percent capacity factor with 5 percent grid loss delivers about 5 TWh per year.

3 100 TWh x 0.66 (two-thirds of lamps converted) x 0.62 (savings per lamp) = 40.1 TWh savings. Commercial lighting represents about 56 GW of summer peak load. Incandescents use about 45 percent of commercial sector lighting energy use. If we assume that downlights represent two-thirds of that incandescent share, or 30 percent of commercial sector lighting energy use and a comparable 30 percent share of commercial lighting peak load, they represent about 17 GW of peak load. A 40% reduction [17 GW x 0.66 (two-thirds of lamps converted) x 0.62 peak reduction per lamp] yields a saving of 7 GW.

4 Assumes a two-lamp ballast for $20, two T8 lamps for $5, a $50 reflector kit, $5 for a new lens, 20 minutes of labor at $35 per hour, and project management overhead of 20 percent.

5 Mark S. Rea, ed., *Lighting Handbook: Reference & Application*, 8th Edition (New York: Illuminating Engineering Society of North America, 1993); and Evan Mills and Nils Borg, "Trends in Recommended Lighting Levels: An International Comparison," *Proceedings of the Second European Conference on Energy-Efficient Lighting*, Arnhem, The Netherlands (September 26–29, 1993).

6 Mills and Borg [5].

7 Assuming typical office-space lighting densities of 2 watts per square foot, 3,000 hours of lighting operation per year, 7¢ per kWh, and a typical worker costing $30,000 in salary and benefits while occupying 200 square feet of space.

8 "Special Section on Lighting: Are You Using the Proper Bulbs?" *Better Buildings* (April 1987), p. 25.

9 Nancy Clanton, personal communication (April 1994), Clanton Engineering, Boulder, CO, tel 303-530-7229, fax 303-530-7227.

10 H. Luirink, "Practical Possibilities for Energy Saving in Lighting," Energycon 80 (Amsterdam) lecture reprinted as Philips Lighting Division/Eindhoven marketing brochure "Light 00.48."

11 Pierre Villere, with Intersource Technologies, used this phrase frequently in presentations during that period.

2 LIGHTING FOR PEOPLE

Lighting is an art as well as a science. Efficient lighting can and should be beautiful too. It feeds esthetic, emotional, and spiritual as well as economic and functional appetites. Efficient but ugly lighting will neither satisfy its users nor reward its providers. Yet surprisingly few programs or practitioners go beyond catalogs and calculations to integrate superlative engineering with magical design. That harmony is not only a goal to strive for; it is ultimately the wellspring of customer success.

2.1 LIGHTING DESIGN BASICS

This report is mostly about the technology of energy efficient lighting, but we feel compelled to include a rudimentary discussion of lighting design principles. Without good design, the best technology will fail. We summarize here some key design principles that relate most closely to the efficiency orientation of this volume. For more detail on lighting design, see the sources listed in the sidebar on the next page. Following this summary of design principles, we cover some of the fundamental units used to measure lighting.

Too many lighting efficiency programs focus exclusively on lighting technology. To be sure, technology is important, but it is only one part of a comprehensive approach. The designer of high-quality, efficient lighting should approach the task in logical sequence, as follows:

HIGHLIGHTS

- Because the eye can function well over a ten-trillion-fold (10^{13}) range of lighting levels, lighting quality is more important than lighting quantity.
- The eye is more sensitive to contrast and differences in luminance than to absolute lighting levels, so good lighting design controls variations in luminance.
- In most cases, lighting quality will be much better served by removing sources of glare and veiling reflection than by supplying more light.
- Don't confuse requirements for ambient lighting with the light needed on tasks. It is often more attractive, and clearly more energy efficient, to maintain ambient lighting at relatively low levels and to use task lighting to raise light levels on work surfaces to recommended levels.
- Use the highest practical CRI (color rendering index) lamp whenever possible. Objects and people not only look better, but object clarity is actually improved under such light. Using high-CRI lamps can sometimes make up for a light level otherwise considered marginally sufficient.
- Lamps with high color temperatures at low light levels make spaces appear cold and dim. Conversely, lamps with low color temperature at high levels of illumination make spaces look overly colorful. Sources at about 3,500 K look good over a wide range of illuminances.

■ *Improve the visual task.* This could entail color coding the objects on an assembly line to make them easier to distinguish, or fixing the photocopier to provide crisp rather than muddy copies, or designing easily read signs to guide travelers through an airport. The lighting designer often has no control over such factors, but should consider and act on them whenever possible.

■ *Improve the space.* The color and configuration of room surfaces and furnishings have a great deal to do with the success or failure of any lighting design.

■ *Optimize lighting quality.* A little attention to fundamental issues—the reduction of glare and veiling reflections, proper luminance ratios, the best mix of direct and indirect sources, appropriate color temperature and color rendering—can have a huge impact on the efficiency and effectiveness of any lighting design.

■ *Consider lighting quantity.* Only after issues of quality have been addressed should the amount of light be considered. As is discussed further in this chapter, less light properly delivered is preferable to more light poorly delivered.

■ *Capture daylight.* The first place to look for fulfilling lighting needs is the sun. See **CHAPTER 5** for more on daylighting.

■ *Select and implement efficient electric lighting technology.* This is all too often considered the first (or only) step in the process. The right lighting technology depends on the task, the level of quality desired, and the amount of light required.

■ *Training for operation and maintenance.* All the hard work and talent that goes into developing a good lighting design can be undone if the people who run and maintain the system aren't properly trained.

TO LEARN MORE ABOUT LIGHTING DESIGN

Lighting Handbook (1993) and the *Lighting Education Intermediate Level Manual* (IESNA ED-150)
Published by:
Illuminating Engineering Society of North America (IESNA)
120 Wall Street
New York, NY 10005
tel 212-248-5000
fax 212-248-5017

Architectural Lighting Design (1990)
Author: Gary Steffy
Published by:
Van Nostrand Reinhold
tel 800-842-3636

Advanced Lighting Guidelines (1993, Revision 1)
Available from two sources:
Electric Power Research Institute
tel 415-855-2000
fax 415-855-2954; and
California Energy Commission
tel 916-654-3298
fax 916-654-4304

With this general sequence in mind, we turn to some specific issues in lighting design.

■ The eye is more sensitive to contrast and differences in luminance than to absolute lighting levels. The eye can adapt comfortably to nearby objects and surfaces whose luminance varies by about a factor of 3. Luminance ratios above 3 to 1 nearby and 10 to 1 for more distant objects can cause eyestrain and fatigue, as the eye continually tries to adjust to higher and then lower levels of illumination.

■ Glare and veiling reflections are the greatest enemies of quality lighting design. In most cases, lighting quality will be better served by removing sources of glare than by supplying more light. In fact, so many spaces are overlit and full of glare that quality can often be improved by delivering less light in a superior manner. It is easier to see under 20 footcandles of indirect light, bounced off a light-colored ceiling, than under 100 footcandles of glary downlight. Not only do the 20 footcandles of indirect light deliver better service, they use far less energy.

■ Don't confuse requirements for ambient lighting with the light needed on tasks. It is often more attractive, and clearly more energy efficient, to maintain ambient lighting at 10 to 30 footcandles and use efficient (compact fluorescent, for instance) task lighting to raise levels on work surfaces to 50 or more footcandles.

2.1.1 Minimize variations in luminance

Paramount to understanding good lighting design is the fact that the eye sees *differences in luminance*, not absolute footcandle levels. A paper task could be adequately lit at 45 footcandles, but if areas in the visual field are much darker (for example, wood paneling at 5 footcandles) or much brighter (such as large unshaded windows), the space will not be comfortable, and the lighting on the task may not be adequate. Try to attain reasonably even luminances (footcandles x reflectance) on areas in the visual field (walls, furnishings, tasks). Ratios of 3:1 nearby and 10:1 far away are considered by the IES to be good goals in areas containing video display terminals (VDTs). For reflectances of typical surfaces, see the IES *Lighting Handbook*.[1]

Consistency within a space reduces the need of the eye to adapt continuously to variations. When consistent within a small range (about 20 percent), the illumination level can be reduced in many cases to 50 percent of IES recommended levels without creating discomfort or hindering performance. This opens the possibility for dramatic energy savings *if* the lighting is properly designed so as not to create a gloomy atmosphere. This requires careful attention to many details, including the relative light levels in adjoining spaces.

The eye adapts to predominant light levels, and subconscious expectations exist regarding levels in different areas. A lobby space at 10 footcandles may be acceptable in a building's core where most people have recently experienced 50 footcandles in nearby rooms, but that same space would seem extremely dark if entered from the outdoors where even on a gloomy day the level is at least 400 footcandles. Try to make levels comparable to nearby spaces, and follow logical progressions. For example, corridors are usually kept at 10 to 20 footcandles while adjacent offices are often at 30 to 60 footcandles, yielding ratios of less than 6:1 (assuming the surfaces are similarly reflective) in order to avoid major changes in perception as one enters or leaves such areas.

Light-colored room surfaces and furnishings can be a critical component of good lighting design.

Luminance ratios are greatly reduced when lighting is provided indirectly by bouncing it off highly reflective surfaces, such as white acoustic ceiling tiles. In fact, light-colored room surfaces and furnishings can be a critical component of good lighting design. Light-colored surfaces spread light around, smooth out differences in luminance, and create a more spacious and less cavelike feel. They also absorb less light than dark surfaces, which means that the space needs less illumination.

Note, however, that people have an expectation of glare from a lamp that localizes the source of light. Some occupants may feel indirect light isn't bright enough (even at levels above 60 footcandles) because the usual visual cues of light source and shadow have been deleted. This suggests that people don't like completely uniform lighting; they want some degree of variation.

Subtle variation can be accomplished by mixing direct and indirect light to adjust luminance ratios. Some newly available (and in some cases, portable) indirect fixtures distribute light toward the ceiling or walls and can potentially allow overhead lighting levels to be reduced. Doing this properly requires some understanding of light distribution and, perhaps, professional assistance using a simulation program. Peerless and Orgatec, among others, make fixtures that can be mounted on the top or walls of modular furniture in open-plan offices to bounce light off the ceiling. The Peerless units use 50-watt compact fluorescents, while Orgatec uses metal halides.

Architectural elements in the space may also present opportunities to adjust luminance ratios, particularly by hiding lamps so they cannot create large differences in luminance between fixtures and their surroundings.

Some basic techniques to improve lighting quality are as follows:

- Choose fixtures that recess and/or cover the lamp and distribute its light by reflection, refraction, or diffusion.
- Place the light source behind or inside a cornice, cove, valance, or trough.
- Use indirect fixtures and bounce the light off the wall and/or ceiling.
- Use lamps or fixtures with narrow beam-spreads to highlight only small areas.
- Raise the light source, or recess it into the ceiling.
- Add dimming controls to allow for reductions in fixture light output.

There are general rules for spacing fixtures which help avoid major footcandle variations within a space. Do not, for example, place a ceiling fixture closer than one foot to a wall; doing so creates a "hot spot" on the wall. Unless there is a logical reason to vary lighting levels (such as merchandising), avoid variances at task height directly below and between fixtures by observing the proper spacing methods.

Determining fixture spacing. There are two alternatives to calculate fixture spacing:

- *Spacing-to-mounting-height (S/MH) ratio,* as applied parallel to the lamp's (and usually fixture's) length, and perpendicular to the lamp's length, to assure an even footcandle level between and below the fixtures, preferably less than 20 percent variation. Typical 2 x 4-foot fixtures have S/MH ratios of 0.9 to 1.2 going across them, meaning that the center-to-center spacing of the fixtures should not be more than 1.2 times farther apart than their height above the task (**Figure 2-1**). A similar number exists for spacing end-to-end. These ratios are the most common method to check maximum fixture spacing. Manufacturers list the recommended S/MH ratios for their fixtures. For retrofits where the placement of fixtures is already determined, users can ask reflector manufacturers for reflector kits with the appropriate S/MH ratio for the space.
- *Spacing criterion (SC)* is a single number that indicates the diagonal spacing needed to

Figure 2-1

Spacing-to-mounting-height ratio

The recommended spacing-to-mounting-height ratio of most 2 x 4-foot fluorescent fixtures is between 0.9 and 1.2. That means the center-to-center distance (d) measured across the fixtures should be 0.9 to 1.2 times the distance between the bottom of the fixture and the height of the desk or other task surface (h).

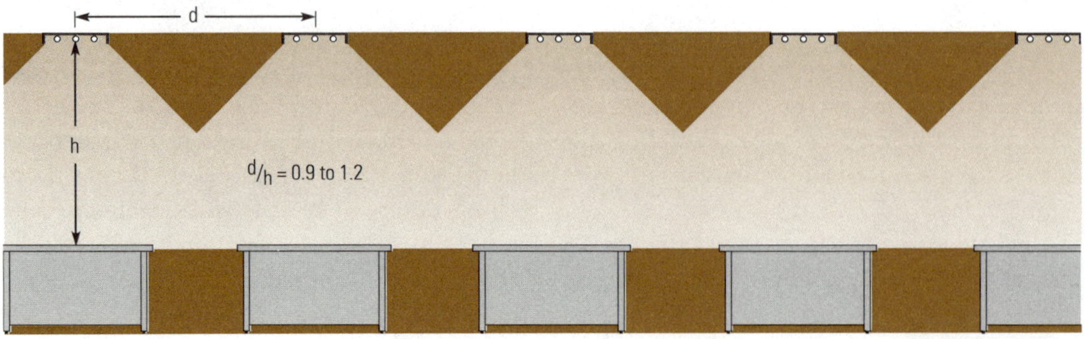

assure that light at the cross point of diagonals connecting the centers of four fixtures will be the same as directly below the fixture. SC and S/MH are often numerically close for typical fixtures.

These ratios are actually fairly conservative because they assume no surface interreflectances (for example, off walls), which tend to soften variations. Some variation in levels between fixtures is allowable since the eye generally doesn't notice variations under 50 percent (i.e. 40 to 60 footcandles). Special reflectors and lenses are available that can spread light farther, thus mitigating the need to add fixtures between those found to be installed farther apart than justified by their catalog-value spacing ratio. Such devices usually add cost to a retrofit but greatly improve light level consistency.

2.1.2 Glare and veiling reflections

Glare and veiling reflections come in many forms. Direct glare is familiar to the layman, and can be defined as luminance so much greater than that to which the eye has adapted that it causes annoyance, discomfort, or loss in visual performance and visibility. This type of glare is relatively easy to address. It might involve adjusting window shades to block direct glare from the sun or installing reflectors with tighter "cut-off angles" in overhead fixtures to prevent light from shining directly into an occupant's eyes.

All lighting designs should pass the baseball cap test, meaning that a person using the space shouldn't feel like they need a visor to shield glare from overhead lights. If shading your eyes with your hand makes your face muscles relax, there is too much direct glare in the room.

Veiling reflections—glare from reflected sources—is a more challenging issue for the lighting designer to deal with. Before computers became ubiquitous, the most common visual problem associated with tasks on a horizontal work plane was veiling reflections. This occurs when light strikes a task and produces shiny reflections that reduce or veil contrast between details of the task and the background, such as print or photos on paper (**Figure 2-2**). As shown in **Figure 2-3**, lower contrast reduces visibility, which can cause eyestrain and impaired productivity.

Most people have experienced veiling reflection when light from a ceiling fixture reflects on the page of a glossy magazine and makes it difficult to read. Veiling reflections can be more subtle than this, however. Less shiny reflections from matte paper or a light-colored desktop can cause continual eyestrain of which the user is not even aware.

There are many potential ways to minimize veiling reflections, including moving the light source, installing different reflectors or lenses, and moving or reorienting the task.

Figure 2-2

Common sources of glare

Veiling reflections commonly occur when the light source is directly above and in front of the viewer.

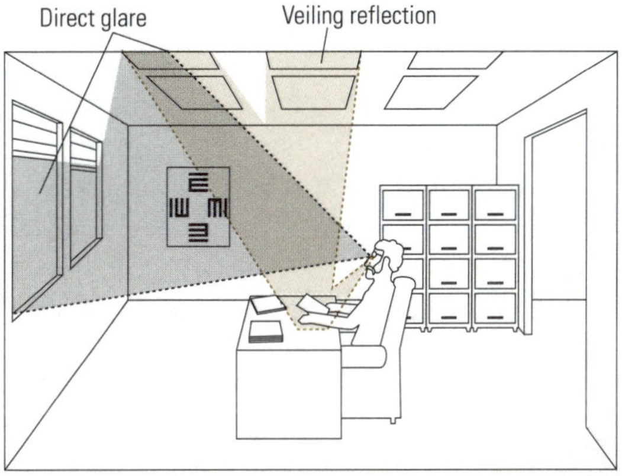

Figure 2-3

Visibility diminishes as contrast is reduced

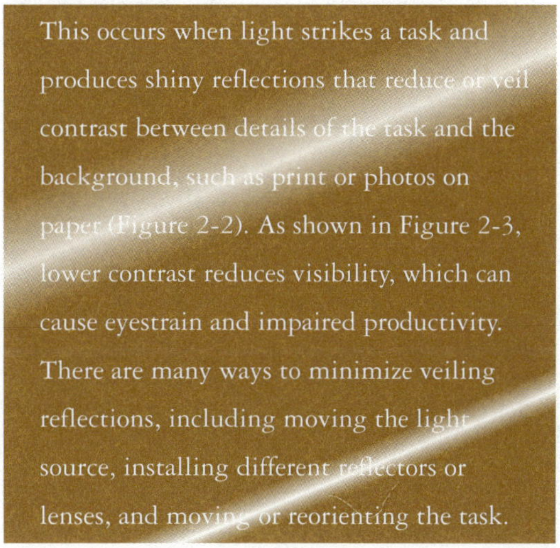

There are many potential ways to minimize veiling reflections, including moving the light source, installing different reflectors or lenses, and moving or reorienting the task. One can also use the brute force approach of adding more light to wash out the veiling reflection, but this is a wasteful solution that uses more energy and may create eyestrain from putting too much light on the task. Ideally, the lighting designer should eliminate the sources of veiling reflection so that the user need not continually squint or adjust tasks to avoid such discomfort. The designer can place fixtures in a way that avoids veiling reflections, and delivers more usable light using fewer footcandles. As noted above, it is easier to see under 20 glare-free footcandles of indirect light than 100 glary footcandles of direct light. A proper understanding of veiling reflections and how to mitigate them is central to the goal of making lighting that is both more effective and more efficient.

The IES has recognized the importance of lighting for VDT spaces with a specific publication:

RP-24
Illuminating Engineering Society of North America
120 Wall Street
New York, NY 10005

Glare problems are more complex in environments with video display terminals (**Figure 2-4**). Good designs for VDT environments may offer indirect light to reduce screen glare, options for raising and lowering light levels, and adjustable chairs, workstations, and VDT stands.

Figure 2-4

Sources of glare in VDT environments

The worker/task/environment relationships associated with VDT tasks are significantly different from those associated with traditional tasks.

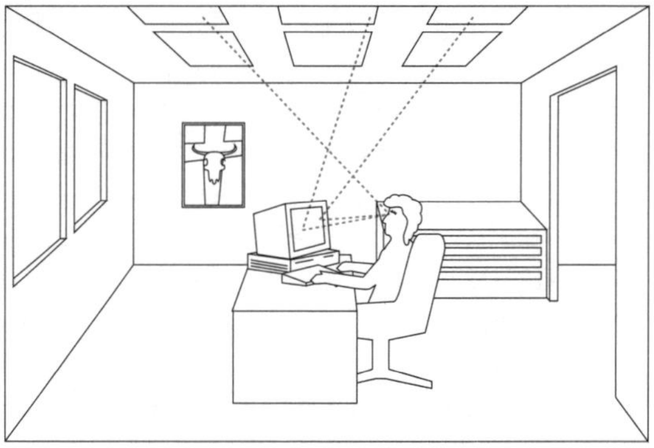

2.1.3 How much light for the task?

So far we've covered items that affect the *quality* of lighting. With those items in mind, let's now look at the *quantity*, or footcandle/lux illuminance levels, that make up an acceptable design. (Footcandles and other lighting measurement units are defined in **2.2**). All such analyses start by examining the tasks that are involved. **Table 2-1** gives a sampling of the task lighting levels recommended by the Illuminating Engineering

DESIGN TIPS

Respect the space

Whenever possible, the decor of a space should be respected. Don't install "high-tech" looking fixtures where traditional or "old style" fixtures previously existed. If there is a predominant color scheme and fixtures are exposed (for example, pendant-mounted), consider having them painted to match or "disappear" within that scheme. Trim rings and reflectors on downlights deserve attention, and many acceptable options exist. A little care in this area can make a better looking job, and elicit more cooperation from the occupants and the building architect/designer. Such nonstandard options may, however, require more detailed specification, longer lead times, and better job coordination to avoid foul-ups.

Go for high CRI

Use the highest practical CRI (color rendering index) lamp whenever possible. Objects and people not only look better, but object clarity is actually better under such light. Using high-CRI lamps can sometimes make up for a light level otherwise considered marginally sufficient.

Society of North America. IESNA lists recommended values for hundreds of activities in indoor and outdoor spaces.[2] If you're ever wondering how much light is recommended for drafting on mylar with India ink, or for tanning leather, or for illuminating the infield in AAA-league professional baseball, IESNA has a recommendation. Keep in mind that these are consensus-based guidelines that vary among countries and over time. Nevertheless, they are a good starting point.

The IESNA recommendations are expressed in a range of three numbers. Which one to use in a given application depends on a combination of factors, including the details to be seen, the age of the observers, the importance of speed and accuracy in the task, and background reflectance. IESNA specifies a methodology for weighting these factors in the selection of illuminance levels.[3]

In most typical office, institutional, and educational applications, the tasks being performed are sufficiently uniform so that ambient levels between 20 and 60 footcandles are acceptable. Extra attention is needed for spaces in which laboratory equipment, machinery, or computer screens are used, where higher (or lower) levels may be more appropriate, or where additional light sources are needed to provide high localized levels. Such sources provide what we call "task" lighting.

In most typical office, institutional, and educational applications, ambient levels between 20 and 60 footcandles are acceptable.

Aged occupants often require higher light levels due to diminished vision. Tasks involving poor contrast, rapid identification, or small letters or objects may require higher-than-normal light levels. Such tasks are often best served by adding compact fluorescent task lighting attached or built into the desk or work area, rather than raising the general light level.

Table 2-1

Currently recommended illuminance categories and illuminance values for generic types of activities in interiors

The types of activities are broken into 9 general categories, with recommended illuminance ranges from 2 to 2,000 footcandles, designated by the letters A through I. IESNA then lists hundreds of specific tasks and assigns to each one of the letter codes. (1 footcandle = 10.76 lux.)

Type of activity	Illuminance	Range of illuminances Lux	Range of illuminances Footcandles	Reference work-plane
Public spaces with dark surroundings	A	20–30–50	2–3–5	General lighting throughout spaces
Simple orientation for short, temporary visits	B	50–75–100	5–7.5–10	
Working spaces where visual tasks are only occasionally performed	C	100–150–200	10–15–20	
Performance of visual tasks of high contrast or large size	D	200–300–500	20–30–50	
Performance of visual tasks of medium contrast or small size	E	500–750–1,000	50–75–100	Illuminance on task preferably aided by task lighting
Performance of visual tasks of low contrast or very small size	F	1,000–1,500–2,000	100–150–200	
Performance of visual tasks of low contrast and very small size over a prolonged period	G	2,000–3,000–5,000	200–300–500	Illuminance on task obtained by a combination of general and high-intensity task lighting
Performance of very prolonged and exacting visual tasks	H	5,000–7,500–10,000	500–750–1,000	
Performance of very special visual tasks of extremely low contrast and small size	I	10,000–15,000–20,000	1,000–1,500–2,000	

Sample commercial, institutional, residential, and public assembly interiors

Auditoriums		Banks		Office		Conference rooms	
Assembly	C	Lobby:		Typical	D	Conferring	D
Social activity	B	General	C	VDTs	Add task lights	Critical seeing (refer to the individual task)	
		Writing area	D				
		Teller stations	E				

Source: IESNA

In most cases, ambient and task lighting both contribute to lighting on the task, so it is the *sum* of the task source and the ambient source that should deliver the required task illuminance. Thus, if the task requires 60 footcandles, and ambient provides 30 footcandles on the work surface, the task source need only provide an additional 30 footcandles. Installing a 60-footcandle task lamp would yield a total of 90 footcandles on the work plane, which is more light than is required.

In most cases, ambient and task lighting both contribute to lighting on the task, so it is the sum of the task source and the ambient source that should deliver the required task illuminance.

2.1.4 How much light for the space?

Once the task-related factors have pointed to the required footcandle level, the space-related factors must be considered to determine the amount of fixture light output needed to attain that level. The surface reflectances of walls, ceilings, floor, and furnishings all influence the amount of light needed from fixtures to do the job. Similarly, the placement of fixtures in the room and the room's shape can both affect the amount of usable light they can supply to a task. Light losses due to lamp degradation and dirt buildup should also be considered. All of these factors are combined in a procedure called the "lumen method" or "zonal cavity method."

Zonal cavity method. The basic formulas used in the zonal cavity method spring from the definition of illuminance: 1 footcandle (fc) = 1 lumen/ft^2. That is, to maintain an average of 40 fc in an area of 100 ft^2, we need 40 x 100 = 4,000 lumens coming out of our fixtures. But several modifying factors must be considered: the fixture is not absolutely efficient in dispensing light, much of that light may be lost while being reflected off of various surfaces before it arrives at the task, and light sources degrade with age and dirt buildup. To complete the zonal cavity calculation, three fundamental quantities must be known: the room cavity ratio (RCR), the coefficient of utilization (CU), and the light loss factors.

■ *Room cavity ratio.* The RCR characterizes a room by its shape, using the formula below. RCRs of typical spaces are shown in **Table 2-2** and a sample calculation is illustrated in **Figure 2-5**.

$$\text{RCR} = \frac{5(\text{H})(\text{L}+\text{W})}{\text{L} \times \text{W}}$$

where:
L = room length,
W = room width, and
H = height between fixture bottom and task height (typically at 30 inches above the floor).

Figure 2-5

Sample calculation of room cavity ratio (RCR)

$$\text{RCR} = \frac{5(\text{H})(\text{L}+\text{W})}{\text{L} \times \text{W}} = \frac{5(10-2.5-1.5)(20+10)}{20 \times 10} = \frac{5(6)(30)}{200} = \frac{900}{200} = 4.5 \text{ room cavity ratio}$$

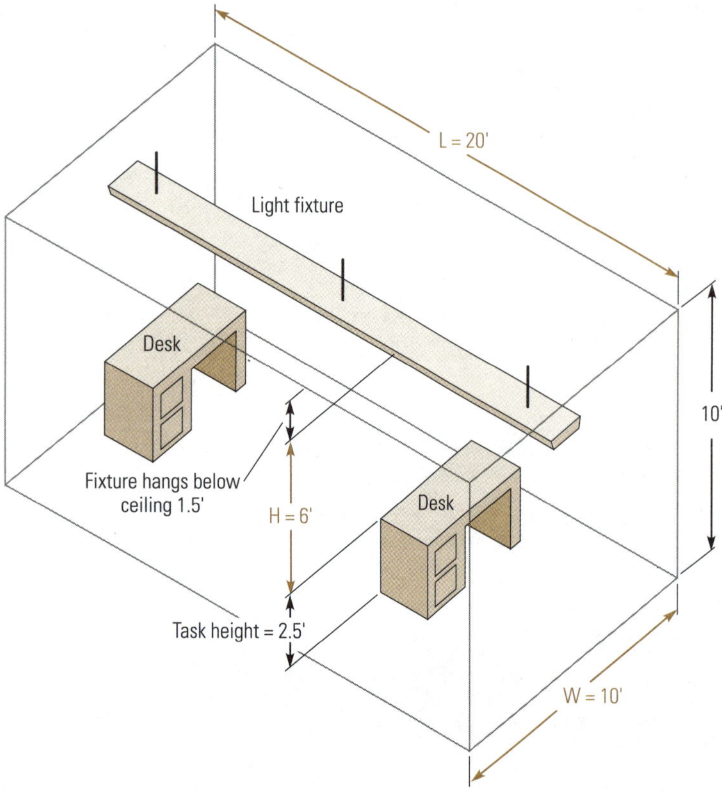

Table 2-2

Room cavity ratio (RCR) of typical spaces

Space type	Height (ft)	Length (ft)	Width (ft)	RCR
Private office	6.5	10	8	7.31
Hallway	6.5	60	6	5.96
Lobby	13.5	40	60	2.81
Classroom	6.5	40	20	2.44
Dining/lecture	8.5	40	60	1.77
Showroom	8.5	80	60	1.24

■ *Coefficient of utilization.* The CU is a measure of the fixture's ability to distribute light down to the work plane (the imaginary surface which includes all desktops), using the RCR value and the surface reflectances of the walls, floor, and ceiling. Some refer to this as the fixture "efficiency," but that is incorrect. Fixture efficiency is the proportion of lamp light that *escapes* the fixture at any angle, while CU is the proportion of lamp light that *reaches* the work plane. The two are calculated differently and are not interchangeable. The CU is supplied with a fixture's photometric data, either in a catalog or by request. **Table 2-3** shows how this information is typically presented.

Note how rapidly the CU drops as wall reflectance decreases or as RCR increases. In private offices (RCR = 7) with dark wallpaper (reflectance = 30 percent), for example, only about 42 percent of the lamps' light output actually reaches the work plane. Similar situations can occur in spaces with modular furniture using darker fabric dividers.

■ *Light-loss factors.* Light distribution is not only affected by the color and reflectance of room surfaces and furnishings, but also by change in lighting output over time, which is principally a function of lamp lumen depreciation and fixture dirt buildup (see **Figures 2-6** and **2-7**). Lumen depreciation data can be found in technical information supplied by the lamp manufacturer, while dirt depreciation values can be taken from IES graphs for various types of fixtures and dirt environments. In most cases, 0.75 is a good conservative number for combined loss factors for fluorescent lighting in clean, nonindustrial environments. Lamp and fixture maintenance is discussed in **CHAPTER 12**.

In most cases, 0.75 is a good conservative number for combined loss factors for fluorescent lighting in clean, nonindustrial environments.

Table 2-3

Typical table of coefficient of utilization values

Assuming specific reflectances for room surfaces, and room cavity ratios ranging from 1 to 10, the coefficient of utilization for the fixture in this example ranges from 67 to 34 percent, which is the proportion of the lamp light reaching the work plane.

Reflectance (wall)		50		30		10	
Reflectance (ceiling)		80	50	80	50	80	50
RCR	1	67	56	65	53	53	51
	2	66	54	63	51	51	49
	3	65	52	61	50	50	48
	4	64	50	59	48	48	46
	5	63	48	57	46	46	44
	6	61	46	53	44	44	42
	7	59	43	51	42	42	40
	8	56	41	49	40	40	38
	9	54	40	47	38	38	36
	10	51	39	45	36	36	34

Note: Floor reflectance is 20%.

■ *Zonal cavity formula.* Once these light loss factors have been taken into account, one has a more realistic picture of the "maintained footcandle level." The number of lamps (or fixtures) needed to attain that sustained minimum light level over a lamp's lifetime can then be determined by the zonal cavity formula:

$$\text{Number of fixtures} = \frac{\text{Maintained footcandle level} \times \text{Room length} \times \text{Room width}}{\text{Lumens/lamp} \times \text{CU} \times \text{Loss factors} \times \text{Lamps/fixture}}$$

Note that this method is heavily dependent on several assumptions: that surface reflectances are reasonably accurate (one common error is to look at a floor reflectance but ignore the dark desktops that are covering it); that the fixtures are evenly distributed in the room; and that other concerns, such as voltage, room temperature, lampwall temperature, and ballast factor, are normal and will not affect lamp lumen output. The basic RCR calculation assumes that the fixtures are mounted on the ceiling; variations in the RCR calculation method can account for direct/indirect fixtures mounted on pendants.

Depending on the needs of the task and occupants, it may also be necessary to consider the "visual comfort probability" (VCP) of a fixture. This factor is taken from a table available from the fixture manufacturer and indicates the proportion of people who will experience very little glare from a group of fixtures when those people are seated in the least desirable location in the room. A VCP of 70 or greater is considered acceptable by IES, and is usually achieved when light from a fixture emanates below an angle of about 65°, as measured from the nadir. The candela distribution curve of the fixture will show the intensity of light at such angles. VCP can be adjusted by changing a troffer-type fixture's lens or louver. Doing so may also affect the consistency of the light level between fixtures, or may result in darkening of the walls.

Fixtures with higher VCPs usually are deeper (to conceal the lamp) and usually have lower CU values, meaning more light will be need-

Figure 2-7

Dirt depreciation for a typical fluorescent fixture in a clean environment

Even in clean office environments, 18 months of dirt buildup can reduce light output by 10 percent.

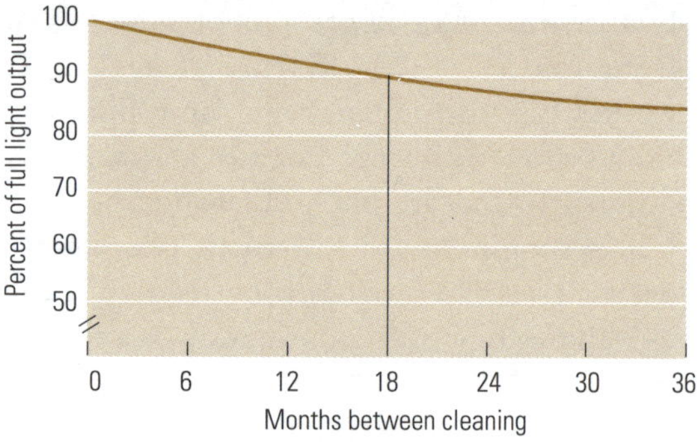

Figure 2-6

Lamp lumen depreciation

The light output of lamps declines over time.

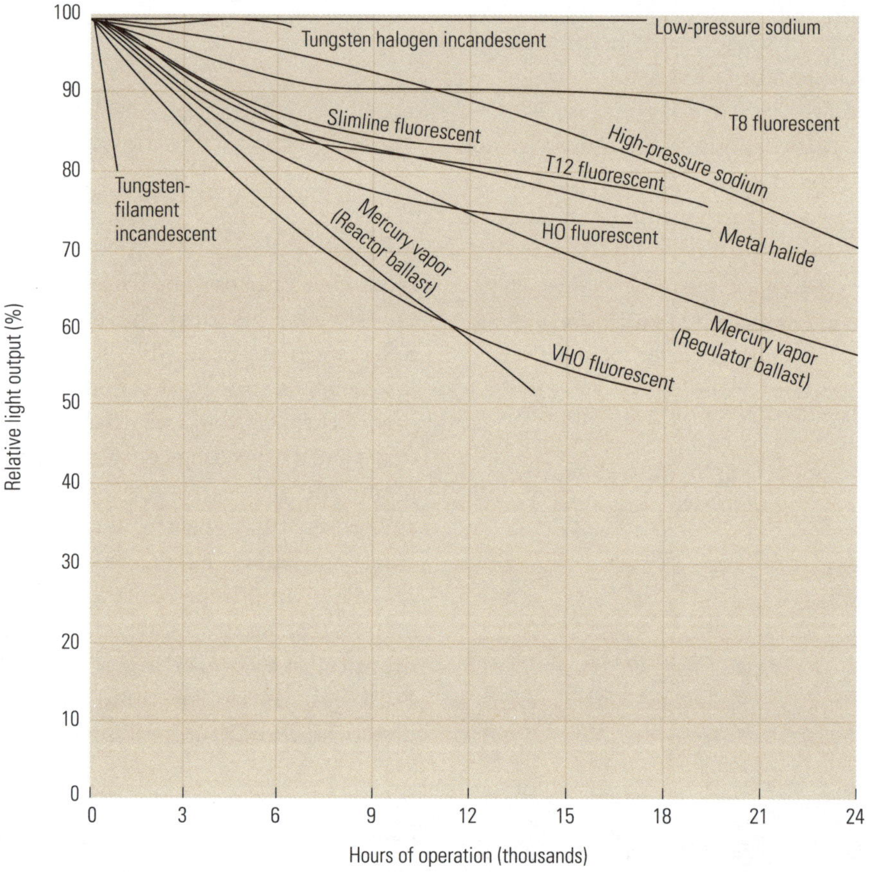

Source: National Lighting Bureau

ed to attain a given footcandle level. More watts will also be needed, so care is essential in weighing the need for higher VCP in any upgrade situation.

Nonuniform conditions. So far we have covered situations involving uniform light levels. There are cases when nonuniform levels are better, even if existing levels are uniform. This typically occurs in merchandising, where we want products to stand out (known as "highlighting" or "accent" lighting), or where vertical surfaces need special illumination (as in conference rooms). A general rule of thumb: for anything to stand out, it should reflect at least three times as much light as its surroundings. That means consideration of both illuminance and reflectance. Separate lighting systems, such as tracks and wall washers, are best suited for this task. Here are several tips when considering such equipment:

- Get good data from the manufacturer, preferably charts showing footcandles at regular heights and distances from the fixture, along an adjacent wall, and on the floor.

- Set up a mockup of the installation and verify that the fixture performs properly, taking special note of glare at the fixture's periphery.

- Be sure proper placement of the fixture will not conflict with existing HVAC, electrical, or structural items in the ceiling.

- Verify that the display will remain reasonably unchanged. Otherwise, determine the maximum degree of change and design systems that can accommodate that range.

- Place these accent lighting systems on separately switched circuits.

When confronted by oddly shaped spaces, specialty tasks (such as rapid assembly of small parts), or other factors not easily handled by this simplified approach, consider involving a professional lighting designer. Doing so may cost extra, but will usually avoid expensive errors and disappointment that could jeopardize future sales or installations.

For anything to stand out, it should reflect at least three times as much light as its surroundings.

To summarize some of the key points of this section, high-quality efficient lighting must:

- Start with the task.

- Avoid large contrasts due to uneven light distribution, and/or surfaces with low reflectance.

- Avoid glare and veiling reflections.

- Allow for the adaptation of the eye to bright or dark surroundings.

- Use optimal combinations of ambient and task lighting to achieve efficiency and enhanced visual performance.

- Create variations in light levels for special purposes.

Having summarized some of the lighting design basics, we turn now to some fundamentals on the units used to measure light.

2.2 MEASURING LIGHT

Light is measured in a variety of ways. Some of the most important attributes for readers concerned with energy efficient lighting design are intensity, color, distribution, and efficacy. One important distinction to keep in mind is that most of the light we see is reflected off some surface. Light that has been reflected or passed through a modifying material such as a lamp shade is referred to in terms of *luminance*. Light coming directly from the source is measured in terms of *illuminance*.

2.2.1 The electromagnetic spectrum

Vision depends on light, that band of electromagnetic radiation with wavelengths ranging from about 380 nanometers (one nanometer = 10^{-9} meters) to 770 nanometers. We sense different wavelengths of light as color (**Figure 2-8**). Most light sources emit energy in the non-visible ultraviolet and infrared ranges as well.

2.2.2 Intensity and quantity of light

The fundamental photometric unit is the *candela* (cd), a measure of the intensity of a light source in a given direction. The total amount of light emanating from a light source is measured in *lumens* (lm). *Footcandles* (fc) indicate the density of light landing on a surface and are equivalent to lumens per square foot. The metric (SI) unit for lighting density is the *lux* which equals one lumen per square meter (10.76 square feet). One footcandle thus equals 10.76 lux. The relationship of these parameters is illustrated in **Figure 2-9**.

Candelas and lumens are measured by special photometric machines in laboratories and are used primarily for comparing light sources independent of site conditions. Footcandles are measured in the field with a meter, and may be very dependent on site conditions because, unlike candelas or lumens, they are influenced by surface reflectances.

All of these units measure illuminance, or direct light from the source, not the reflected light that we normally see. Light that is purely reflected, or transmitted through a diffusing medium, is measured in footlamberts (or, in SI units, candelas per square meter, which are also called "nits"). Since most objects reflect light from other sources, we almost always see footlamberts. Footlamberts are equal to footcandles multiplied by the

Figure 2-8

The electromagnetic spectrum

The shortest wavelength radiation our eyes can detect is violet light at about 380 nanometers (nm); the longest is red light at about 770 nm.

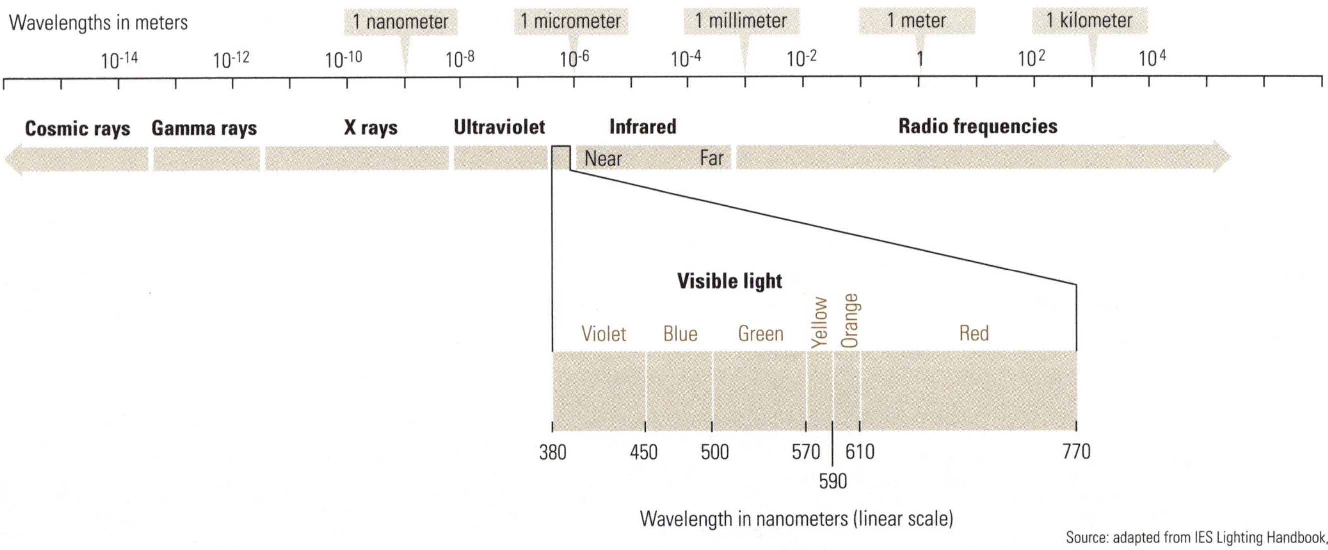

Source: adapted from IES Lighting Handbook, 1984

reflectance or transmittance of the surface being lit.

As with basic units of electricity, the garden hose provides a useful analogy for understanding lighting units (**Figure 2-10**).

- Pressure in the hose is analogous to the brightness of a light source (candelas).

- The flow rate of water spraying out of the hose (gallons per minute) is analogous to the lumen output of a source. Just as one can have a tiny hose with high pressure that doesn't deliver much water, or a huge hose with low pressure that delivers large amounts of water, one can have tiny lights that are very bright (lots of candelas) but don't deliver a large quantity of light (lumens), or big lights that aren't very bright (low candelas) but deliver a lot of light (lumens).

- The amount of water spray landing per square foot of surface is analogous to the footcandle level.

- The amount of water bouncing off the one square foot surface is analogous to the footlambert level.

footcandle = lumens per square foot

lux = lumens per square meter

footcandle = 10.76 lux

How do these numbers get used? Since no light source provides light evenly in all directions, candelas (which measure intensity in a given direction) are used to describe the distribution of brightness of a light source. Lighting fixtures are often compared on the basis of

Figure 2-9

Relationship of light measurement terms

A one candlepower light source delivers a luminous intensity of one candela in all directions. Assuming that the sphere has a radius of one foot, the light source will deliver one lumen of light to each square foot of surface of the sphere, for an illuminance of one footcandle.

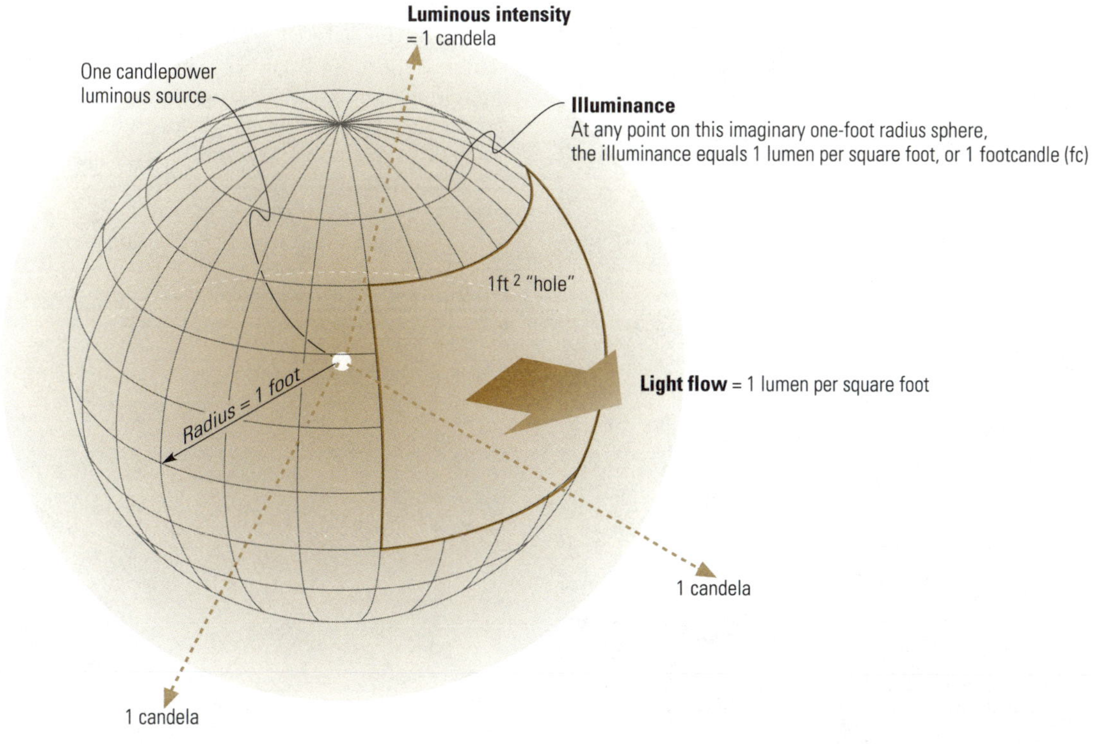

how they distribute light, as displayed on a candlepower distribution curve (**Figure 2-11**). These imposing looking diagrams aren't all that complicated once you get used to them, and they are fundamental to good lighting design.

Lamps are rated in lumens, which is the total amount of light they emit, not their brightness, and not the light level on a surface. To determine how many lumens are arriving on a given surface, we measure the footcandles at that surface. We generally define a desired footcandle level based on the type of task being performed. In reality, the specification of a task (such as reading black type on white paper) also often roughly defines the reflectances and contrast involved, such that the footlambert level is implicit in the footcandle specification.

Figure 2-10

Understanding lighting units through analogy to the garden hose

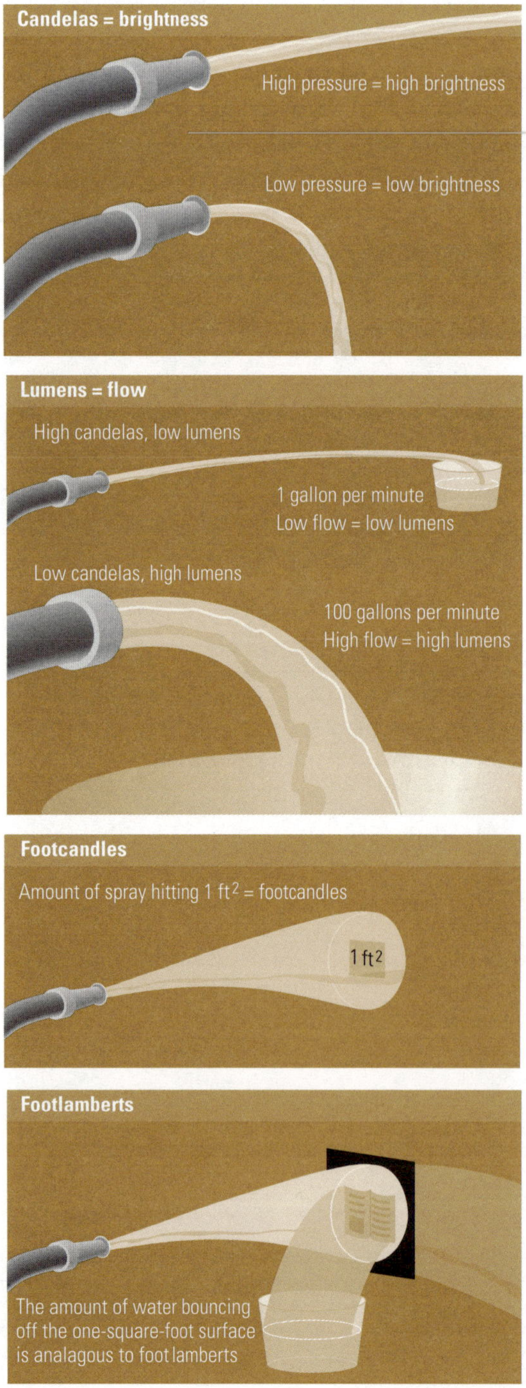

Figure 2-11

Candlepower distribution curve

This chart conveys the brightness of light at any angle from the point directly below the light source, or the nadir (0°). The light source is at the center of the radiating lines. The concentric circles represent constant brightness levels in candelas with the values increasing with greater distance from the center of the chart. In this case, the highest intensity reached is about 17,200 candelas at 26° from the nadir. If you imagine this as a ceiling-mounted fixture delivering light downward into the space it becomes clear why no light is being delivered up into the ceiling cavity, between 90° and 270° from the nadir.

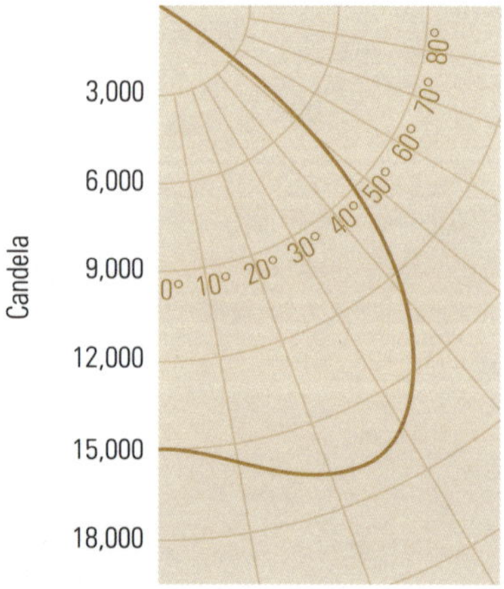

2.2.3 Color

We rarely see pure colors; most of the light we see is a mixture of colors that has reflected off surfaces that absorb some colors and reflect others. The eye is more sensitive to some colors than others, which opens opportunities for more efficient lighting that targets those zones of heightened sensitivity (see **2.2.6**).

To demonstrate the mix of color coming from a light source or being absorbed by a surface, we use a graph called a spectral power distribution (SPD) curve, which shows the intensity of a color on the vertical axis and the wavelength (color) on the horizontal axis (**Figure 2-12**).

Figure 2-12

Typical spectral distribution curve

Most lighting sources emit varying levels of light at different wavelengths. This example shows the emission from a cool-white (4,100 K) F40 fluorescent lamp.

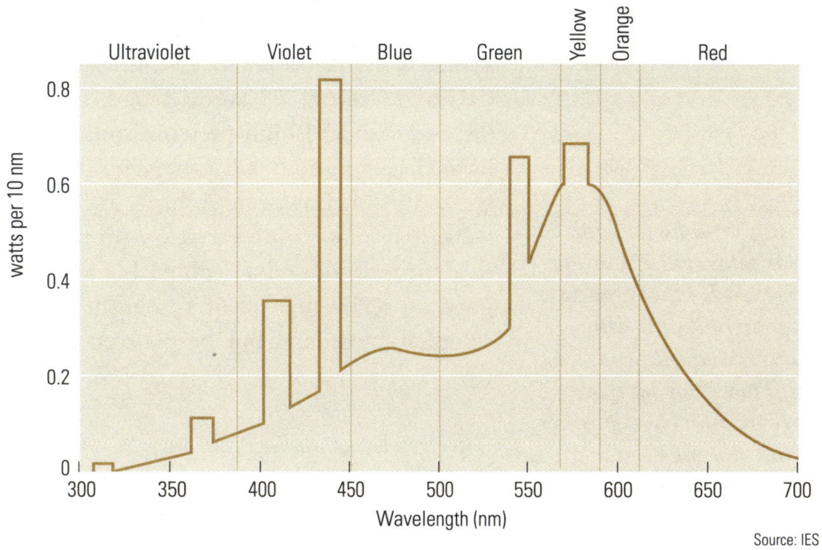

Source: IES

There are three basic ways to describe the color mix of light from a source:

- color temperature
- color rendering index
- chromaticity color chart

Color temperature. Different light sources emit light that is perceived as having different "whiteness." Incandescent lamps have a yellowish-white light, while many fluorescent sources and midday sunlight are perceived as having a bluish-white appearance. Color temperature is the metric used to describe these qualities. It is expressed in kelvins (K), a unit that has no relation to the actual temperature of the light source. Rather, it corresponds to the temperature of a blackbody radiator emitting light of comparable color.

For example, an electric stove element will glow red if it gets hot enough. If one could turn it up even higher, it would get whiter and then bluish-white as it gets hotter, and eventually it would get hot enough to emit a blue light. The color temperature scale essentially says that a lamp emits the color of light that would be emitted by a blackbody radiator (the electric cooking element in our example) heated to this temperature.

The scale is physically correct but may seem intuitively backward; the higher the color temperature, the "cooler" or bluer the light. Incandescent lamps have a color temperature of about 2,700 K. Fluorescents generally range from 3,000 for the "warm white" lamps to about 4,000 for "cool white" lamps. Daylight typically ranges from 5,000 to 10,000 K.

The "right" color temperature for an application may depend on the footcandle level being maintained. Using lamps with high color temperature at low light levels makes spaces appear cold and dim. Conversely, using lamps with low color temperature at high levels of illumination will make a space look overly colorful. This phenomenon, known as the Kruithof effect, is illustrated in **Figure 2-13**. It is particularly important to consider this in dimming applications because a light source that looks good at 50 footcandles may not look as good at higher or lower illuminances. Sources at about 3,500 K are a good middle ground that look good at a wide range of illuminances. This issue needs to be considered in retail display, and restaurants, where the appearance of objects, people, and food is very important.

The higher the color temperature, the "cooler" or bluer the light

Color rendering index (CRI). Measured on a scale of 0 to 100, CRI conveys the ability of a light source to convey accurately a sample of eight standard colors, relative to a standard source with excellent color rendering that has the same color temperature. The CRI averages the rendering of the eight colors; the higher the average value, the better the source presumably is. A cool-white fluorescent has a CRI of 58 to 62; T8 lamps range from 75 to 98; and standard high-pressure sodium lamps have CRIs of about 27. One should use CRI with caution however. A high average CRI value may mask wide variations in the lamp's rendering of individual colors. It may be very good at rendering six of the eight standard colors, for instance, but be poor on the other two.

Using lamps with high color temperature at low light levels makes spaces appear cold and dim. Conversely, using lamps with low color temperature at high levels of illumination may make a space look overly colorful.

Color rendering indices can only be compared for lamps of similar color temperature. It makes no sense to compare the CRI values of an incandescent lamp with color temperature of about 2,700 K with that of a cool-white fluorescent with color temperature of about 4,000 K, because their CRIs are evaluated in relation to two different standard sources.

Figure 2-13

Kruithof's curve

As light levels drop, color temperature should also be reduced.

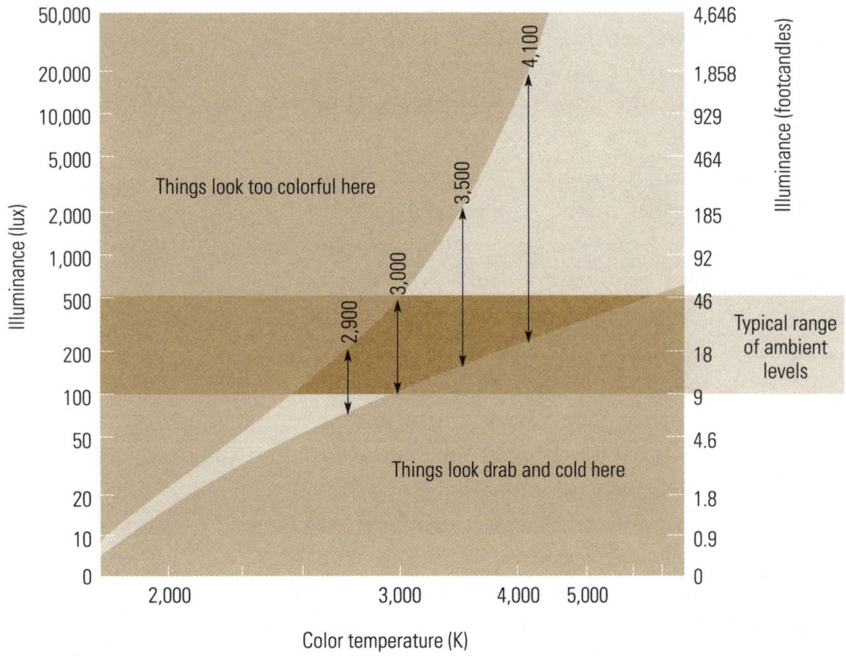

Source: IES

Chromaticity color chart. This is a rather complicated system of chromaticity values on an x, y diagram that is used for precise color specification in specialty applications, encountered most often in the paint and printing industries. It doesn't get used widely in the world of lighting design. For more on this system, consult the *IES Lighting Handbook.*

Color temperature and CRI are important concepts to grasp, because they play key roles in the esthetic acceptability of lighting installations. Several key points are worth noting:

- Use the highest CRI you can afford—go for 75 or higher, but beware that the cost of fluorescent lamps climbs steeply above the CRI of 80. The 1992 U.S. Energy Policy Act created minimum CRI standards for many common fluorescent lamps. See the box on EPACT impacts in **CHAPTER 7**, page 122.

- Avoid mixing sources in near proximity that have different color rendering, because they will impart different appearances to people and furnishings. For instance, lower-CRI lamps will make a wood-paneled wall look worn and dull compared to a similar surface washed by high-CRI light.

- Color temperatures of about 3,500 K are a good average level for most facilities. Values well above 3,500 K may impart a cooler ambiance and, at typical indoor light levels, may make people and furnishings look less attractive. Temperatures below 3,500 K may make people look too pink. In some applications, such as retail display, other color temperatures may be more appropriate, and should be carefully selected by skilled designers in consultation with the client.

- Lighting sources with different color temperatures should generally not be mixed in near proximity to each other. Keeping sources within 400 K of each other is more important than designing around any particular color temperature.

2.2.4 Efficacy

No lamp converts all the electricity it receives into light. Much of it gets turned into heat. The number of lumens that a lamp and its ballast (if there is one) produce from a watt of power is called "efficacy." For example, a 100-watt incandescent lamp that supplies 1,800 lumens has an efficacy of 1,800 lumens / 100 watts = 18 lumens per watt. Be sure to consider whether ballast losses are included in an efficacy value; they can often add 20 percent or more to total losses. Typical efficacy ranges for the major types of light sources are shown in **Figure 2-14**.

Lamp efficacy values are based exclusively on the lamp's performance and do not include ballast losses. For ballasted lamps, lamp efficacy is determined for the lamp when it is operated by a reference ballast that drives the lamp at precisely its full rated output. Many ballasts underdrive or overdrive lamps, however. *System efficacy* values measure the performance of particular lamp and ballast combinations and include the ballast losses.

Why the term *efficacy*, instead of *efficiency?* Because there's a lot more than the efficacy of the lamp and ballast that determines the efficiency with which electricity gets turned into useful, high-quality light on the task. The most efficacious light source is worthless if it is enclosed in a poor fixture that traps much of the light or reflects it at angles that cause eye-straining glare. Moreover, high-efficacy lamps can produce poor-quality light. High- and low-pressure sodium lamps have the highest efficacies available, but the light they produce generally has poor color and color rendering. Finally, there is the fact that manufacturer-rated lumen output is measured with a defined lamp orientation and temperature. Real-world conditions may differ widely from these laboratory conditions, leading to different—generally less—light output than product ratings may suggest. Compact fluorescent lamps, for example, are rated at 77°F in a base-up orientation. Their efficacy can be 20 percent less in field applications where they are subjected to hotter conditions in a sealed fixture and mounted in a base-down position (see **8.4.5**).

Figure 2-14

Lamp efficacy of major light sources

Actual values in field installations may vary considerably from rated efficacies determined in laboratory conditions.

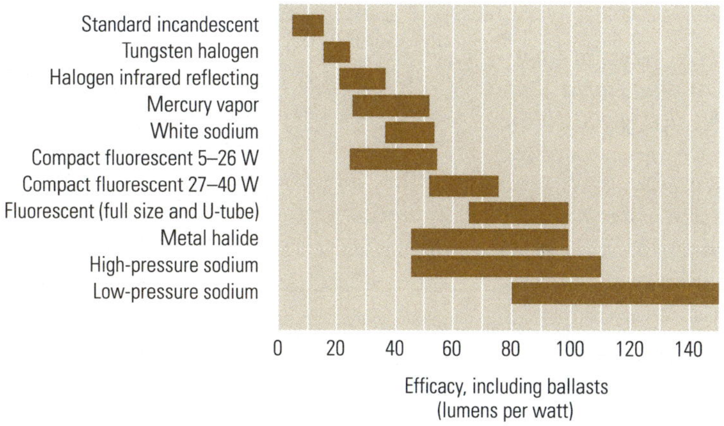

2.2.5 The eye's adaptability

From the dimmest conditions in which we can see to the brightest we can endure, the eye can see over a roughly *ten-trillion-fold* (10^{13}) range of illuminance. That's a remarkable feat of adaptability—with significant implications for energy efficiency—and it is accomplished in three principal ways:[4]

The eye can see over a roughly ten-trillion-fold (10^{13}) range of illuminance.

■ *Change in pupil size.* In response to varying levels of ambient light, the muscles of the iris expand or contract to vary the exposed diameter of the pupil by a range of 8 to 1, which varies the area of the pupil by a factor of 64.

■ *Photochemical adaptation.* The rods and cones of the eye contain pigments which, when they absorb light energy, change composition and release ions that then get translated into an electrical signal to the brain. Under steady lighting conditions, the levels of these pigments remain the same. In the dark they are regenerated. And in very bright conditions, they are depleted.

■ *Neural adaptation.* The fastest change in sensitivity to different light levels comes from variations in the neural components at the back of the eye, which send differing levels of electrical signals to the brain as light levels change.

The huge range of the eye's adaptability is apparent to anyone who has ever attended a movie matinee. Coming in from the bright outdoors, what appears to be an unnavigable darkness gradually unfolds over several minutes into a readily discernible room full of seats, aisles, and people. Once fully adapted to the darkened interior of a theater, one can make out remarkable levels of detail. This phenomenon has important ramifications for the levels of light we need in buildings for good visual performance. Given proper transition spaces that "step down" from the outdoors to building cores, the eye can function very well in relatively low levels of light.

Our perceptions of light levels are not linear, but roughly logarithmic. The nonlinear response of the eye means that a dimmer may have to drop to 10 percent of full light to create a perception of 75 percent reduction in brightness.

Another aspect of the eye's adaptability is that our perceptions of light levels are not linear, but roughly logarithmic.[5] When a typical three-way 50-100-150-watt incandescent lamp is turned on to the 50-watt setting, the perceived change from a dark room to a 50-watt light is dramatic. Switching to the 100-watt setting adds an equal amount of additional light, but the apparent change in brightness is not as great. Going to the highest setting creates an even less perceptible change in light level.

AGING AFFECTS VISION

The eye's lens starts to yellow in individuals over 50, making blue tones look darker and causing more confusion between blue and green colors. In addition to the world appearing more yellow, the lens of the eye gets thicker and less transparent with age. This increases the scattering of light within the eye, which makes older people more susceptible to glare. Since less light makes it through the lens to the retina, older people need higher illumination levels.

This phenomenon relates to the effectiveness of dimming systems. If we want to dim a room's lighting for an audio-visual presentation so that it seems 75 percent darker, the actual light levels need to be reduced by considerably more than 75 percent. As discussed in **11.2**, architectural dimming controls are designed to reduce light levels to 5 or 10 percent of full output. But some step dimming and so-called "energy dimming" controls only go down to 20 or 25 percent of full light output. While this sounds like a significant drop, the nonlinear response of the eye means that a dimmer may have to drop to 10 percent of full light to create a perception of 75 percent reduction in brightness.

In addition to the eye's nonlinear response to varying light levels, researchers are discovering that the eye responds differently to varying spectral wavelengths, in ways that raise questions about the way lumens and thus efficacy are defined.

2.2.6 Photopic vs. scotopic light

The eye is a lot like a camera. It has a lens that focuses light, an aperture (pupil) that opens or closes to adjust the amount of light entering the eye, and a surface onto which the light projects (retina). In the retina, about 120 million photoreceptive cells called rods and another 8 million cells called cones absorb light and convert it to electrical signals for processing by the brain to form the images that we call sight.

Rods and cones are tuned to different wavelengths of light. Cones enable us to sense color and fine details. They are most receptive to light centered around the yellow-green portion of the spectrum called the *photopic* region, at 555 nanometers, and less sensitive to the red and blue ends of the spectrum. Rods, which give us night vision, but also function during the day, are most sensitive to a range of light centered at about 507 nanometers. This range of sensitivity is called the *scotopic* region (**Figure 2-15**).

The photopic vision curve shown in Figure 2-15 is used to determine the lumen output of a light source. Light output is first measured by a meter in terms of light-watts, but these light-watts are then weighted to account for the fact that the human eye is more sensitive to light-watts in the yellow-green range than to light-watts in the red or blue ranges. Light-watts in the yellow-green band are given greater weighting in this calculation. Consequently, a source with most of its light-watts in the yellow-green range—like low-pressure sodium—will be said to have a high number of lumens, even if it produces fewer light-watts than another source that generates most of its light in the blue portion of the spectrum.

Research pioneered by Dr. Sam Berman at Lawrence Berkeley Laboratory suggests that this method of calculating lumens may be flawed. Light sources that are deficient in blue light cause the pupil to open wide, which in turn reduces depth of field and the ability to focus on tasks, particularly in the low light levels typical of building interiors. Bluer light that is richer in scotopic content will lead to smaller pupil apertures, which in turn lead to better depth perception and enhanced ability to determine fine detail.

In response to such findings, Berman and his colleagues are proposing a new way of measuring light-watts that accounts for the eye's scotopic response.[6] The new unit is the "pupil lumen," which factors in the importance of focusing ability on visual performance. When measured in pupil lumens, the relative efficacy of different light sources is very different from standard lumen measurements. High- and low-pressure sodium sources, which have the highest lumen-per-watt ratings under conventional measurements, have about the same number of pupil-lumens per watt as incandescent sources, which are traditionally viewed as the source with the lowest lumens per watt. Sources rich in blue light—including triphosphor fluorescents—get higher ratings under the pupil lumens approach. Table 2-4 contains correction factors for converting conventional lumens to pupil lumens for various light sources.

Figure 2-15

Photopic and scotopic ranges of eye sensitivity

Cones are most receptive to light in the photopic region, rods to the scotopic region. Light sources that deliver more light in the scotopic region may be able to deliver the same level of visual service while reducing energy use.

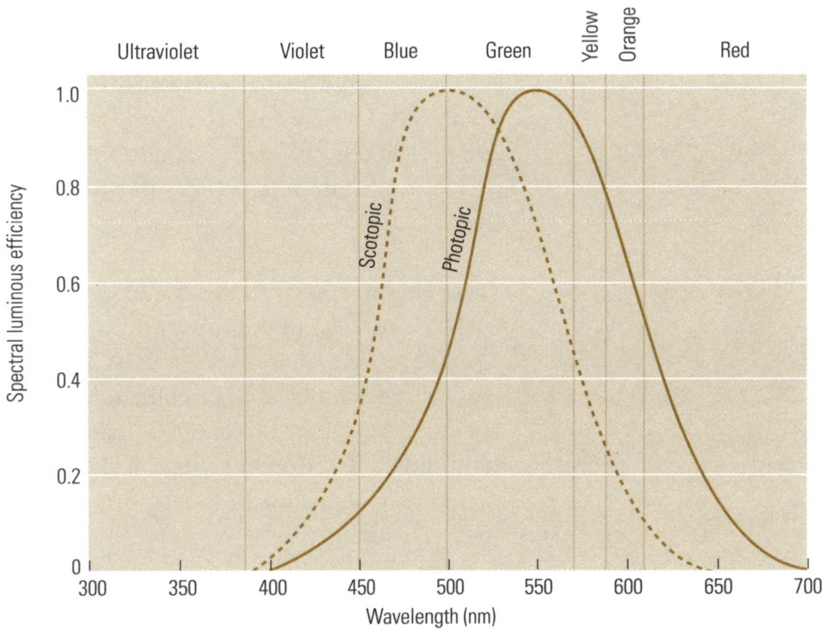

Table 2-4

Conversion factors for lumens to pupil lumens[a]

When the scotopic eye functions are factored into efficacy ratings, sources rich in blue light are big winners.

Source	Conventional lumens/watt	Correction factor	Pupil-lumens/watt
Standard incandescent	17	1.26	21
Tungsten halogen	22	1.32	29
2,900K WW fluorescent[b]	65	0.98	64
2,850K WWDX fluorescent[b]	48	1.09	52
3,500K triphosphor fluorescent (thick coat)[b]	69	1.24	85
5,000K, 90 CRI fluorescent (C-50)[b]	46	1.70	78
Vitalite fluorescent[b]	46	1.71	79
5,000K pure triphosphor fluorescent[b]	70	1.58	111
Daylight fluorescent[b]	55	1.72	95
Deluxe mercury vapor	40	0.86	34
35-watt high-pressure sodium	55	0.57	31
50-watt high-pressure sodium	65	0.76	49
Low-pressure sodium	110	0.38	42

Notes: a: Preliminary values from LBL research.
b: fluorescent efficacies based on two 40-watt lamps operated on an energy-saving magnetic ballast.

Source: Benya [6]

This work—which is still quite controversial—has the potential to shake up the century-old science of photometry and with it many established canons of lighting design and technology. As Sam Berman puts it, "present day photometry may be okay for photography but it isn't very good for lighting." He believes that spectral shifts to account for scotopic sensitivity could significantly reduce energy use in some applications while maintaining the same level of visual performance. Replacement of the ubiquitous cool-white lamp by a high-color-temperature, narrow-band (NB) lamp would elicit the same pupil size and hence the same visual effectiveness with 24 percent less power, Berman maintains.[7]

These findings need to be balanced, however, with a number of factors, including the Kruithof effect discussed in **2.2.3**, which suggests that bluer, higher color temperature sources at low light levels make spaces appear cold and drab. Moreover, the link between visual acuity and actual human performance is not well established.

The international lighting research community is beginning to take notice of this issue. The Commission Internationale de l'Eclairage (CIE), a group of international lighting experts that sets lighting science standards and protocols worldwide, is holding a special technical session in Vienna in early December 1994 to discuss the implications of scotopic sensitivity. While CIE is unlikely to move quickly, if convinced by the findings of Berman and his colleagues, it could set the stage for a reexamination of the photometry guidelines and practices that undergird lighting design.

What should lighting designers do in the meantime? According to noted lighting designer James Benya, designers should make only minor adjustments for now:[8]

> Use sources known to be blue-rich, such as metal halide and triphosphor fluorescent, including compact fluorescent lamps. You may wish to select a slightly higher color temperature than before, keeping in mind that psychologically, a high color temperature creates a cold and unusual feeling interior space (especially at low light levels). In a space where a person might not need to focus at multiple distances, such as a warehouse, high-pressure sodium sources might still be preferable to others for a number of reasons.
>
> This research also raises an interesting issue with night vision. At night, the rods provide greater sensitivity than the cones, and rely heavily on blue light. Yet we measure exterior light sources in lumens corrected for day vision, and use light sources that are blue-poor. Here it also appears that blue-rich sources are needed to achieve the greatest visual sensitivity for exterior lighting at low levels.

1 Mark S. Rea, ed., *Lighting Handbook: Reference & Application*, 8th Edition, Chapter 11 (New York: Illuminating Engineering Society of North America, 1993).

2 Rea [1].

3 Rea [1], p. 476.

4 Rea [1], pp. 73–74; and *Lighting Education ED-150* (New York: Illuminating Engineering Society of North America, 1993), pp. 1–5 to 1–10.

5 *Lighting Education ED-150* [4], p. 1–15.

6 Much of this discussion is drawn from James Benya, "Bluer Light, Better Sight," *Architectural Record Lighting* (February 1991), pp. 63–64.

7 Sam Berman, "Energy Efficiency Consequences of Scotopic Sensitivity," *Journal of the Illuminating Engineering Society*, v. 21, no. 1 (Winter 1992), p. 8.

8 Benya [6].

3 Connections

3 CONNECTIONS

The richest benefits of lighting upgrades come from whole-system integration, not piecemeal application of components. Lighting is intimately connected to the function of other building systems, to the health and productivity of the building occupants, and to the environment at large. These connections need to be considered at all stages, from design through disposal.

3.1 LIGHTING SYSTEM INTEGRATION

Many lighting efficiency efforts are oriented towards the installation of specific pieces of equipment, such as electronic ballasts or compact fluorescent downlights. But as with many other types of complex systems, the interactions between system elements in lighting equipment create energy and power savings that can be greater than the sum of their parts. Following are some examples of these effects:

- Reducing lighting power in ceiling-based fixtures (through delamping or switching to more efficient equipment) also reduces the temperature in the fixture, which in turn increases light output.

HIGHLIGHTS

- The full potential that efficient lighting holds for energy and dollar savings as well as productivity gains can only be achieved through whole-systems design. Integrated packages can pay back just as fast as "cream-skimming" strategies and deliver savings of 80 percent or more.

- Efficient lighting often yields an additional benefit in the form of reduced space conditioning. The hotter the climate and the larger the building, the more likely that the net result is a cooling bonus. Even if there is a heating penalty in energy terms, however, the net economic result may be positive if the heating fuel is less costly than the electricity it is displacing.

- Efficient lighting may allow the cooling system to be downsized, yielding capital as well as energy savings. The checklist for any lighting retrofit should include readjustment of reheat coils or other HVAC controls to ensure that occupants are not overcooled in the summer.

- Increasing numbers of building operators are engaging in proper disposal or recycling of used lamps and ballasts. This can add about a dollar to the lifecycle cost of a lamp and $1 to $8 to the lifecycle cost of a ballast. Factoring these often-ignored costs into the economic equation on lighting provides added support for the use of daylighting and electric lighting designs that minimize the number of lamps and ballasts. Reduced risk of environmental liability alone may justify preemptive lighting retrofits to remove old magnetic ballasts and replace them with electronically ballasted systems.

- Increasing numbers of case studies are emerging that demonstrate measurable productivity benefits of high quality lighting. A few percent productivity gain can be worth more to the bottom line of a building owner or operator than eliminating all of the energy bills.

■ Electronically ballasted T8 lamps not only have better initial efficacy than magnetically ballasted cool-white T12 systems, but better maintain their light output over time. This means that fewer initial lumens are required.

■ Switching from a 70-percent-efficient diffuser to a 93-percent-efficient acrylic lens reduces the amount of required lumens in the fixture, which in turn lowers temperatures and improves efficacy.

3.1.1 Performance comparison of cream-skimming and whole-system retrofits

Integrated lighting retrofit packages save two or three times as much as "cream-skimming" retrofits and pay back just as fast.

So how can these kinds of interactions be accounted for in lighting system retrofits or new designs? While every situation has individual eccentricities, one illustration of the integration of lighting efficiency strategies is shown in **Table 3-1**.[1] In this case, we examine retrofit options for a room with twelve fluorescent fixtures dating from 1970 (**Figure 3-1**), and look at seven incremental steps of upgrading the system.

Case 1: base case. We assume a 1970-vintage lighting scheme with four T12 lamps per fixture, driven by two standard magnetic ballasts, a white-painted metal luminaire, and a flat diffuser. This system maintains an average of 28 footcandles on the work surfaces.

Case 2. So-called "energy-saving" T12s (see **7.1.1**) are substituted for the normal T12 lamps. They use less energy but produce less light, and while the payback is just over two years, the savings are quite small.

Case 3. This and Case 4 are very common "cream-skimming" retrofits. Four T8 lamps and two, two-lamp electronic ballasts are substituted for the existing lamps and ballasts in each fixture. The existing fixture and diffusers are left in place. The new system uses 35 percent less energy and pays back in just under five years.

Case 4. This retrofit is similar to Case 3, except that a single four-lamp electronic ballast is used in each fixture. The four-lamp ballast has a lower ballast factor and dissipates less energy than two of the two-lamp ballasts, plus it cuts installation labor costs relative to Case 3. This approach reduces energy use by 42 percent relative to the base case, and pays back in 2.7 years.

Case 5. This option begins to capture some system interactions. Each fixture is equipped with a specular reflector and a new acrylic flat prismatic lens, which are so much better at getting light out of the fixture than the old white-painted luminaire and aged diffuser that the fixture can be delamped by 50 percent, to two T8s and a single two-lamp electronic ballast, while delivering virtually the same amount of light to the task. About one-third of the improvement in light delivery (coefficient of utilization) is due to the better lens and two-thirds is due to the reflector and cooler operating temperature. This approach reduces annual lighting energy use by 67 percent with a 2.7-year payback.

Case 6. Similar to Case 5 with the addition of a ceiling-mounted occupancy sensor, which reduces the annual burn hours from 4,000 to 2,500 and yields energy savings of 80 percent compared to the base case, with a payback just under three years. The bill savings are slightly less in percentage terms because the sensor saves energy but does not reduce demand charges.

Case 7. Now we're getting somewhere, but there's more. Suppose in addition to all the technical changes in Case 6 we institute a maintenance program to group relamp and clean the lenses at 70 percent of lamp life. Because there will be less lumen depreciation from the lamps and less dirt depreciation on the lenses, low-power-factor ballasts can be used, which save additional energy by underdriving the lamps. The total energy savings from this approach is an impressive 83 percent, at a 3.2-year payback. Note that by going to this level of system integration the savings are twice what they were in the "cream-skimming" Case 4 and it takes just a few more months than Case 4 to pay back. Surely with 83 percent of the lighting energy saved there's nothing left to cost-effectively capture. Wrong. Enter the dimming ballast.

Case 8. Now we add to Case 6 dimmable ballasts with their own photocells, installed in the two rows of lamps closest to the windows (zone A). By harvesting daylight the burn hours of the electric lighting in zone A drops to 1,293 hours per year, compared to 2,500 hours in zone B, which doesn't receive enough daylight to cost-effectively utilize dimming. The result is a whopping 86 percent lighting

Table 3-1

Performance comparison of fluorescent retrofit options

Packages of efficiency measures achieve deep savings with attractive economics. Cases 6, 7, and 8 save twice as much energy as the "cream-skimming" Cases 2, 3, and 4, and pay back about as fast.

	Base case **Case 1**	"Energy saving" lamps **Case 2**	T8 + 2-lamp electronic ballast **Case 3**	T8 + 4-lamp electronic ballast **Case 4**	T8 + 2-lamp EB + reflector lens + 50% delamping **Case 5**	Same as Case 5 + sensor **Case 6**	Same as Case 6 + maintenance **Case 7**	Same as Case 6 + dimming ballasts **Case 8**
Design criteria								
Minimum maintained illumination level	25	25	25	25	25	25	25	25
Light loss factors:								
Phosphor degradation	15%	18%	11%	11%	11%	11%	8%	11%
Thermal effects (included in CU)	11%	10%	8%	8%	5%	5%	5%	5%
Dirt depreciation (at lamp changeout)	30%	30%	30%	30%	30%	30%	25%	30%
Luminaire data								
Reflective surface	white paint	white paint	white paint	white paint	spec. reflec.	spec. reflec.	spec. reflec.	spec. reflec.
Distributing media	opal diffuser (aged)	opal diffuser (aged)	opal diffuser (aged)	opal diffuser (aged)	acrylic lens	acrylic lens	acrylic lens	acrylic lens
Coefficient of Utilization (aged)	0.40	0.42	0.45	0.45	0.80	0.80	0.80	0.80
Lamp data								
Description	F40WW T12	F40WW T12/ES	FO32 T8/35	FO32 T8/35	FO32 T8/35	FO32 T8/35	FO32 T8/35	FO32 T8/35
Quantity	48	48	48	48	24	24	24	24
Initial lumens (at rated conditions)	3,150	2,750	2,900	2,900	2,900	2,900	2,900	2,900
Lumens at changeout	2,678	2,255	2,581	2,581	2,581	2,581	2,668	2,581
Ballast data								
Description	2-lamp standard magnetic	2-lamp standard magnetic	R5, 2-lamp E B	R5, 4-lamp EB	R5, 2-lamp EB	R5, 2-lamp EB	R5,2-lamp, low pwr EB	2-lamp EB, dim. EB
Arrangement	2/fixture	2/fixture	2/fixture	1/fixture	1/fixture	1/fixture	1/fixture	1/fixture
Ballast factor	0.88 warm, 0.93 startup	0.88 warm, 0.93 startup	0.88	0.84	0.88	0.88	0.80	0.88
Input watts per fixture	184	156.4	120	106	60	60	50	60
Lighting system data								
Fixture quantity	12	12	12	12	12	12	12	12
Delivered fc (after 1 week of burn)	47	43	49	47	43	43	39	47
Average maintained footcandles (fc)	**28**	**25**	**30**	**29**	**27**	**27**	**27**	**27**
Control method	wall switch	wall switch	wall switch	wall switch	wall switch	ceiling mount ultrasonic	ceiling-mount ultrasonic	sensors + photocells
Maintenance method	relamp at burnout	relamp at burnout	relamp at burnout	relamp at burnout	relamp at burnout	relamp at burnout	relamp at 70% life	relamp at burnout
Total kW	2.208	1.877	1.440	1.202	0.720	0.720	0.600	0.720
Annual burn hours	4,000	4,000	4,000	4,000	4,000	2,500	2,500	1,293/2,500
Annual energy use (kWh)	8,832	7,507	5,760	5,088	2,880	1,800	1,500	1,221
Energy savings (%)	N/A	**15%**	**35%**	**42%**	**67%**	**80%**	**83%**	**86%**
Annual operating cost for energy ($)	883.70	750.74	576.00	508.80	288.00	212.40	177.00	171.87
Operating cost savings (%)	N/A	15%	35%	42%	67%	76%	80%	81%
Installation cost ($)	N/A	312	1,440	1,020	1,620	1,970	1,970	2,300
Simple payback (years)	N/A	**2.4**	**4.7**	**2.7**	**2.7**	**2.9**	**3.2**	**3.2**

Note: See endnote [1] for assumptions used in the analysis of retrofit options.

energy savings for the room, with an overall payback of 3.2 years. Note that with enhanced design daylight might be transmitted even farther back into the space and could yield enough savings in zone B to be cost-effective. On the other hand, we assumed a relatively inexpensive and unproven dimming ballast technology for this example. Other systems we could have assumed, such as the Lutron or Advance Mark VII systems, would involve more control wiring, a controller, and separate photocells around the room, leading to a higher cost than the incremental $330 we assumed in this example.

As dimming ballasts become more widely available, their price is expected to fall. If and when that occurs, they may become cost-effective even for applications in nondaylit core zones. Why? First, they will allow "task tuning" to the various light levels needed in a space. Walkways in an open-plan office, for instance, can be adjusted for 10 or 20 percent less ambient light than work areas. Dimming ballasts will automatically adjust the power provided to the lamps according to the lumen depreciation status of the lamps. That means lamps don't have to be overdriven early in their life to ensure that they will produce sufficient light late in their life. In addition, dimming ballasts can be tied to scheduling controls that would dim light levels slowly, slightly, and generally imperceptibly during periods of peak demand, when power is most costly.

Figure 3-1

Illustrative open-plan office

We assume an open-plan office area 39 feet long, 29 feet wide, with 9 foot, 2 inch ceilings. The space receives natural south light from extensive glazing on one of the 39-foot walls. Electric lighting is provided by 12 ceiling-mounted 2 x 4 foot fluorescent fixtures. In Case 8, dimming ballasts are used with photocells and we assume that daylight reduces electric lighting use for the two most southerly rows of fixtures, 20 feet into the room (Zone A).

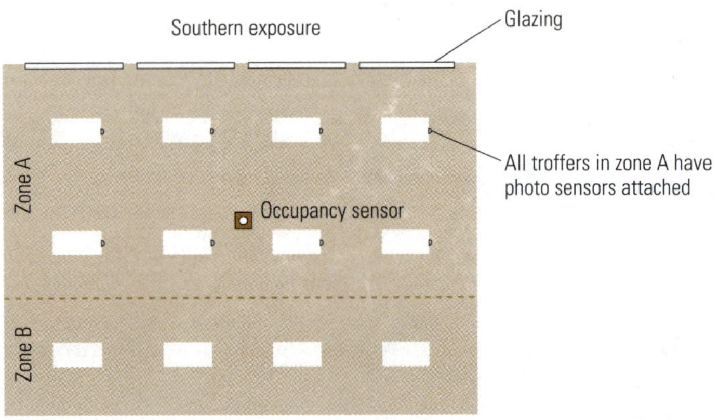

The deepest savings come from complete packages of measures that also include non-technology options such as improved maintenance practices. An analysis that failed to take advantage of interactions by reducing design margins would show substantially less savings.

On the other hand, the incremental economics of successive measures, if viewed separately, may look relatively poor because the remaining energy use gets smaller and smaller, making a smaller "pie" from which to remove further slices of savings. This kind of marginal analysis will lead to smaller savings than analyses of complete measures. For example, adding occupancy sensors to Case 5 costs an additional $350 for this room, but only saves an additional $76—a simple payback of 4.6 years. However, the complete package economics only slip from a 2.7-year payback to 2.9 years, so the option would still fit within a 3-year payback criterion.

Fluorescent dimming systems frequently fall victim to this kind of economic analysis, because the equipment is currently expensive, and other highly cost-effective measures are usually contemplated first. In Case 8, the dimming equipment costs $330 compared to Case 6 and saves only $40 per year, for an incremental payback of 8.1 years. When bundled with all the other measures, however, the package has a respectable 3.2-year payback.

High incremental costs should thus not dissuade the lighting user from considering the savings that can be had with dimming systems, whether they are used with perimeter daylighting, for core-zone occupancy- and schedule-controlled dimming, or for lumen maintenance. The cost of dimming ballasts has been long predicted to fall, but has yet to do so. If and when it does, dimmable efficiency upgrade *packages* may be the ticket to the deepest energy savings.

The analysis above doesn't include some of the nonenergy benefits that can accrue to whole-system lighting solutions. Counting these kinds of benefits can make a marginally cost-effective lighting package look a lot more attractive. Some of the most important nonenergy benefits are summarized here and discussed more completely later in this chapter:

- *Indirect savings of cooling energy use and plant downsizing.* Many building owners are replacing or converting their chiller systems because CFC refrigerants are being phased out. By reducing the cooling load, efficient lighting can allow replacement chillers to be smaller and hence less costly to buy and operate. Even in cases where no changes in HVAC equipment are envisioned, lighting upgrades in most commercial buildings are likely to reduce cooling costs more than they increase heating costs.

- *Environmental risk mitigation and reduced disposal costs.* Replacing older (pre-1980) ballasts makes sense purely from a risk management point of view because many of them contain PCBs. Going to integrated reflector retrofits that allow the number of lamps and ballasts to be reduced has the often-ignored benefit of reducing future lamp and ballast disposal or recycling costs.

- *Productivity benefits.* If it is done right, efficient lighting can also improve lighting quality. If that improvement has even a tiny beneficial impact on worker productivity, the economic benefit can easily exceed the entire cost of the lighting improvement (see **3.5**).

3.2 Effects on Other Building Systems

Lighting influences other building systems in a variety of ways, ranging from changes in heating and cooling loads to effects on power quality.

3.2.1 Interactions with heating and cooling systems

Except for the small percentage of light that escapes through windows, all energy used by interior lights ends up as heat inside the building. Although some of this heat may be picked up by the return air stream and directly exhausted, and a small fraction may be absorbed by the building mass and later released during unoccupied times, lighting typically constitutes the largest internal heat gain in commercial buildings. This internal heat gain is useful when the building requires heat, but is unwanted when the building requires cooling. Energy efficient lighting adds less heat to a space per unit of light output than inefficient lighting. The net effect on HVAC systems can be expressed in terms of annual site or source energy, peak kW demand for cooling, required cooling capacity, energy bills, or environmental impacts.

Large buildings in hot climates may have net annual site energy HVAC bonuses of 40 percent of more.

The net annual effect on site HVAC energy use depends on the type of building, the climate, the heating and cooling systems, and the relative size of the heating and air-conditioning loads. Large buildings that are dominated by internal loads and use far more air conditioning than heating can experience a site energy HVAC bonus of 40 percent or more, meaning that each kWh of reduction in annual lighting energy use yields an additional 0.4 kWh of annual reduction in HVAC energy, by reducing cooling energy more than it increases required heating energy. For small, envelope-dominated buildings—especially those in cold climates—the net impact is often an HVAC penalty, meaning that each kWh in lighting energy use increases HVAC energy by increasing the annual heating energy use more than it reduces cooling energy. The precise impact on any given building is best determined by computer simulation, although some simplified methods for approximating the effect are discussed below.

Even if a building has an HVAC penalty in terms of annual site energy, it may experience a net reduction in energy bills, source energy, or pollutant emissions. If the building is heated with natural gas that costs a lot less than the electricity used for cooling, a small reduction in cooling energy can save more money than a relatively larger increase in heating energy costs. Moreover, the fuel-to-delivered-electricity losses of the electric power system mean that each kWh saved by the customer saves three to four units of source energy at the power plant, with a correspondingly leveraged reduction in pollutant emissions, particularly if the power plant burns coal, which emits more pollution than natural gas. Thus, for example, a small commercial building in a cold climate that has a net site HVAC penalty of 25 percent, may have a net economic HVAC bonus of two percent, a source energy bonus of 10 percent, and perhaps an even higher bonus in terms of reduced pollutant emissions.[2]

The reduction in summer peak cooling demand is often greater than the net HVAC energy bonus, if there is one. In the predecessor edition to this volume, Competitek used parametric simulations to determine the HVAC/lighting interactions for a 60,000-square-foot office building and a 12,000-square-foot retail store in Austin, Texas. This analysis found a net annual electricity bonus of 35 percent, and a net summer peak electricity demand bonus of 52 percent.[3]

There are several reasons why the summer peak benefits often will be greater than the cooling energy bonus. Because one doesn't heat in midsummer, no lost contribution to useful space heating need be subtracted from the peak cooling bonus (except with some reheat systems—see Figure 3-2). If daylight dimming is part of the lighting system, some of the lights won't be on much in the morning and early afternoon, thus reducing heat buildup in the hours leading up to the afternoon peak. In systems equipped with variable-speed drives, delivering less cooling means slower operation of pumps and fans, whose power consumption depends roughly on the cube of the flow they must deliver. Air conditioners are less efficient in the hottest weather, so using them less during peak periods saves proportionately more power than reducing their use at other times. Whether the peak demand reductions will lead to corresponding reductions in the demand charges on the electricity bill depends on whether the reduction occurs consistently throughout the period upon which the demand charges are based.

Harvested Cooling Capacity

Saving 1 kW of lighting removes up to 3,412 Btu/hour of heat. With 1 ton of cooling capacity equal to 12,000 Btu/hour, each saved kW of lighting reduces the cooling load by:

$$\frac{1\ \text{kW} \times 3{,}412\ \text{Btu/hour per kW}}{12{,}000\ \text{Btu/hour}} = 0.28\ \text{tons}$$

Assuming that a lighting retrofit costing $2/ft^2 reduces lighting demand from 3 watts/ft^2 to 1.2 watts/ft^2. This yields 1.8 kW of lighting savings per 1,000 square feet of upgraded space. That in turn equals a 1/2-ton reduction in required cooling capacity:

$$1.8\ \text{kW} \times 0.28\ \text{ton/kW} = 1/2\ \text{ton}$$

Assuming that new cooling capacity and its associated ducts, electric service, drainage, and so on costs at least $3,000 per ton installed, the lighting retrofit virtually pays for itself in freed-up or forgone expansion of cooling capacity:

$3,000/ton x 1/2 ton per 1,000 ft^2 = $1.50 per ft^2, *which is nearly as much as the total cost of the lighting upgrade.*

While the net impact on HVAC energy and summer peak cooling demand may or may not be significant in a given building, more efficient lighting will always reduce the amount of cooling capacity needed to cool the building. Since cooling capacity is very expensive ($3,000 or more per ton installed) lighting retrofits can virtually pay for themselves by stretching the capacity of the existing cooling system, counterbalancing capacity reductions that can accompany conversion to non-CFC refrigerants, foregoing the need to expand cooling capacity to serve new loads, or allowing a new cooling system to be made smaller than it otherwise would be (see sidebar on previous page for a sample calculation).

In buildings with electric resistance heating, improving lighting efficiency will only switch the job of heating from one electric heater—the lights—to another—the electric resistance coils. However, if the building is heated by an electric heat pump, the heating system will have a better efficiency than the lights, or if the building is heated by natural gas, the cost of heating with the main system may be far less than the cost of heating with the lights. In the worst-case scenario, the building heating system cannot satisfy the load without the aid of the heat provided by lights. While this scenario is unusual, improving lighting efficiency in this case would result in the need to increase the heating system capacity.

Many HVAC systems deliver cold air throughout the building and use reheat coils or heating ducts to warm air back up to comfort conditions in particular spaces. Care should be taken to adjust reheat coils or increase the supply air temperature after lighting retrofits to maintain comfort and avoid energy waste. This should be part of the checklist for any lighting retrofit. Otherwise, people who were previously comfortable in summer may be uncomfortably cold after the lighting retrofit.

Consider the constant-volume system with electric reheat shown in **Figure 3-2**—an inefficient but not uncommon type of air-conditioning system. Without increasing the supply air temperature after a lighting retrofit, the reheat coil will have to use more energy to make up for the reduced internal load. In this situation, there is no HVAC cooling bonus, and some of the electricity savings from the lighting retrofit is actually negated. To avoid this problem, reset the supply air temperature to a warmer temperature, or, better yet, combine the lighting retrofit with a switch to a variable-air-volume air-conditioning system that avoids simultaneous heating and cooling (see section 5.4 of the *Space Cooling Technology Atlas*).

The decrease in cooling load from efficient lights may offset the reduction in cooling capacity that can accompany the switch to some non-CFC refrigerants.

Quantifying HVAC/lighting interactions. Researchers have made various attempts to quantify the interactions between HVAC systems and internal gain reduction resulting from lighting and other efficiency improvements. The HVAC system effects make it difficult to develop a single simplified method

Figure 3-2

Interaction of lighting and HVAC systems with electric reheat

Cooling coil discharge temperatures may need to be raised after a lighting retrofit to maintain comfort and ensure that electric reheat systems don't negate the lighting savings.

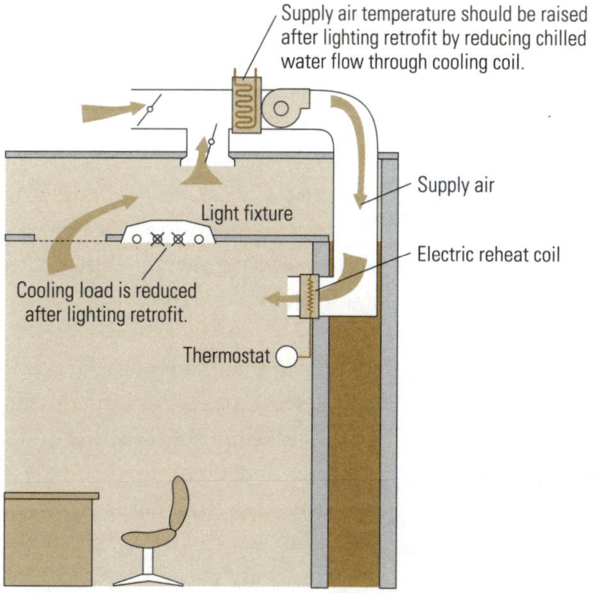

that will quantify the interactions for all building types. Detailed building simulation, which can account for system effects including potential cube-law fan and pump energy savings, is the best method for predicting the HVAC/lighting interaction in a given building. Some of the results of research to date are presented below.

In one study, weather data from 17 major cities was used to simulate the energy consumption of a 46,000-square-foot, two-story office building in different climates using different internal gains.[4] The building used a closed-loop water-source heat pump system, divided into sixteen zones with two heat pumps in each zone, and employed a 40,000-gallon storage tank to minimize the need for supplementary heating or cooling.

The total annual HVAC energy consumption for the building (**Figure 3-3**) shows that in hot climates like Miami and Houston, an HVAC energy savings bonus is achieved when reducing the internal gains in this building. In cold locations like Minneapolis, however, this building's HVAC energy consumption increases substantially when the internal gains are reduced to a minimum. This example accentuates the heating penalty because it is an electrically heated building and does not have the potential for cube-law cooling savings with pumps and fans that a more conventional commercial building HVAC design would experience.

Other studies have found HVAC bonuses of 38 percent in Fresno, California and Chicago office buildings,[5] and 42 percent in a Ventura, California video store.[6]

Researchers at Lawrence Berkeley Laboratory have used simulations to develop a method for calculating the heating and cooling load and demand changes due to a reduction in lighting internal gains.[7] The LBL method does not include the effects of the HVAC system, but only the changes to load. In order to calculate changes in energy use, assumptions must be made about the total system COP or efficiency. Further adjustments would be necessary to account for an economizer cycle.

Using the LBL approach and assuming an average cooling system COP of 2.4 (including all auxiliary and distribution systems) and a seasonal heating efficiency of 0.75, the annual cooling energy would be reduced by 34 to 38 percent of the annual lighting energy savings in a large office building in a hot climate, and 20 to 25 percent of the annual lighting energy savings in a cold climate.[8] The annual heating energy use would be increased by 0 to 5 percent of the annual lighting energy savings in a hot climate, and 33 to 47 percent of the annual lighting energy savings in a cold climate. This results in a net site energy effect of a 29 to 37 percent bonus in hot climates and a 13 to 27 percent penalty in cold climates. Similarly, the cooling peak demand

Figure 3-3

Commercial building HVAC energy use vs internal gains for six cities

Computer simulation of a 46,000-ft² commercial building shows that changes in internal gains can either increase or decrease the total HVAC energy required, depending on the climate. Because this is an electrically heated building, the heating penalty is larger and the cooling bonus smaller than would be the case in a similar building heated with gas.

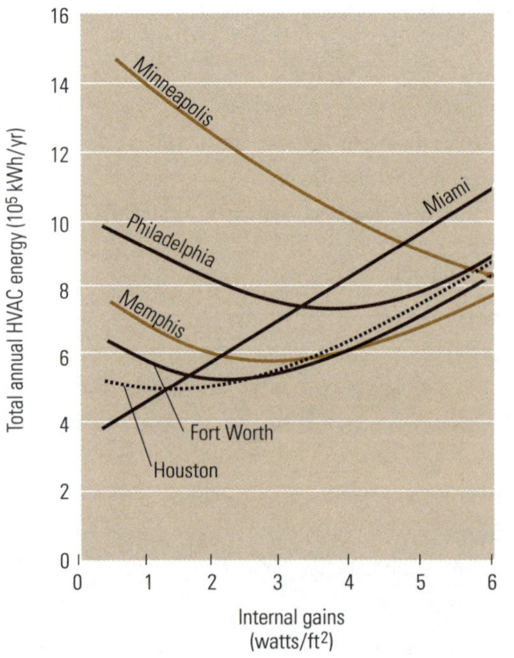

Source: Howell [4]

would be reduced by 33 to 35 percent of the lighting power reduction in both hot and cold climates, and the peak heating demand would be increased by as much as 18 percent of the lighting power reduction in a hot climate and 38 percent of the lighting power reduction in a cold climate.

Another simplified method for determining HVAC and lighting interactions was developed by R.A. Rundquist and others.[9] The Rundquist paper includes worksheets and cooling and heating fractions for 153 cities across the United States. According to this method, again assuming a total cooling system COP of 2.4 and a seasonal heating efficiency of 0.75, the cooling energy in a large office building with economizer would be reduced by 23 to 30 percent of the lighting energy reduction in a hot climate, and about 13 to 15 percent of the lighting energy reduction in a cold climate. In a large office building without economizer the cooling bonus is larger—30 to 36 percent of the lighting energy savings in a hot climate and 28 percent in a cold climate. The heating energy use would increase 0 percent in a hot climate and 18 to 21 percent in a cold climate. By this method, the net site energy effect ranges from a 23 to 36 percent net cooling bonus in hot climates, and from a 7 percent penalty to a 10 percent bonus in cold climates.

The HVAC bonuses or penalties presented above are representative of typical large office buildings. Various building and HVAC system particulars can drive these numbers higher or lower. As shown in the Rundquist calculations, the cooling bonus will be significantly higher if the building does not utilize an economizer cycle. Similarly, if the total cooling system COP were lower, as is the case with many older rooftop cooling units, the cooling energy bonus would be greater, or if the heating efficiency were higher, as is the case for electric resistance heating, the heating energy penalty would be lower. Furthermore, buildings that are more envelope-dominated, like small offices, supermarkets, or schools, will have greater heating penalties and smaller cooling bonuses than large office buildings.

Regardless of these assumptions, if the heating is supplied by natural gas rather than electricity, the HVAC interaction usually results in a net economic, source energy, and environmental bonus even if a site energy penalty is calculated. For example, another study from LBL found that for a 160,000-square-foot, eight-story office building, heated with natural gas and cooled with electricity, the HVAC interaction based on economics varied from a 2 percent penalty to a 16 percent bonus, even though the energy penalty was as great as 30 percent.[10]

Common sources of error in the calculation of HVAC effects. Finally, LBL researchers also offer some caveats regarding the calculation of HVAC bonus or penalty. HVAC interactions depend heavily on the HVAC system configuration and operation, and part-load plant efficiencies. LBL researchers found that simplified calculations of HVAC interactions often overestimate the cooling savings and underestimate the additional heating requirements resulting from improved lighting efficiency. They found that one of the most significant contributors to error was the use of rated input lighting power rather than an application-specific value that takes into account thermal and component interaction effects.[11] For example, a typical lighting troffer containing four 40-watt T12 fluorescents and two magnetic ballasts has a rated power draw of about 192 watts, but in the overheated conditions in which many such fixtures operate they will draw closer to 170 watts. Using the rated value as the "base case" leads to significant overestimates of energy savings in many efficiency retrofits. This common error has more to do with misunderstanding of lighting energy input than calculations of HVAC interactions, but it apparently occurs quite often.

The assumption that all of the heat from the lights contribute to zone internal gains is another source of error. Researchers at the National Institute of Standards and

Technology (NIST) found, for example, that if lights are recessed and unvented, about half of the heat generated is picked up by the return air stream. In many buildings, about 90 percent of the return air is recirculated. In those buildings that use economizers to provide cooling with outside air, however, a significant amount of the lights' heat may be directly exhausted rather than adding to the internal gains of the space.[12]

THE LIGHTING AND HVAC INTERACTION TEST FACILITY

The National Institute of Standards and Technology in Gaithersburg, Maryland has constructed a full-scale test facility to simulate an office space. The test facility is extensively instrumented and can be used to test numerous lighting and HVAC systems configurations. A detailed computer model of the facility has been developed to extend the measurement results.

Over the past few years, the test facility has been used to determine energy transfer from lighting systems to building spaces and to evaluate the performance of various lighting and HVAC equipment configurations under typical operating conditions.[13] For more information on the work of this facility, contact:

Stephen Treado
Building Environment Division
National Institute of Standards and Technology
Gaithersburg, MD 20877
tel 301-975-6444

3.2.2 Power quality effects

Lighting systems can influence and be influenced by the power factor and harmonics characteristics of the electrical environment in the building.

Power factor. Power factor is an indicator of how much of a power system's capacity is available for productive work—making it an important and constant concern for utilities and large power users alike. There are two types of power factor: phase power factor and total or true power factor.

Phase power factor is the classic definition commonly referred to in power system texts and is often referred to as "displacement" power factor. Phase power factor examines only the fundamental (60 Hz or 50 Hz) voltage and current and measures to what degree the two waveforms are out of phase, and is equal to the cosine of the phase angle difference between voltage and current. Voltage and current that are perfectly in phase (as shown in **Figure 3-4**) are said to have a phase difference of zero degrees, and therefore a phase power factor of 1 (or "unity"). Purely reactive loads will have phase angles of 90 degrees (as shown in **Figure 3-5**) and will carry phase power factors of zero.

Phase power factor is also equivalent to the ratio of watts to volt-amperes (or of real power to total apparent power), and reflects how much of the power drawn by the load is "real"—in phase with voltage—and thus able to do work. It is the ratio of the actual power being used in a circuit to the power which is apparently being drawn from the supply line.

The higher the power factor, the larger the in-phase component of the current and voltage. The lower the power factor, the greater is the out-of-phase component of reactive current. This reactive power cannot do work but requires sufficient capacity to carry it throughout the grid and the local distribution system, and incurs its share of system losses. Though out-of-phase current cannot do real work, it has real and large costs.

Total power factor (also called true power factor) reflects the fact that electrical power has a shape component, indicating waveform distortions such as harmonics which, like out-of-phase current, dissipate energy against any resistance but cannot do work. The irregular waveforms commonly found in power grids can be considered as the sum of the fundamental frequency and of various harmonics, along with nonperiodic noise. Because the harmonics and noise are generally out of phase with the fundamental-frequency current, they too cannot do useful work and thus degrade power factor. Total power factor thus embraces two distinct kinds of out-of-phase current—the phase component, which is recurrent and easy to measure, and the shape component, which is just starting to be widely appreciated, and can be quite difficult to measure.

Harmonics. Harmonics are voltage and current frequencies (usually in integral multiples of the fundamental frequency) riding on top of the normal voltage and current waveforms. Current harmonics are induced onto the power line by nonlinear electronic equipment and systems (including lighting ballasts). Voltage distortion typically results from current distortion reacting with system impedance. These waveforms, usually lumped together as total harmonic distortion (THD), perform no useful work and can be a significant nuisance.

Many utility DSM programs limit their incentives to lighting products with high power factor (>0.9) and low harmonic distortion (<0.32), and with good reason. Products with low power factor and high levels of harmonics can cause problems in the utility and customer facilities. Low-power-factor equipment requires more current to perform the same work and can overload wiring or cause greater resistive losses in wires. More transformer capacity gets used, eliminating its availability for new equipment or higher system loads. In extreme cases, overall system voltage may drop, affecting the operation of other equipment supplied by the same transformer. And if the utility has a separate charge for low power factor, the customer's bill will be higher. High harmonics can cause interference with other sensitive equipment and may induce hazardous currents in the neutral wiring of the building which can increase the risk of fire and electrical shock.

Many utility DSM programs limit their incentives to lighting products with high power factor (>0.9) and low harmonic distortion (<0.32).

A related problem can occur when partially altering a lighting system. Lighting uses single-phase power, so a building's lighting load is typically divided equally among all three phases of an electric service. If the lighting load is reduced on one phase but not the others, current may be induced in the neutral line, creating a hazard and greater line losses.

Figure 3-4

Voltage and current for a purely resistive load

In-phase voltage and current waveforms.

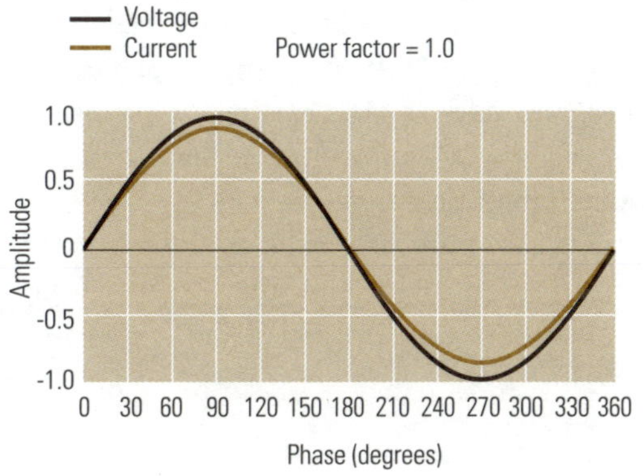

Figure 3-5

Voltage and current for a purely reactive load

Voltage and current waveform are 90° out of phase in this inductive load.

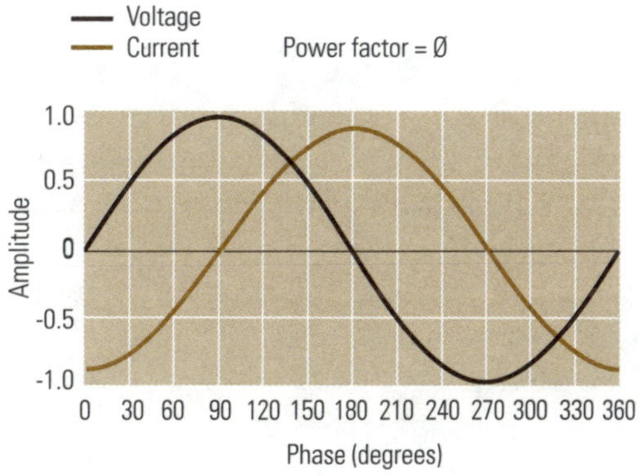

As an example of what can go wrong by ignoring these issues, consider the impact of a lighting retrofit that raises light levels by replacing incandescents with a power factor of 1.0 with compact fluorescents or HID lighting that have a low power factor of 0.35. The base case is a dimly lit lobby with twelve 100-watt incandescent fixtures, each drawing about 0.8 amps, for a total draw of 9.6 amps. One could double the light levels with very little increase in kWh use by replacing the incandescents with 50-watt high pressure sodium lamps with magnetic ballasts that each dissipate 8 watts and have power factors of 0.35. From an energy point of view, this seems like a good strategy. But each of the HPS lamps will draw 1.38 amps; twelve of them will draw 16.5 amps, for a 7-amp (72 percent) increase in wire load, which may lead to overloaded circuit breakers and possibly burned-out wall switches.

To avoid such situations, discuss pending renovations with a facility's electrical maintenance supervisor to be sure loads can be balanced once the job is done. Use low-harmonic (<0.32) and high-power-factor (>0.90) electronic ballasts. Note that so-called normal-power-factor magnetic ballasts typically have power factors of 0.4 to 0.6 while high-power-factor magnetic ballasts have power factors of 0.8 to 0.9. Many electronic ballasts have power factors of 0.95 or higher. Be particularly aware of these issues when replacing incandescents, which have a power factor of 1.0. In these cases one should consider using fixtures or retrofit devices that have built-in power-factor correction capacitors, or high-quality electronic instead of magnetic ballasts.

Increased concern from utilities and users sensitive to power quality are creating a stronger market for equipment with "clean" power quality characteristics. The European Community recently adopted a regulation requiring all electronic equipment over 50 watts sold in the EC to carry power-factor correction components.[14]

HARVESTED ELECTRICAL CAPACITY

Upgrading to higher efficiency and higher-power-factor equipment can often pay for itself by freeing up expensive electrical service capacity. For example, replacing two T12 F40 lamps driven by a 0.85-power-factor magnetic ballast with one T8 lamp and reflector driven by a 0.97-power-factor electronic ballast saves 81 volt-amperes in ballast and lamp savings:

$$\left(\frac{96\text{ W}}{0.85\text{ PF}}\right) - \left(\frac{31\text{ W}}{0.97\text{ PF}}\right) = 81\text{ VA}$$

Since electrical service (transformers, switchgear, feeders, circuit breakers, etc.) costs \$100 to \$200 per kVA, each such ballast and lamp changeout at 81 VA is worth \$16.20, or nearly as much as the electronic ballast costs:

$$\left(\frac{81\text{ VA}}{1{,}000\text{ VA}}\right) \times \$200/\text{kVA} = \$16.20$$

In a recent real-world example, Columbia University recently found that lighting upgrades in two adjacent buildings freed up sufficient electrical capacity to allow ganging them onto one building's electric service. Purchasing the transformers for that one system then became a cost-effective way to place both structures on high-tension service, saving another 29 percent on the two buildings' electric bills. When the costs and savings of this option were combined with that of the lighting upgrades, overall payback remained at about 2.5 years, but total savings nearly doubled, demonstrating the synergistic effects that can result by using lighting upgrades for electric load management. This option was previously unavailable because purchasing the transformers for *both* buildings yielded an unacceptably high payback.

3.3 ENVIRONMENTAL ISSUES

Fluorescent lamps and high-intensity discharge lamps contain mercury, and pre-1980 ballasts contain polychlorinated biphenyls (PCBs) as an insulating material. People usually don't consider lamps and ballasts as potentially hazardous to the environment, and most discarded lamps and ballasts end up in landfills.

3.3.1 Lamp disposal

More than 500 million fluorescent lamps and a smaller but significant number of high-intensity discharge lamps are discarded in the U.S. each year.[15] Proper disposal or recycling of these lamps is important since they represent the second largest source of mercury in the municipal solid waste stream (5 percent), behind batteries (88 percent). As mercury is phased out of alkaline batteries over the next several years, lighting products will become the largest source of mercury in the municipal waste stream.

Although lamps are a significant share of the municipal waste stream's release of mercury, other sources are collectively larger, including industrial processes, mining, emission from fossil-fuel-fired power plants, and golf course fungicides. For this reason, proper lamp disposal has historically been a low priority for regulatory agencies. That is beginning to change as state and federal regulators are focusing increasing attention on lamp disposal issues in hopes of heading off environmental and health problems.

MERCURY FROM POWER PLANTS EXCEEDS RELEASE FROM LAMPS

Incandescent lamps contain no mercury, but fossil fuels burned at power plants do. Because incandescents are less efficient and require more fuel at the power plant, they lead to the release of 2 to 10 times as much mercury as using a fluorescent lamp serving the same task, even assuming that all of the fluorescent lamp's mercury is released to the environment.[16] As discussed in this section, proper disposal or recycling of fluorescents can ensure that the mercury contained in the lamp is not released to the environment.

Typically, problems occur when solid mercury leaches into groundwater or airborne mercury comes down with precipitation. Fish are then exposed to contaminated waters and are eaten by other animals and people. Certain forms of mercury are highly toxic and can cause permanent neurological and kidney damage in wildlife and humans. Even small releases of mercury can cause problems because it becomes highly concentrated as it passes up the food chain. In some northern Minnesota lakes with average concentrations of 2 nanograms per liter, the average concentration in the tissues of mature northern pike reaches 450 nanograms per gram—a bioconcentration factor of 225,000.[17] This creates a situation in which it may be safe to drink the water from a lake but not safe to eat the fish. This phenomenon has caused some states with large areas of wetlands and lakes to establish strict mercury disposal regulations. Twenty states have issued health warnings banning consumption of fish from mercury-contaminated lakes.

Because incandescents are less efficient and require more fuel at the power plant, they lead to the release of 2 to 10 times as much mercury as using a fluorescent lamp serving the same task, even assuming that all of the fluorescent lamp's mercury is released to the environment.

Mercury poses no threat as long as it is contained in the unbroken lamp. When the lamp gets broken during collection, transportation, or disposal (where many lamps are crushed), the mercury is released. Even this release may not pose an immediate toxicity problem because by the time a lamp reaches the end of its useful life most of the mercury vapor has been absorbed into the phosphors and the lamp glass. This means that accidental breakage is unlikely to expose people to significant quantities of free mercury vapor. It is the leaching in the landfills and conversion by bacteria into volatile forms like methyl mercury, or improper incineration of the lamps which create potential environmental problems. For these reasons, proper handling and disposal or recycling of lamps is becoming a major issue.

Fluorescent lamps are the second largest source of mercury in the municipal solid waste stream.

Current EPA regulations require fluorescent lamps to be disposed of in a hazardous waste landfill or mercury reclamation facility. Enforcement is left to the states, however, and is ignored in many jurisdictions.

Regulation in the U.S. and Canada. In both the U.S. and Canada, there are no federal guidelines specifically tailored for lamps, which means all lamps must be treated as any other waste that contains mercury. In the U.S. some controversy exists over lamp disposal due to the current testing procedures for determining toxicity levels. According to current EPA regulations, most fluorescent lamps fail a standardized leaching test and must therefore be disposed of in either a hazardous waste landfill or a mercury reclamation facility. Enforcement is left to the states, however, and is ignored in many jurisdictions.

The EPA is presently determining whether or not lamps will be exempted from the hazardous waste laws or placed in a special regulatory category. Two basic options were proposed in a Notice of Proposed Rulemaking (NOPR) in July 1994: (1) categorization as a class of "universal wastes" that although hazardous, can be collected at a centralized location and then taken to a hazardous waste or reclamation facility, or (2) exclusion of fluorescents from consideration as a hazardous waste if disposed of in a municipal landfill, but with incineration in municipal combustors prohibited.[18] There is no federal regulation specifically addressing the transport of lamps to disposal sites.

In 1990, the Resource Conservation and Recovery Act (RCRA) determined a toxicity testing procedure for mercury-contaminated wastes. This "leachate test" is intended to mimic the leaching of wastes into the groundwater from a landfill. A concentration of more than 0.2 milligrams of mercury per liter of water indicates a failing grade. Most fluorescent, mercury vapor, and metal halide lamps fail this test, and subsequently must be considered potential hazardous waste. There may be exceptions, but it is wise to assume that all fluorescent and HID lamps will fail. Many waste generators are unaware of this, and simply throw away lamps like any other solid waste.

EPA's 1994 NOPR generated considerable controversy among lamp manufacturers, and EPA has not yet moved forward with this process. As a result of actual field tests at leaching landfills, both EPA and its critics have also raised questions about the seriousness of the leaching problem. According to EPA, "the available data on landfill leachate suggests that mercury-containing lamps may not pose a threat to groundwater when placed in a state-controlled municipal landfill due to the low levels of mercury found in landfill leachate." On the other hand, EPA estimates that incineration accounts for more than 99.999 percent of the airborne mercury from discarded lamps. It would therefore make more sense to regulate lamp incineration to a much greater degree than landfilling. EPA estimates that about 14 percent of lamps are presently being incinerated.[19]

A facility that produces less than 100 kilograms (220 pounds) per month of mercury wastes is considered a small-quantity generator and will avoid much of the regulation regarding storage and treatment of that waste. The maximum quantity of waste that a generator is allowed to stockpile on site is 1,000 kilograms (2,200 pounds), which translates to about 4,000 broken four-foot T8 lamps.The EPA has a handbook entitled, "Understanding the Small-Quantity Generator Hazardous Waste Rules: A Handbook for Small Business" which has an excellent explanation of the lengthier "Code of Federal Regulations," as well as names and addresses of federal and state authorities.

Having put lamps into the category of potential hazardous waste, the U.S. federal government then allows individual states to define stricter rules. It is also the states' responsibility to enforce proper disposal practices. Some states have developed specific guidelines for lamps, but most states allow lamps to fall under the hazardous waste umbrella. Usually, the rules are based on a minimum quantity generated per day or per month.

California allows up to 25 lamps per user per day to be disposed of in landfills. If discarded during normal spot relamping, such an exemption would eliminate most commercial facilities with areas under about 200,000 square feet. Additional lamps must be recycled or sent to a mercury reclamation facility. While this approach clearly leaves room for abuse (25 to the landfill, one slipped in the neighbor's dumpster), many businesses in California that generate fewer than 25 spent lamps per day are going the recycling route.[20] More than 600,000 lamps per month are being recycled in California.[21]

In Minnesota, fluorescent and HID lamps from businesses are banned from landfills and household lamps will be banned after August 1994.[22] Massachusetts, Oregon, and Vermont are now considering laws specifically for lamp disposal. Florida is developing a code dedicated to lamp disposal and in the interim has a small-generator exemption of 10 lamps per month. Since the laws are changing, and each individual state has the authority to have stricter mercury waste laws than the federal government, it is important to check with state authorities to determine what the rules are in a given location.

Facility managers should also carefully consider the on-site impacts of lamp disposal. Some facilities have tried to reduce their disposal costs by crushing lamps and sending the results to a hazardous waste landfill, but the federal Occupational Safety and Health Administration (OSHA) has not certified any lamp crushers as meeting its airborne mercury standards. As a result, employee contamination could result in an OSHA (not EPA) fine.

In Canada, the federal authority, Environment Canada, has similarly chosen to leave the provinces in charge of regulating the disposal of mercury waste. The provinces in turn have formed a Canadian Council of Ministers which is presently standardizing the rules for wastes in order to retain some consistency throughout the country.[23] The term "special waste" is used for any waste which requires special treatment. Again, as in the U.S., there are no rules specifically for lamp disposal.

At the provincial level there are two testing procedures to determine if there are sufficient amounts of mercury to classify a waste as special waste. The first determines the total amount of mercury by weight. If there is more than 100 parts per million of mercury in a lamp, then the lamp fails the test. The other method of testing is a leachate extraction procedure similar to that used by RCRA in the U.S. For the present time, the quantities of mercury in fluorescent lamps is still below the allowable levels for both of these tests.[24] This means lamps are considered municipal solid waste. There is a practical side to this policy. There are no functioning lamp recycling facilities in Canada, and in some provinces there are no special waste landfill sites.

Most provinces do not have a lamp disposal policy. Ontario and Quebec are currently reviewing the classification of lamps as hazardous waste, but they now are considered nonhazardous. There may be limits set by disposal companies regarding the number of lamps allowed in a dumpster, but there do not appear to be limits set for municipal landfills. A desire to be a good corporate citizen may be the primary factor in prompting generators to transport lamps to the nearest recycling facility. There is no problem transporting lamps across the international border to the United States, because a hazardous waste manifest is not necessary.[25] The cost of transportation, however, can be equal to or greater than the cost of the recycling (see recycling section on next page).

An additional wrinkle is the debate over whether lamps should be considered hazardous waste or hazardous products. This is more than an issue of semantics. If lamps are hazardous *products*, then the *manufacturer* is the originator of the waste. If they are hazardous *waste*, then the *building owner* is the originator. While this debate continues, used lamps are viewed as waste, and building owners and managers must take responsibility for them.

It is up to the building owner to ensure that lamps are disposed of properly.

Disposal and recycling options. Unfortunately, once materials leave a job site, the owner may no longer be aware of how the waste is treated. Merely knowing where the materials are going allows companies to be more accountable in their waste disposal practices. Current confusion with regard to the classification of lamps as hazardous waste means that many states lack any serious enforcement regulations, and lamps can easily end up being incinerated or buried in the municipal landfill. At the present time it is up to the building owner to ensure that lamps are disposed of properly. Some building owners are taking the initiative and having their lamps recycled even though there is no immediate economic justification for doing so. This is largely because they wish to avoid potential costs in the future associated with cleaning up mercury-tainted landfill sites. Here are the options:

■ *Incineration.* Incineration can be the most environmentally damaging method of dealing with mercury-containing waste, because it produces highly volatile mercury vapor. Incinerators range widely in their ability to capture mercury vapor. Incinerators now used in Minnesota capture only 10 to 50 percent.[26] EPA has found that activated carbon injection technology can reduce mercury levels by 80 to 90 percent.[27] Mercury in a vapor state is the most dangerous for assimilation into the environment and human tissues.

■ *Landfilling.* Disposal of lamps in standard landfills without leachate collection systems may allow the oxidized mercury to seep into the ground water. Disposal of lamps in a landfill suitable for hazardous waste prevents the unconfined release of mercury into the rest of the environment.

■ *Recycling.* The process of breaking down each lamp into its constituent parts varies according to the recycler. Some facilities use recycling as a method to lower the volume of mercury-tainted waste. This means that the glass tubes and metal endcaps will be recycled, while the small volume of remaining mercury and phosphors will be stored or disposed of as hazardous waste. Other facilities take this process one step further by removing the mercury so that no heavy metal byproducts remain. Mercury Technologies Corp. has a process which separates the mercury from the phosphors. The phosphors are then reusable, and the mercury undergoes a purification process. There are markets for both the phosphors and the mercury, and little waste remains. Two firms in the U.S. accept fluorescent and HID phosphors for further refining: Mercury Refining in Latham, New York and Bethlehem Apparatus in Hellertown, Pennsylvania. Before enlisting a recycling company, it is wise to know what happens to the waste, to avoid future clean-up responsibilities. **Table 3-2** lists North American lamp recycling companies.

With any of the recycling processes there is some controversy over the residual mercury left in the glass itself as well as the aluminum endcaps. Different testing methods produce varying results for the degree of glass contamination, but this glass is not generally accepted for food and beverage containers. Surprisingly, one of the markets for the recycled glass is in wine bottle manufacturing which has different standards for its glass.[28] Even though this is one of the few beverages which has the potential to be aging for many years in the bottle, the manufacturers are confident that mercury is not a problem. In the process of forming bottles, the glass reaches temperatures far above the point where mercury becomes volatile. This means that the mercury will vaporize and the resulting glass product will have no residual mercury. (One wonders, however, if the people making the wine bottles are being exposed to dangerous levels of mercury vapor.) In lamp manufacturing, it is unfortunate that the resulting glass from this process is not suitable to make more lamps. This is not a result of residual mercury, but of lead content. Leaded glass is used for the lamp ends, while unleaded is used for the main tubes. These two types of glass are inevitably mixed when the glass is crushed.

Most recyclers require that lamps be received unbroken. One facility has this requirement because the recycling process itself needs the lamp to be whole in order to properly evacuate the tube. Other recyclers prefer unbroken lamps for assurance that all contaminants remain in a contained form, as well as avoiding the immediate hazard of handling broken glass.

At present, the least hazardous and most economical way to handle discarded lamps may be to ship them in the cardboard boxes they originally came in. Some lamp recycling firms, including Lighting Resources, can provide new boxes designed for shipping lamps to a recycling facility, but this will add to the expense. These containers have a reinforced slot for each lamp, and may be stacked or strapped onto pallets for shipment by truck. Such packaging avoids breaking during transit (a possible violation of federal regulations) or storage (a potential violation of state laws).

The cost for collecting, transporting, and recycling used lamps varies between 10¢ and 15¢ (U.S.) per linear foot of lamps with discounts available for higher volume customers. This amounts to an average of 50¢ per four-foot lamp. By comparison, it costs about $64 to operate that lamp for its rated life of 20,000 hours (including ballast losses and assuming 7¢/kWh power). While the disposal cost of 50¢ per lamp is small in relation to operating costs, it will add up over thousands of lamps, and could affect a facility's waste disposal budget (which is rarely connected to its energy budget). This cost should be considered in the economic evaluation of lighting design options and favors the use of reflectorized fixture designs that reduce the number of lamps (and perhaps ballasts) that will need to be disposed of in the future.

Lamp production trends. The mercury contained in fluorescent and most high-intensity discharge lamps is essential for light produc-

Table 3-2

Fluorescent lamp recycling firms in the United States

Advanced Environmental Recycling Corporation
2591 Mitchell Avenue, Allentown, PA 18103
tel 800-554-2372 fax 610-797-7696
web www.aerc-mti.com

Bethlehem Apparatus Company, Inc.
890 Front Street, Hellertown, PA 18055
tel 610-838-7034 fax 610-838-6333
e-mail bacohg@aol.com
web www.thomasregister.com/bethco

DYNEX Industries, Inc.
4751 Mustang Circle, Moundsville, MN 55112
tel 612-784-4040 fax 612-717-3161

Full Circle Recyclers
509 Manida Street, Bronx, NY 10474
tel 800-775-1516 fax 718-328-4462
e-mail fullcircle@evtc.com
web www.evtc.com

Global Recycling Technologies
P.O. Box 651, Randolph, MA 02368
tel 617-341-6080 fax 617-341-6088

Lamp Recyclers of Louisiana, Inc.
P.O. Box 2962, Hammond, LA 70404-2962
tel 504-345-4147 fax 504-345-4775
web www.i-55.com/lamprecyclers

Light Cycle, Inc.
1222 University Avenue, St. Paul, MN 55104
tel 612-641-1309 fax 612-641-1422

Lighting Resources, Inc.
1522 East Victory Street, Suite 2
Phoenix, AZ 85040
tel 800-57-CYCLE fax 602-376-5432

Mercury Recovery Services
2021 South Myrtle, Monrovia, CA 91016
tel 818-303-2053 fax 818-358-2703

Mercury Refining Co., Inc.
1218 Central Avenue, Albany, NY 12205
tel 518-459-0820 fax 518-459-2234
web www.mercuryrefining.com

Mercury Technologies of MN Inc.
Pine City Industrial Park
P.O. Box 13, Pine City, MN 55063-0013
tel 800-864-3821 fax 320-629-7799

Mercury Technology International
30677 Huntwood Avenue, Hayward, CA 94544
tel 510-429-1129 fax 510-429-1498

Midwest Recycling & Mercury Recovery
860 White Street, Dubuque, IA 52001
tel 319-556-8037 fax 319-582-3072

Northeast Lamp Recycling, Inc.
250 Main Street, East Windsor, CT 06088
tel 860-292-1992 fax 860-292-1114
e-mail nlr@eci.com

NSSI, Inc.
5711 Etheridge Street, Houston, TX 77087
tel 713-641-0391 fax 713-641-6153

P-3, Inc.
4515 Oak Fair Boulevard,
Suite 103, Tampa, FL 33610
tel 813-628-8888 fax 813-628-8889
e-mail onesource@p-3.com

Recyclights, Inc.
401 West 86th Street, Bloomington, MN 55420
tel 800-831-2852 or 612-948-0627
fax 612-948-0627

Salesco U.S.A.
5736 West Jefferson, Phoenix, AZ 85043
tel 602-233-2955 fax 603-278-0608
web www.ssusa.com

Superior Lamp Recycling, Inc.
P.O. Box 1323, Fond du Lac, WI 54936-1323
tel 800-556-5267 fax 414-923-9010

USA Lights
2007 County Road C-2, Roseville, MN 55113
tel 612-628-9370 fax 612-628-9371

Source: EPA Green Lights Program and E SOURCE

tion. When energized, the vaporized mercury produces electromagnetic wavelengths which either directly or indirectly produce light. There are no substitutes for mercury which would create equivalent light output and create less of an environmental hazard. Some manufacturers have chosen to consciously reduce the amount of mercury in their lamps, but it is unlikely that building owners will base their specification on mercury content alone.

For more information on lamp and ballast disposal, read "Lighting Waste Disposal," an 18-page summary, available from: EPA Green Lights Program: tel 202-775-6650 fax 202-775-6680

It is unrealistic to predict that these discharge lamps as we know them will eventually be produced with no mercury. As more efficient manufacturing technologies have evolved, however, the quantity of mercury embodied in lamps has decreased. In fluorescent technology, a typical four-foot T12 lamp produced in 1985 contained 48.2 milligrams of mercury fill, while the same lamp type produced in 1990 contains 41.6 milligrams. These averages mask considerable variation, however. Recent measurements conducted for EPA found a range of 14 to 45 milligrams in the four-foot fluorescents of just one manufacturer.[29] The more efficient T8 lamp has less mercury than the older T12, while of the T12 technology family, high output (HO) and very high output (VHO) lamps are particularly heavy in mercury concentration.

Philips introduced its "Alto" series of low-mercury T12 lamps in 1995, and began shipping T8 models in 1996. Both Osram Sylvania and General Electric have produced their own models and plan to have T8 versions available in 1997. Each utilizes processes involving reduced mercury content and encapsulation to control releases during both manufacturing and disposal. All three firms claim that these new lamps will pass the leachate test discussed above. As competition drives down the cost of these presently more expensive lamps, the need to recycle linear fluorescent lamps may be significantly reduced. As of early 1997, no such low-mercury options have been offered for compact fluorescent, HO, or VHO lamps.

Other types of lamps that don't rely on mercury are also being developed. One concept is a sulfur lamp which is energized by microwaves to produce light (see **10.6**). This technology is now available as a commercial product.

3.3.2 Ballast disposal

Fluorescent and HID ballasts manufactured in the U.S. through the late 1970s used polychlorinated biphenyls (PCBs) as an insulating material. Because exposure to PCBs causes skin, liver, and reproductive disorders and is suspected of causing cancer in humans, production of PCBs was banned in the United States in 1976. About half of the one billion ballasts in service in the U.S. as of 1993 were manufactured prior to 1979 and contain about one ounce of PCB per ballast, which means that about 30 million pounds of these highly toxic compounds are suspended in the ceilings of American buildings.[30]

A similar problem exists with ballasts produced between 1979 and 1984 which contained diethylhexyl phthalate (DEHP). This solvent was used in dielectric fluids in ballasts having wet capacitors. Like PCB ballasts, DEHP ballasts cannot be discarded in sanitary landfills.

Federal and state regulations. There is no federal regulation of non-PCB ballasts built after 1979; they can go to the local landfill unless state law requires otherwise. Regulation of PCB-containing ballasts is another story. In the United States, two federal laws are relevant and sometimes contradictory. The Toxic Substances Control Act (TSCA) states that it is permissible to dispose of nonleaking PCB ballasts in a sanitary landfill. The Comprehensive Environmental Response, Compensation, and Liability Act (the Superfund law) requires building owners and waste generators to notify the National Response Center (tel 1-800-424-8802) when disposing of a pound or more of PCBs (12 to 16 ballasts) in a 24-hour period and says that generators of such waste could be liable in subsequent Superfund cleanups at disposal sites. In practice, the Superfund provisions regarding ballasts are not enforced, and many nonleaking PCB ballasts are disposed of in landfills, with no recordkeeping. Without proof that a building owner or operator has disposed of PCB ballasts in a landfill and that those ballasts subsequently became the source of any leakage of PCBs from the landfill, it would be extremely difficult for the government to bring a successful Superfund claim against a building owner or operator.

If a ballast is leaking PCBs or if the asphalt potting material inside the ballast contains more than 50 parts per million of PCBs, TSCA requires that the ballast be incinerated in an approved facility.

Many states have developed regulations governing the disposal of nonleaking PCB-containing ballasts that are more stringent than federal regulations. The status of state regulations as of January 1994 is summarized in **Table 3-3**. Because new regulations are adopted periodically, readers should consult with their state environmental agency or regional EPA office to get up-to-date guidance. EPA's pamphlet on "Lighting Waste Disposal" lists regional EPA and state environmental offices.

FIRE RISK JUSTIFIES REMOVAL OF ALL PCB BALLASTS

In 1981 fire broke out in the basement of a new state office building in Binghamton, New York. Transformers in the basement contained PCBs, which got dispersed through the ductwork and throughout the building. After more than a decade of cleanup *at costs approaching the entire cost of the building*, the building is still not open.

Building owners and operators that are considering leaving disconnected PCB-containing ballasts in the ceiling of their building so as to avoid disposal costs should think again. A fire that disperses those PCBs could essentially render the building uninhabitable for many years and lead to massive cleanup costs.

Disposal options. A flow chart for dealing with ballasts is presented in **Figure 3-6**. There are seven basic options for handling *nonleaking* PCB ballasts:[31]

- *Disconnect and leave them in the ceiling.* While there is a natural tendency to "leave well enough alone," it makes sense to remove PCB ballasts because of potential risks of contamination in the event of a building fire (see box).

- *Dispose of in a sanitary landfill.* This is what many people do. It's a legal option, though many building owners and operators feel that it is not a moral option because sanitary landfills

Figure 3-6

Ballast disposal options

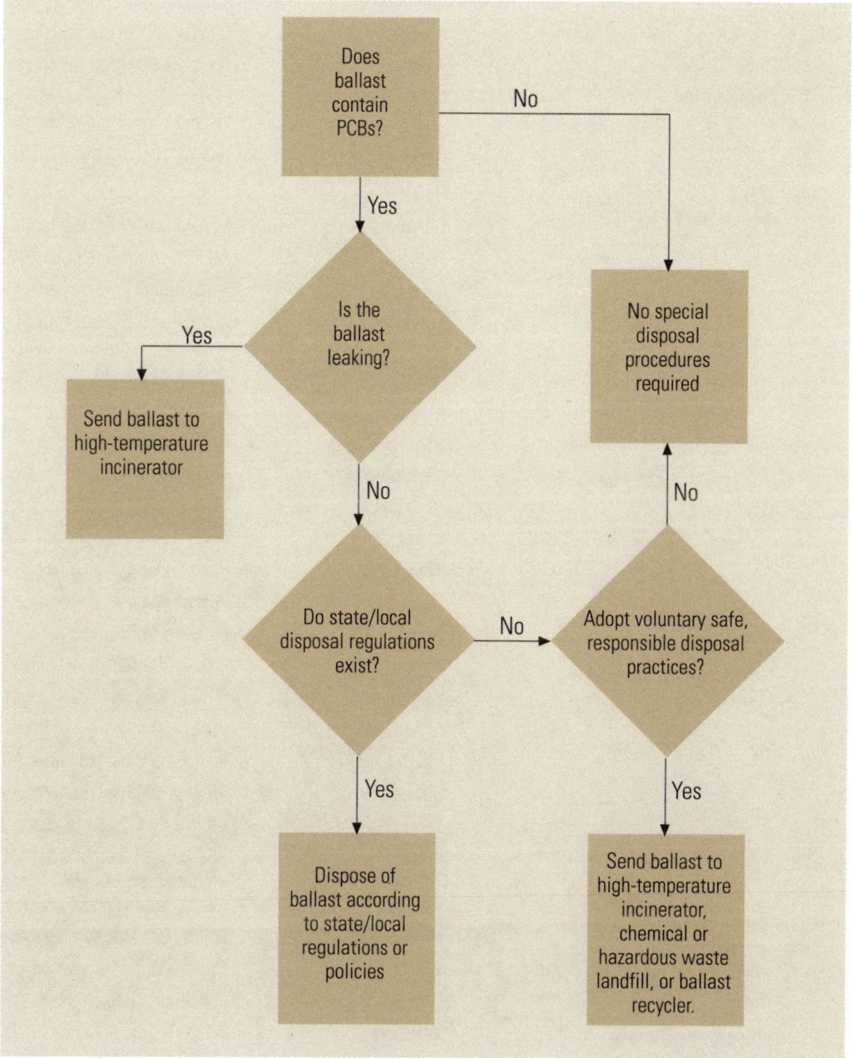

Source: EPA

Table 3-4

U.S. hazardous waste landfills

For an update to this list, call the Toxic Substances Control Act Hotline at 202-554-1404 or the Green Lights Hotline at 202-775-6650.

Chemical Waste Management Call 1-800-843-3604 for information on CWM disposal facilities nationwide e-mail cwmi@ix.netcom.com
CWM Chemical Services Control, Inc. 1550 Balmer Road Model City, NY 14107 tel 716-754-8231 fax 716-754-0211 web www.wmx.com
Envirosafe Services Inc. of Idaho P.O. Box 400 Grandview, ID 83624 tel 800-274-1516 fax 208-834-2997
U.S. Ecology, Inc. Box 578 Beatty, NV 89003 tel 702-553-2203 fax 702-553-2742 web www.americanecology.com

Source: EPA

are not designed to prevent PCBs from leaking into surrounding soils and groundwater.

■ *Send to a municipal incinerator.* This is not a desirable option because most of these facilities are not designed to destroy PCBs completely so the ash is likely to contain PCBs.

■ *Send to a chemical waste landfill.* This is a common practice because it is the least expensive option that provides high levels of assurance that the PCBs will not contaminate the environment. A list of commercially permitted hazardous waste landfills is presented in **Table 3-4**.

■ *Send whole ballasts to a PCB incinerator.* This is a fairly expensive option because the entire ballast, 80 percent of which is metal, is incinerated. Permitted PCB incinerators are listed in **Table 3-5**.

■ *Recycle the ballast.* PCBs can be removed and sent to an approved incinerator and the metal (80 percent of the ballast by weight) can be recycled. Ballast recycling firms are listed in **Table 3-6**.

Table 3-3

State regulations regarding disposal of PCB-containing ballasts as of early 1994

State	Comments
Alabama	In-state landfill requires prior approval.
Arizona	Can be sent to municipal landfill if packed in approved drums.
Arkansas	Regulates transportation of PCBs > 50 ppm.
California	PCBs > 50 ppm are hazardous waste and must either be placed in tab packs and disposed of in hazardous waste landfill or incinerated.
Connecticut	PCB ballasts must be incinerated or sent to a chemical waste landfill.
Florida	Follow EPA Region 4 policy, which recommends chemical waste landfill.
Georgia	Must check with local landfill and see if it will accept the waste.
Illinois	All PCB-containing ballasts meet definition of special waste (35 IAC).
Indiana	Need approval to dispose of more than 25 small capacitors of ballasts per day.
Kentucky	Solid waste PCBs ≥ 1 ppm cannot be placed on the land; waste containing < 50 ppm can be placed in a contained landfill. Residual landfills may dispose of PCBs according to their permit.
Louisiana	PCBs > 50 ppm considered hazardous waste.
Maine	All PCBs > 50 ppm regulated as hazardous waste.
Maryland	Based on entire weight of ballast. Average 1–2 ballasts limit.
Massachusetts	Policy on disposal and handling more stringent than federal legislation.
Michigan	PCB disposal prohibited in Michigan landfills with no small quantity exemptions.
Minnesota	All PCBs > 50 ppm regulated as hazardous waste.
Mississippi	PCBs > 25 ppm must be disposed of in hazardous waste landfill.
New Jersey	PCBs > 50 ppm is hazardous waste, unless meets certain conditions.
New Mexico	Follow EPA Region 6 policy.
Oregon	Follow EPA Region 10 policy (> 5 ballasts/year must be incinerated or sent to chemical waste landfill).
Pennsylvania	If PCBs > 50 ppm, then waste is regulated by DER.
Rhode Island	PCBs > 50 ppm regulated as hazardous waste.
South Carolina	In-state disposal requires prior approval.
Tennessee	In-state disposal requires prior approval.
Vermont	All PCBs > 50 ppm regulated as hazardous waste.
Virginia	PCB-containing materials are regulated as Special Solid Waste. PCBs > 50 ppm may not be disposed or stored without EPA approval. PCBs between 1 and 50 ppm restricted to disposal in sanitary landfills or industrial waste landfills.
Washington	Follow EPA Region 10 policy (> 5 ballasts/year must be incinerated or sent to chemical waste landfill).
Washington, DC	Recommends incineration or chemical waste landfill.
West Virginia	Follow EPA Region 3 policy.
Wisconsin	All PCBs > 50 ppm regulated as PCB waste.

Source: EPA

■ *Cut open the ballast.* Some firms dispose of the portion containing the capacitor as hazardous waste and sending the rest to a sanitary landfill. The key issue here is how carefully the company is cutting open the ballast to make sure that they don't rupture the capacitor or release PCB-contaminated asphalt dust.

There is no one right answer on how to deal with PCB ballasts. EPA encourages incineration, chemical or hazardous waste landfilling, or recycling as the most responsible options. Some concern has been raised about whether incinerators are closely enough monitored and kept in compliance. In one instance, a hazardous waste incinerator in North Carolina was shut down in 1989 after it was determined that the facility had operated unsafely for a number of years.[32] Anyone faced with making such a decision should consider the liabilities of the various options and involve the organization's risk management staff in the final choice.

Economics of ballast disposal. The costs of disposal options vary widely. Sending them to the local sanitary landfill or municipal incin-

Table 3-5

U.S. PCB incineration facilities

For an update to this list, call the Toxic Substances Control Act Hotline at 202-554-1404 or the Green Lights Hotline at 202-775-6650.

Aptus, Inc.
P.O. Box 1328
Coffeyville, KS 67337
tel 316-251-6380
fax 316-252-1483

Chemical Waste Management
P.O. Box 2563
Port Arthur, TX 77643
tel 409-736-2821
fax 409-736-1636

Rollins Co.
P.O. Box 609
Deer Park, TX 77536
tel 713-930-2300
fax 713-930-2316

Roy F. Weston, Inc.
One Weston Way
West Chester, PA 19380
tel 610-701-3000
fax 610-701-3124
web www.rfweston.com

Source: EPA

Table 3-6

U.S. ballast recycling firms

For an update to this list call the Toxic Substances Control Act Hotline at 202-554-1404 or the Green Lights Hotline at 202-775-6650.

Advanced Environmental Recycling Corporation
2591 Mitchell Avenue
Allentown, PA 18103
tel 800-554-2372
fax 610-797-7696
web www.aerc-mti.com

Bethlehem Apparatus Company, Inc.
890 Front Street
Hellertown, PA 18055
tel 610-838-7034
fax 610-838-6333
e-mail bacohg@aol.com
web www.thomasregister.com/bethco

DYNEX Industries, Inc.
4751 Mustang Circle
Moundsville, MN 55112
tel 612-784-4040
fax 612-717-3161

Full Circle Recyclers
509 Manida Street
Bronx, NY 10474
tel 800-775-1516
fax 718-328-4462
e-mail fullcircle@evtc.com
web www.evtc.com

Global Recycling Technologies
P.O. Box 651
Randolph, MA 02368
tel 617-341-6080
fax 617-341-6088

HID Recycling Technologies
32000 Aurora Road
Cleveland, OH 44139
tel 800-200-9716
fax 800-200-9718
e-mail chris_cole@adlt.com
web www.adlt.com/venture

Lamp Recyclers of Louisiana, Inc.
P.O. Box 2962
Hammond, LA 70404-2962
tel 504-345-4147
fax 504-345-4775
web www.i-55.com/lamprecyclers

Light Cycle, Inc.
1222 University Avenue
St. Paul, MN 55104
tel 612-641-1309
fax 612-641-1422

Lighting Resources, Inc.
1522 East Victory Street, Suite 2
Phoenix, AZ 85040
tel 800-57-CYCLE
fax 602-376-5432

Mercury Recovery Services
2021 South Myrtle
Monrovia, CA 91016
tel 818-303-2053
fax 818-358-2703

Mercury Technologies of MN, Inc.
Pine City Industrial Park
P.O. Box 13
Pine City, MN 55063-0013
tel 800-864-3821
fax 320-629-7799
e-mail ecmis.mtm@norsol.com

Mercury Technology International
30677 Huntwood Avenue
Hayward, CA 94544
tel 510-429-1129
fax 510-429-1498

Midwest Recycling & Mercury Recovery
860 White Street
Dubuque, IA 52001
tel 319-556-8037
fax 319-582-3072

Northeast Lamp Recycling, Inc.
250 Main Street
East Windsor, CT 06088
tel 860-292-1992
fax 860-292-1114
e-mail nlr@eci.com

NSSI, Inc.
5711 Etheridge Street
Houston, TX 77087
tel 713-641-0391
fax 713-641-6153

P-3, Inc.
4515 Oak Fair Boulevard, Suite 103
Tampa, FL 33610
tel 813-628-8888
fax 813-628-8889
e-mail onesource@p-3.com

Recyclights, Inc.
401 West 86th Street
Bloomington, MN 55420
tel 800-831-2852 or 612-948-0627
fax 612-948-0627

S.D. Myers
180 South Avenue
Tallmadge, OH 44278
tel 330-633-2666
fax 330-633-0756

Salesco U.S.A.
5736 West Jefferson
Phoenix, AZ 85043
tel 602-233-2955
fax 602-278-0608
web www.ssusa.com

Superior Lamp Recycling, Inc.
P.O. Box 1323
Fond du Lac, WI 54936-1323
tel 800-556-5267
fax 414-923-9010

Transformer Service, Inc.
74 Regional Drive
Concord, NH 03301
tel 603-224-4006
fax 603-228-2430
e-mail tsisales@tsi1.mv.com

Source: EPA

erator costs on the order of 10¢ to 20¢ per ballast, but these options don't reliably keep the PCBs out of the environment. Chemical waste landfilling costs 50¢ to $2 per ballast. Whole ballast incineration costs $5 to $8 per ballast. Recycling costs $3 to $5 per ballast.

These estimates are averages for high-volume disposal. They don't include the cost of packaging the ballasts in 55-gallon drums, transportation, or profile fees levied in some cases by incinerators or landfills to document incoming hazardous waste. One also needs to consider the cost of getting the ballasts out of the ceiling; this can often be more than the cost of recycling. One way to minimize the cost of disconnecting and removing ballasts is to use cheaper labor for that task than the trained electricians needed to install new ballasts. Leaking ballasts should of course be handled by specialists. Because of the time and expense involved in determining whether a given ballast does or does not contain PCBs some building operators find it more economical to dispose of them all as if they contained PCBs. This costs more in disposal fees but eliminates the time needed for ballast removal crews to even think about the classification of each ballast.

One key implication of the ballast disposal issue is that the environmental risk mitigation benefits alone may justify immediate retrofit of lighting systems to electronically ballasted equipment. Avoiding the potentially massive cost of PCB leakage by preemptive retrofit is a huge and generally uncounted benefit of lighting retrofits.

Ballast disposal in Canada. The options for PCB ballast disposal in Canada are limited in comparison to the United States. The federal authority, Environment Canada, has given the individual provinces jurisdiction over waste management, but the provinces all have similar rules for PCB ballasts. PCBs have been banned from landfill sites, and the American border is closed to shipments containing PCBs.[33] Provinces will only occasionally allow transport of PCBs across their borders. The generator of the waste, usually the building owner, owns the PCBs until they are destroyed.

The generator of the PCBs has the option of storing the ballasts, or ensuring that the PCBs are destroyed. There is an operating PCB incinerator in Alberta, but it is currently accepting only limited numbers of PCB ballasts from other provinces. Alberta is currently the only province with a decreasing inventory of PCB ballasts.[34] The current recommendation made by the Ministries of Environment is to provide secure storage for these ballasts until other solutions are found. Some limited provincial PCB storage sites exist, but generally owners must provide their own storage facilities.

It is possible to recycle these ballasts, but the PCB portion of the ballast remains as waste. Recycling ballasts reduces the volume of PCB to be stored, which can save on costly storage space. Before accepting the services of a PCB ballast recycler, however, it is wise to research their recycling procedure. The owner of the waste is responsible for the PCBs should any problems arise in the recycling process. The owner must provide storage for any ballast components which contain PCBs after the recycling is complete.

The provinces realize that indefinite storage of PCBs is not a solution. British Columbia's Ministry of Environment is working on a PCB elimination strategy and will attempt to have a solution by 1999. Until that time, it is necessary to follow the provincial rules for handling and storing special wastes. These guidelines detail the necessary labeling, handling, storage containers, and stacking of containers. The guidelines are available from the Provincial Environment Ministry Offices. To determine if your ballasts contain PCBs, Environment Canada has published a report entitled, "Identification of Lamp Ballasts Containing PCBs," Report EPS 2/CC/2, revised August 1991.

3.4 LIGHTING AND HEALTH

Natural light and artificial light directly affect people's health and well-being. Both sources of light can be used as therapy for depression. Flicker and patterns of light, from either natural or artificial sources, can set off seizures in some people. Both sources also emit ultraviolet light, which is known to cause skin cancer. The following sections explore these issues.

3.4.1 Lighting and depression

Research in recent years has shown that cycles of light and darkness exert powerful biological effects on humans. These daily rhythms affect hormone levels and various metabolic processes, which in turn can affect mood. About 1 in 20 people suffer depression in the dark months of winter, with symptoms including lethargy, weight gain, withdrawal, increased need for sleep, and difficulty sleeping.[35] Shift workers who work at night often suffer similar symptoms. Doctors have found that these symptoms can be relieved for many individuals by exposing them to bright light—typically 250 to 1,000 footcandles—for 30 to 60 minutes several time a week.

While the effectiveness of such phototherapy (**Figure 3-7**) for some people suffering from severe depression is well documented, this phenomenon raises questions about the effects of interior lighting levels on the mental state and productivity of the majority of people who do not suffer clinical symptoms. Humans evolved outdoors, where even on a very gloomy day the light levels are on the order of 500 footcandles, which is much higher than the 30 to 60 footcandles typical of indoor spaces. Might some indoor workers be suffering subtle levels of depression or other negative symptoms from spending most of their waking hours in environments considerably dimmer than that in which human beings existed throughout history? What effects might occur from lowering lighting levels even further in response to concerns over energy efficiency?

Figure 3-7

Keeping happy with light therapy

Regular treatments with elevated light levels can alleviate depression for some individuals.

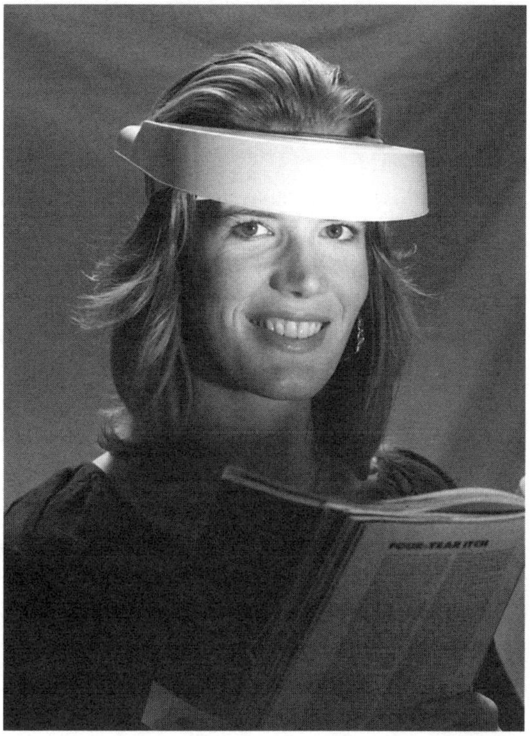

Source: EPRI Journal
Photo by Robert Neroni

Researchers are grappling with these questions but no firm answers have yet emerged. Tests are being conducted to determine the effects of wavelength, duration, and time of exposure, as well as light intensity. Even if it is determined that large numbers of people suffer ill effects from low interior light levels, efficiency goals may still be achieved by using task lighting to raise illumination on the task plane and in the worker's immediate environment.

3.4.2 Lighting and epilepsy

Light affects human beings in a vast number of ways. One of the more unusual and disturbing examples is that certain levels or patterns of light or rates of lighting flicker can stimulate seizures in about 5 percent of persons suffering from epilepsy. Such photosensitivity is most common in epileptics under 20 years of age and is more common among females than males.

> Numerous environmental stimuli may be implicated—including the sun seen through trees or reflected on moving water, discotheque lighting and arcade games—but [one study of 358 individuals suggests that] television is most often the precipitant, in some 50 percent of the patients.[36]

Elements generally associated with good lighting design are likely to minimize ill effects in vulnerable epileptics. That means flicker-free systems, care not to overilluminate spaces, and designs that avoid sharp, regular patterns of light and dark.

Questions have arisen as to whether lighting retrofits could be associated with seizures. In one incident at the Massachusetts Association for the Blind rehabilitation center in Boston, a partially blind patient experienced an epileptic seizure upon entering a space that was undergoing a lighting retrofit. The patient had been in this space previously without experiencing problems. The retrofit involved switching from standard 2 x 4 four-lamp T12 fixtures with magnetic ballasts to a reflector, two T8 lamps, and an electronic ballast in each fixture. Clinic officials asked Boston Edison to suspend the retrofit while they explored whether the new equipment, particularly the electronic ballasts, could be to blame. E SOURCE provided Boston Edison with the information presented in this section, which indicated that the electronic ballasts were less likely than magnetic units to precipitate seizures. The retrofit went forward and no further incidents have been reported, suggesting that the lighting was not responsible in this particular case.[37]

The studies of epilepsy and light cited above suggest that raising the brightness or amount of light and introducing alternating patterns of light and dark could trigger responses in a small percentage of epileptics. It is conceivable, for instance, that a space that previously had flat lenses and was retrofitted with louvers could create such striped patterns. If the retrofit also raised lighting levels, which is often the case, this combination of factors could be to blame. In the case in Boston, the retrofit was anecdotally reported to raise lighting levels initially, although no data is available.

As for flicker, magnetically ballasted systems are more likely to produce detectable flicker than high-frequency electronic systems. Tests suggest that higher frequencies are less likely than low-frequency flicker to induce seizures. Two sets of controlled experiments exposing photosensitive epileptics to stroboscopic light sources found maximum sensitivity to flash rates in the range of 15 to 25 flashes per second, with sensitivity dropping off rapidly at flash rates of 60 per second (**Figure 3-8**). However, 5 to 20 percent of the photosensitive epileptics did have a reaction from 60-flash-per-second strobing. This suggests that conventional fluorescent lighting, which flickers off and on 120 times per second (twice per cycle), could conceivably trigger a response in some individuals. Since the human eye is not sensitive to flicker in the 20,000-cycle and higher frequencies characteristic of electronically ballasted lighting systems, it seems unlikely that high-frequency fluorescent lighting is a contributor to seizures. More research is needed to test this hypothesis. Flicker is discussed further in **7.2**.

Figure 3-8

Sensitivity of photosensitive epileptics to different flash rates

Data from two studies shows the highest degree of sensitivity in the range of 15 to 25 flashes per second.

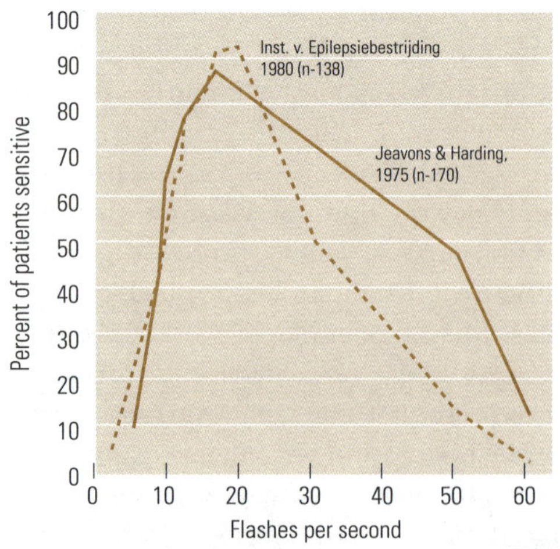

Source: Laidlaw [34]

The bottom line is that elements generally associated with good lighting design are likely to minimize ill effects in vulnerable epileptics. That means flicker-free systems, care not to overilluminate spaces, and designs that avoid sharp, regular patterns of light and dark, which are undesirable in any event.

3.4.3 The effects of ultraviolet light

As many a sunbather knows but chooses to ignore, there is a direct correlation between the intensity of ultraviolet light and certain types of skin cancer. Since every kind of electric light source produces some ultraviolet light, concerns have been raised about the potential risks of UV exposure from electric lighting.

E SOURCE treated this topic in some detail in an August 1993 *Tech Update* and the main points of that report are summarized here.[38]

- UV from the sun is generally 100 to 1,000 times as intense as that from interior lighting.

- A few light sources, such as very high output (VHO) fluorescent lamps, emit enough UV to warrant precautionary measures, but the health risks from the vast majority of lamps are negligible.

- Filters and lenses are available for lighting fixtures to cut UV emissions dramatically and are recommended when there is a need to minimize UV exposure of people or artifacts.

- Misleading information has been published that suggests compact fluorescents emit four times the UV of incandescents and thus should not be used for illuminating sensitive objects in museums or retail stores. These claims were based on inaccurate measurements. When more accurate meters are used, many CFL retrofit kits, especially those equipped with acrylic lenses, are shown to put out comparable or lower amounts of UV as incandescents.

3.5 Lighting and Productivity

Measuring how lighting quantity and quality affect worker productivity is very difficult. Because worker salaries dwarf energy costs, however, even modest effects on productivity resulting from lighting can far outweigh the concurrent changes in energy bills. A few percent productivity gain, if actually achieved and sustained, can be worth more to the collective bottom line of the building owner and operator than eliminating *all* of the energy bills.

As interest in the nonenergy benefits of efficiency upgrades grows, links between lighting upgrades and productivity are receiving closer scrutiny. Given the poor quality of lighting in so many spaces—overlit, flickering, and noisy systems with high levels of glare—it seems intuitive that improving their quality will boost productivity by reducing eye strain, minimizing headaches, and improving visual performance. Such improvements may take the form of reducing the frequency of errors and improving quality control, and it may reduce absenteeism, be it fewer trips to the coffee machine or fewer sick days.

A few percent productivity gain, if actually achieved and sustained, can be worth more to the collective bottom line of the building owner and operator than eliminating all of the energy bills.

The literature is somewhat divided. A study conducted in the late 1980s was unable to substantiate a relationship between better lighting and improved productivity.[39] Increasing numbers of case studies are emerging, however, that are finding a generally positive link between better lighting and improved productivity.[40] Sometimes the results are quite dramatic, as the following sampling of case studies attests:

- *Control Data lighting retrofit pays back in 23 days.* A lighting retrofit at Control Data's offices in Sunnyvale, California sought to reduce energy bills and the reflected glare on VDT screens that was slowing performance and causing input errors. The $15,000 retrofit reduced lighting energy use by 65 percent, for a 23-month payback. Far more importantly, however, the firm credits the better lighting with reducing downtime in the operation group's huge computers by three to four hours per year—saving the facility about

$200,000 a year—and allowing the ten workers in the retrofitted space to increase their productivity by six percent, worth an additional $28,000 per year. In total, the company believes that the lighting enhancements saved them about $235,000 per year, for a simple payback of 23 days.[41]

■ *Drafters' output boosted by lighting upgrade.* A $49,000 lighting retrofit in a drafting room of Pennsylvania Power & Light's Allentown facilities reduced lighting energy use by 76 percent and reduced veiling glare, which had been causing eye strain and leading to unacceptable numbers of errors. Before-and-after measurements of work output showed that the workers performed their tasks more than 13 percent faster after the retrofit, although management claimed a more conservative 7.5 percent improvement for purposes of economic analysis on the project. Even at the 7.5 percent rate, the productivity savings were worth $235,000 per year, leading to a payback of two and a half months on the project. In a similar case, a lighting upgrade at the Superior Die Set Corporation in Oak Creek, Wisconsin reduced lighting levels and glare in an overlit drafting area, reducing the time needed to produce a drawing by 11.3 percent. In terms of energy savings the project paid back in about 20 months, but counting the productivity benefits, the project had a simple payback of about 24 days.[42]

■ *Lighting speeds Santa Fe mail sorters.* A mid-1980s retrofit of the mail sorting area in the Santa Fe, New Mexico post office is credited with raising the number of pieces sorted per hour by six percent. The energy and maintenance savings on the $300,000 project would have saved about $50,000 per year, but the productivity gains were worth over $400,000 per year.

■ *Electronic ballasts cut incidence of headaches and eyestrain.* In a double-blind study in the United Kingdom, the weekly incidence of headaches among office workers was studied for a year.[43] Those workers whose lighting was driven by high-frequency electronic ballasts reported half the rate of headaches and eyestrain as their coworkers with magnetically ballasted 50-Hz systems. The researchers attributed the results to the lack of flicker in the electronically ballasted systems. The potential implications of a widespread reduction in lighting flicker are enormous. Even a modest reduction in such symptoms across a large share of the office worker population could have a significant effect on productivity. This phenomenon alone may justify the use of electronic ballasts instead of magnetic units, even disregarding the energy savings and other benefits that electronic ballasts provide.

Recent research at Lawrence Berkeley Laboratory suggests that "beat" frequencies caused by the interaction of magnetically ballasted lighting and the refresh rate of video display terminals can cause considerable eyestrain. This research found that even though we are not visually aware of it, the brain is sensitive to flicker even at frequencies around 120 Hz and that the muscles in the iris of the eye fluctuate very rapidly when subjected to such flicker.

These case studies document results in settings where productivity is easy to measure and the effect of lighting is easy to isolate from other variables. In many settings, it is far more difficult to measure productivity or to isolate the key parameters that affect it. Some experiments indicate that changing *any* noticeable working condition can improve productivity, but only temporarily.[44] Nonrepetitive tasks, with no standard output measure, are especially hard to monitor for productivity changes. Different people, particularly at different ages, and even the same people at different times of day, require different amounts of light for similar tasks. Few researchers properly control for nonlighting variables—holidays, weather (which can also affect mood), flu outbreaks, labor relations, outside economic factors, and so on. Experiments which test relationships between lighting levels and visual performance (speed and accuracy of reading, for example) are often seeking relatively small marginal effects in the midst of strong saturations. Short-term results are often reported without documentation of whether the effect persists over time. Most experiments are more concerned with lighting quantity than quality, yet small and often uncontrolled-for changes in quality may matter more than far larger changes in quantity.

One must use extreme caution then in making sweeping generalizations about the links between lighting and productivity. All too many lighting installations conducted in the name of efficiency save energy but deliver terrible quality. Many a facility manager has delamped to save energy and in the process created a gloomy space with poor light distribution. The canonical efficiency upgrade from T12 lamps to T8s can create serious glare if the T8s are mounted at the same height in the fixture as the T12s. Daylighting, if done poorly, can create glare and overheating, which are more likely to reduce productivity than enhance it.

The key is education that widely disseminates the skills needed to create high-quality lighting. As this knowledge spreads, and as more practitioners learn how to combine better lighting quality with better efficiency, the chances are good that the satisfaction, health, and productivity of the people who spend most of their waking hours under those lights will improve, even if that change is difficult to measure.

1 Assumptions used in analysis of retrofit options:

• The base case is 1970-vintage IES type 41 luminaire, four T12 lamps, flat bottom diffuser, 2 x 4, dirt category V (see *IES Lighting Handbook*, 8th Edition, 1993, p. 438). We assume an RCR of 2, reflectance 80/50/20. The luminaire is unchanged in Cases 1–4, but receives a flat prismatic lens and specular reflector in Case 5, raising the coefficient of utilization to 0.80.

• Initial interior temperature of fixture (not lamps) with four T12 lamps is 100°F; temperature with four T12 34-watt lamps drops to 95°F; temperature with two T8s and a specular reflector drops to 85°F. Light loss calculation is based on 1% loss per 2°F above 77°F (see *IES Lighting Handbook*, p. 209). CU of original fixture includes thermal effects.

• Fixture cleaning occurs at end of rated life base case, assuming burn hours of 4,000 hr/yr and 20,000-hour rated life, yielding 5 years between cleanings for total dirt loss of 30%; when lamps are changed at 70% of rated life (14,000 hours) we get a loss of only 25% at 35 months (see graph V, *IES Lighting Handbook,* p. 398).

• Room temperature near ceiling = 77°F, voltage = 120 V.

• Room contains 12 evenly spaced fixtures in 4 columns, each having 3 fixtures.

• Space air is air conditioned, well maintained, and could be described as "clean" (see *IES Lighting Handbook,* p. 397).

• Specular reflector retrofit kit is designed to maintain same spacing ratio.

• Existing diffuser has yellowed and gathered sufficient adhesive dirt (not easily removed during routine cleaning) to reduce transmittance by another 10%.

• Original ballasts are in base case, 15% losses, circa 1970.

• Ballast factors: 0.92 for magnetic, which (according to Motorola ballast fact sheet) drops to 0.88 once ballast is warmed up; 0.88 for two-lamp EB, 0.84 for four-lamp EB, 0.80 for two-lamp low-power EB.

• Equipment costs: 34-watt lamp = $1.50, T8 = $2.50, relamp labor = $5/lamp, four-lamp EB = $35, two-lamp EB = $25, ballast labor = $20/ballast (regardless of type), reflector/lens kit = $50/fixture + $5/lens + $20 labor = $75 total/fixture.

• No credit taken for increased lamp life due to reduced burn hours due to sensors; increased starts cancel stretched life; same with lumen depreciation.

• Maintenance cost assumptions: *Present relamping* is at the end of rated lamp life, every 5 years (20,000-hour life/4,000 hours per year). In Case 7, *group relamping* is at 70% of rated life, or every 3 to 5 years. This adds $85.70 per year in maintenance cost ($2.50/lamp, $5 labor, $10/fixture cleaning labor), which is deducted from the annual savings.

• Energy costs: demand = $10/kW/month (all 12 months of year), consumption = 7¢/kWh (all times of day).

• Electric peak is reached every month, even when sensors and dimming ballasts are used.

• Assumptions for daylight dimming in Case 8: (a) Specified system is Luxmatic™ variable-arc electronic ballast, from DWA, 7845 Cessna Avenue, Gaithersburg, MD 20879-4109, tel 301-670-7845, fax 301-948-2879. (b) Each ballasts comes equipped with a fiber optic connection, a light sensor connected to a fiber optic cable, and a manually adjustable light level control for setting maximum output. (c) This is a new product with little installed track record. We used it in this analysis because of its simplicity and minimal installation cost relative to standard dimming ballasts available from Advance or Lutron, which would add $200 to $300 to the installed cost in this example. (d) Maximum ballast light output setting is used in this example for the 8 (out of a total of 12) ballasts that are inside the daylight dimming zone. (e) That zone is assumed to extend 20 feet into the 29-foot wide room, covering the 8 fixtures nearest the window. (f) Cost for the installation is based on $80 for materials (Luxmatic ballast, cable, light sensor), $20 ballast installation cost, $30 fiber optic cable installation cost.

2 Assumes natural gas heating at 50¢/therm and electricity for cooling at 5¢/kWh.

3 Amory B. Lovins et al., *The State of the Art: Lighting*, Competitek (1988), pp. 16–18.

4 R.H. Howell, "Energy Requirements for Closed Loop Storage Heat Pump Systems with Different Internal Loads for Various Geographic Locations," *ASHRAE Transactions,* v. 94, part 1 (1988).

5 "Energy Savings Potential in California's Existing Office and Retail Buildings," California Energy Commission (June 1984); and Rocky Mountain Institute's simulation of a 4-story Chicago office building, cited in Amory B. Lovins et al., *The State of the Art: Lighting,* Competitek (1988), p. 18, footnote 40.

6 C. Linn, "Calculating Daylight for Successful Retail Design," *Architectural Lighting* (January 1987), pp. 28–34.

7 A. Osman Sezgen and Y. Joe Huang, "Lighting/HVAC Interactions and Their Effects on Energy Consumption and Peak HVAC Requirements in Commercial Buildings," draft paper for ACEEE 1994 *Summer Study on Energy Efficiency in Buildings*, Energy and Environment Division, Lawrence Berkeley Laboratory, Berkeley, CA 94720.

8 "Hot climate" values are based on Phoenix, Los Angeles, and Miami. "Cold climate" values are based on Chicago, New York City, and Minneapolis. A "large" office building has typically a 20,000 to 25,000 ft^2 footprint.

9 Robert A. Rundquist et al., "Calculating Lighting and HVAC Interactions," *ASHRAE Journal* (November 1993).

10 Ellen Franconi and Francis Rubinstein, "Considering Lighting System Performance and HVAC Interactions in Lighting Retrofit Analyses," submitted for the Conference Record of the 1992 IEEE Industry Applications Society Annual Meeting, Energy and Environment Division, Lawrence Berkeley Laboratory, Berkeley, CA 94720, p. 6.

11 Franconi and Rubinstein [10], p. 7.

12 Franconi and Rubinstein [10], p. 5.

13 Stephen J. Treado, "The NIST lighting and HVAC Interaction Test Facility," *ASHRAE Journal*, v. 33, no. 4 (April 1991), pp. 47–49.

14 "The EC Lobs Another Edict At Outsiders," *Electronic Components* (May 1993), p. A-50.

15 Tom Watson, "Fluorescent Lamps—A Bright New Recyclable," *Resource Recycling* (March 1992), p. 71.

16 Warren Liebold and Lindsay Audin, "Compact Fluorescents, Radioisotopes and Solid Wastes," *Proceedings of ACEEE Summer Study, 1992*, American Council for an Energy-Efficient Economy, Washington DC.

17 Emily Moore, Edward Swain, and Jim Brist, "Recycling Fluorescent and High-Intensity Discharge Lamps," *Proceedings of 9th International Conference of Solid Waste Management*, Philadelphia, PA (November 14–17, 1993).

18 *Federal Register*, v. 59, no. 143 (July 27, 1994), pp. 36288–38304.

19 *Federal Register* [18].

20 Watson [15], p. 74.

21 Watson [15], p. 74.

22 Moore et al. [17].

23 Kevin Hicke, personal communication (February 1994), Manager, Technical Services and Special Wastes, BC Ministry of Environment, Victoria, BC, Canada, tel 604-387-9953, fax 604-953-3856.

24 Wayne Harrick, personal communication (March 1994), Regional Coordinator for Hazardous Materials, Ontario Ministry of Environment, tel 615-549-4000.

25 Mary Janet Ruzika, personal communication (December 1993), Consultant for Lighting Resources Inc., Pomona, CA, tel 602-279-3541

26 Moore et al. [17].

27 *Federal Register* [18].

28 Paul Abernathy, personal communication (February 1994), Mercury Technologies Corporation, Hayward, CA, tel 510-499-1129.

29 Robert Truesdale, Stephen Beaulieu, and Terrence Pierson, "Management of Used Fluorescent Lamps: Preliminary Risk Assessment," Final Report RTI Project No. 94U-5400-010 (May 1993), Research Triangle Institute, P.O. Box 12194, Research Triangle Park, NC 27709.

30 Mitchell Dong and Brian McCagg, "The Practical Guide to PCB Ballast Disposal," FulCircle Ballast Recyclers, 509 Manida Street, Bronx, NY 10474, tel 718-328-4667, fax 718-328-4462.

31 List adapted from Dong and McCagg [30].

32 "A North Carolina Incinerator's Noncompliance with EPA and OSHA Requirements," U.S. General Accounting Office, Washington DC GAO/RCED-92-78 (June 1992).

33 Province of British Columbia, Ministry of Environment, "Fact Sheet, Special Wastes, Handling Waste PCB Ballasts From Fluorescent Light Fixtures," ISSN 0711-8953 (October 1985).

34 Deanna Cymbaluk, personal communication (April 1994), Environment Canada, Industrial Contaminants, Edmonton, Alberta, tel 403-468-8034.

35 Taylor Moore, "Human Performance in the Spotlight," *EPRI Journal* (September 1993).

36 John Laidlaw, Alan Richens, and David Chadwick, *A Textbook of Epilepsy* (New York: Churchill Livingstone, 1993), pp. 314–320.

37 Steve Shelley, personal communication (June 1994), Boston Edison, 800 Boylston Street, P308, Boston, MA 02199, tel 617-424-3603, fax 617-424-3538.

38 Lindsay Audin, "Ultraviolet Radiation from Electric Lighting," *E SOURCE Tech Update TU-93-7* (August 1993).

39 Peter Boyce, Sam Berman, Belinda Collins, Alan Lewis, and Mark Rea, "Lighting and Human Performance: A Review," prepared for NEMA, 2101 L Street NW, Suite 300, Washington, DC 20037, tel 202-457-8487.

40 See for example, Joseph T. Romm and William D. Browning, "Greening the Building and the Bottom Line: Increasing Productivity through Energy Efficient Design," *Proceedings of the 1994 Summer Study on Energy Efficiency in Buildings*, ACEEE, Washington, DC.

41 "Office Lighting and Productivity," National Lighting Bureau, 2101 L Street NW, Suite 300, Washington, DC 20037 (1988).

42 Office Lighting and Productivity [42].

43 A. Wilkins, "Health and Efficiency in Lighting Practice," *Energy,* v. 18, no. 2 (1991), pp. 123–129.

44 See S. Berman and R. Clear, "Can Polarized Lighting Panels Reduce Energy Consumption and Improve Visibility in Building Interiors?" Lawrence Berkeley National Laboratory, LBL-8671 (1979), references 42–47 and Appendix D.

4 END-USE AND MARKET DATA

Lighting directly consumes about 15 percent of U.S. electricity and accounts for about 12 percent of summer peak demand. Electronic ballasts and T8 lamps are rapidly increasing their market share, and recent standards have eliminated production of some of the least efficient sources, including standard incandescent floodlights and spotlights.

HIGHLIGHTS

- Lighting directly consumes 15 percent of U.S. electricity, and accounts for about 12 percent of summer peak demand and 10 percent of winter peak.

- Lighting energy use is increasing in all sectors, but efficiency improvements through federal government programs, utility efficiency efforts, and national codes and standards will considerably slow that growth.

- The commercial sector accounts for two-thirds of lighting energy use, residential lighting about 20 percent, and industrial lighting the remainder.

- Fluorescents' share of lamp production has increased at the expense of incandescents, although more than 90 percent of lamps produced are incandescents.

- Electronic ballasts are gaining production share at the expense of magnetic ballasts, and future federal efficiency standards may contribute to their market growth in years to come. Because of the large installed stock of magnetic ballasts, however, it will be at least a decade before electronic ballasts outnumber magnetic units.

4.1 ENERGY USE

Lighting in the United States directly uses about 500 TWh per year, or 15 percent of U.S. electricity. Nearly two-thirds of all lighting electricity is consumed in commercial buildings, almost 20 percent in the residential sector, and 13 percent in industry (**Figure 4-1**).[1] The U.S. Energy Information Administration (EIA) estimates that lighting's share of residential electricity is nine percent, but Lawrence Berkeley National Laboratory (LBNL) puts its share even higher at 10 to 15 percent.[2] Although lighting electricity use continues to grow, other end uses are growing faster, so its share of total electricity use has been declining gradually in recent years, from an estimated 20 percent in the mid-1980s to a projected 15 or 16 percent in 2010.[3] These percentages exclude interactive effects with space-conditioning equipment (see **3.2.1**).

Lighting's share of total electricity use is declining.

Commercial. In the commercial sector, the rate of growth in lighting energy use will be slowed by remaining utility demand-side management programs, federal lamp and ballast efficiency standards, efficiency improvement trends, a predicted decline in the annual gain of floor space, and a slowing of the annual increase in operating hours.[4] The U.S. EIA predicts annual growth of 0.6 percent for U.S. commercial lighting energy use from 1995 to 2015.[5]

LBNL is analyzing the role of U.S. federal lighting policy options on annual lighting electricity growth in the commercial sector. New LBNL projections are expected out around the end of 1997.[6]

Figure 4-1

Lighting's estimated share of 1996 U.S. direct electricity use, by sector

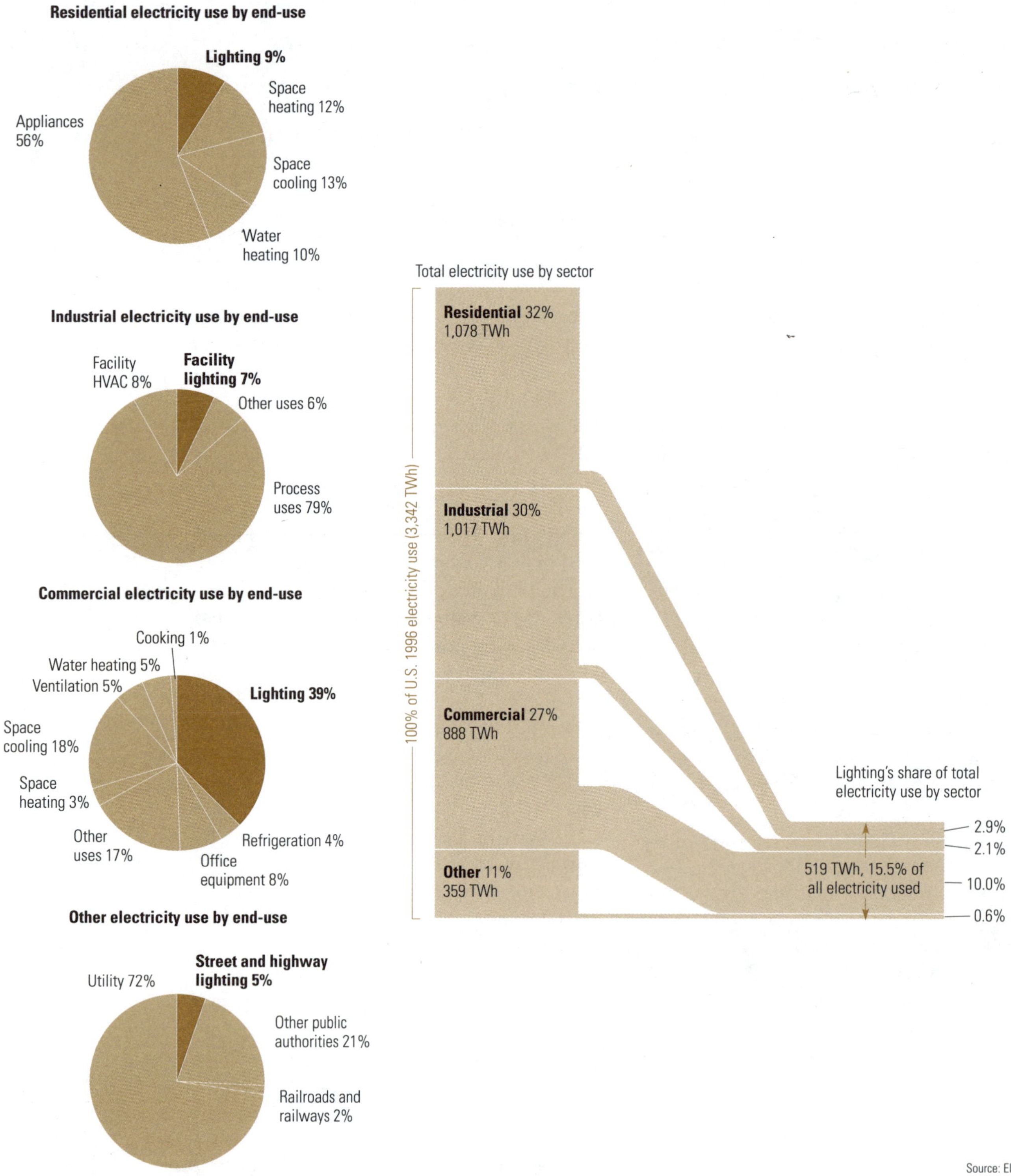

Source: EIA, EEI [1]

Residential. About 2.9 billion light fixtures illuminate 96 million households in the U.S. 85 percent of these are incandescent, while the remainder are fluorescent.[7] From 1970 to 1990, U.S. residential electricity use and peak demand grew rapidly. However, the U.S. EIA predicts annual growth of just 0.7 percent for U.S. residential energy use from 1995 to 2015.[8] According to the Electric Power Research Institute (EPRI), this slowing reflects a decrease in the rate of household growth, efficiency improvements, and utility DSM programs. The resulting moderate growth in residential lighting electricity use will be due to customer growth and increased lighting intensities. National efficiency standards for spot incandescents, the gradual penetration of compact fluorescents in the residential sector, and the remaining DSM programs to encourage the adoption of high-efficiency lighting will help to offset the increases.[9]

Industrial. Lighting accounts for approximately seven percent of all U.S. electricity consumption in the industrial sector.[10] This average value masks wide variations across industries, however. In a 1991 survey of lighting use in Ontario's industrial sector, for instance, Ontario Hydro found that lighting accounts for only one percent of the total load for energy-intensive industries such as mining, pulp and paper, and petroleum refining, but over 40 percent of the total load for less energy-intensive industries, such as tobacco, machinery, and printing.[11]

4.2 PEAK DEMAND

Lighting directly accounts for approximately 12 percent of summer peak demand and 10 percent of winter peak in the U.S.[12] Only space cooling contributes more than lighting to the summer peak, and a significant share of the cooling peak comes from removing waste heat from lighting systems. Commercial lighting accounts for over 80 percent of lighting's total contribution to peak load. EPRI composite load shapes for commercial buildings in 1995 show the components of demand on the winter and summer peak days (**Figure 4-2**).[13]

Figure 4-2

Summer and winter peak day composite load shapes for all U.S. commercial buildings (1995)

Lighting is the second largest contributor to summer peak demand in commercial buildings and rivals heating as the largest contributor to winter peak. Its indirect effects on space cooling increase the cooling and reduce the heating peak.

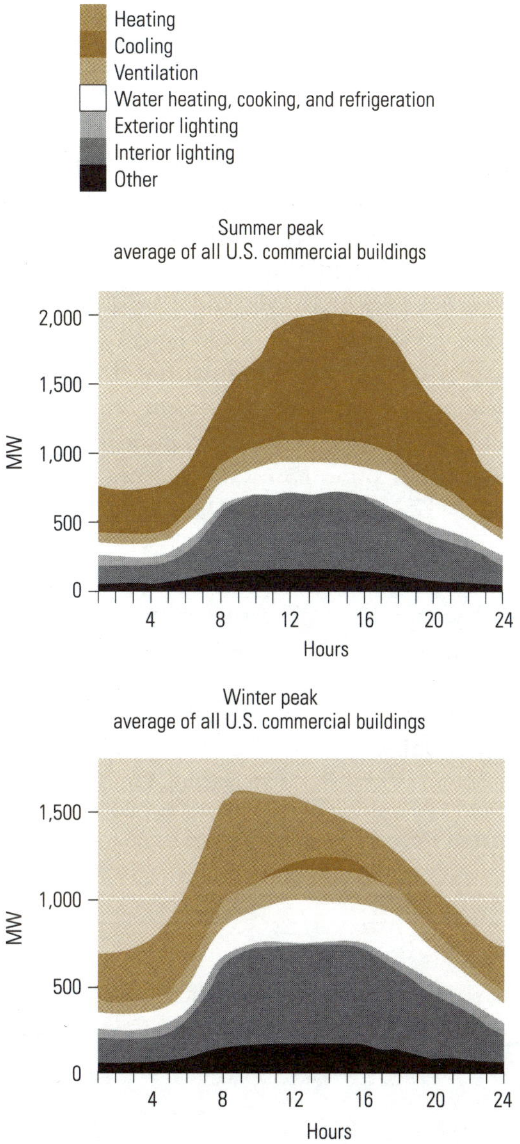

Source: EPRI Commend Database [13]

4.3 Lighting operating hours

How many hours per year are the lights on? LBNL researchers have estimated lighting operating hours for a variety of building types. (**Table 4-1**).[14] These estimates were produced from detailed audit information collected by DOE in 1986 and from several utility studies.

4.4 Lamp production

The U.S. Department of Commerce's Bureau of the Census has recorded data on the U.S. manufacture and sale of electric lamps since 1943. The 1994 data, for example, show that 36 companies produced over 3.5 billion electric lamps worth over three billion dollars. About 70 percent of these lamps were types typically used for illumination of buildings, while the remaining 30 percent was made up of photographic flashbulbs, projection lamps and other photolamps, miniature incandescents, and Christmas tree lamps.[15]

Figure 4-3 shows U.S. lamp production in 1972 and 1994. The vast majority of lamps produced have historically been and still are incandescent. However, fluorescents have gained an increasing share of the total lamp production at the expense of incandescents. HID lamps have maintained a small fraction of the total lamp production.

Graphing the lamp shipment *values* over the same period of time shows a different picture.

Table 4-1

Effective annual lighting hours by building type

These values represent the average hours that a type of lamp is on in buildings that use that type of lamp.

	Fluorescent hours	Incandescent hours	HID hours
Small office	3,624	3,365	3,583
Large office	3,624	3,365	3,583
Restaurant	4,957	5,361	7,223
Retail	4,064	3,867	4,883
Grocery	6,019	5,477	8,601
Warehouse	3,739	3,465	4,711
School	2,462	3,337	2,943
College	3,249	3,186	3,174
Health	7,955	8,086	8,694
Lodging	8,572	8,331	8,502
Miscellaneous	4,005	3,062	5,404

Source: Sezgen et al. [14]

Figure 4-3

History of U.S. lamp production and value

Fluorescent technologies' share of lamp production is increasing. From 1972 to 1994, the market for fluorescent lamps more than doubled, going from 290 million to 599 million units. Fluorescent lamps now account for 46 percent of the lamp market value–though they comprised just over a quarter of the number of lamps produced.

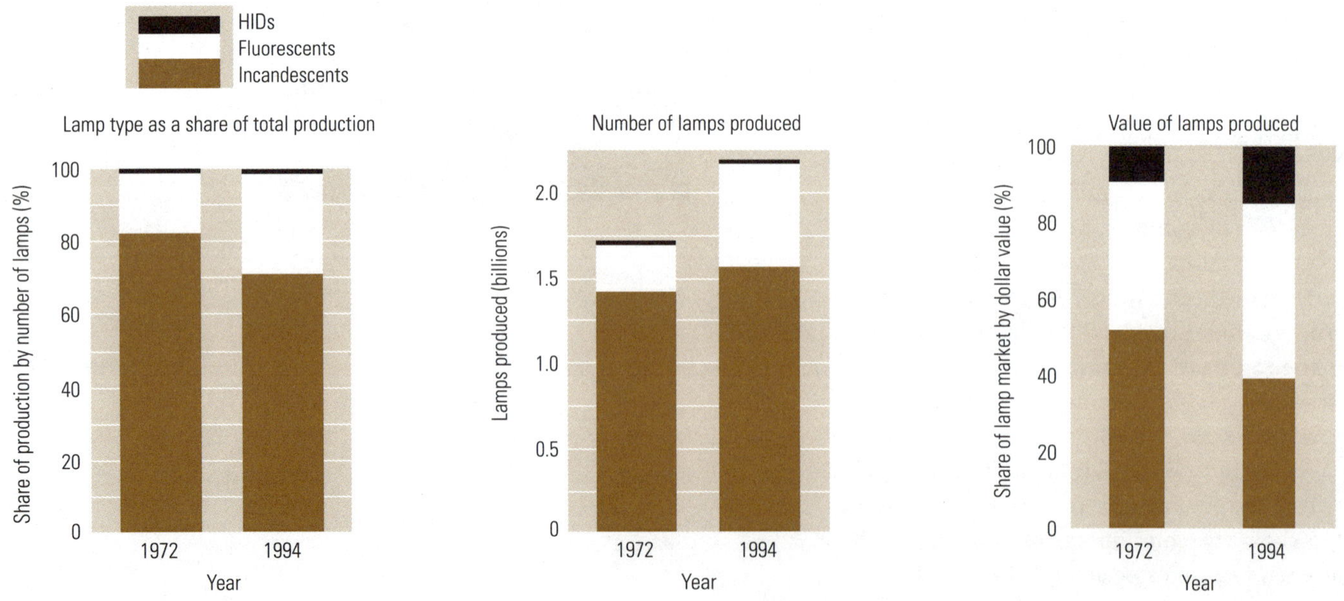

Source: U.S. Dept. of Commerce [15]

Though incandescent lamps outnumber fluorescents, the value of the fluorescents is now greater than the value of incandescents. HIDs also account for a larger percentage of total value than their small numbers reveal.

4.4.1 Incandescents

Just over 90 percent of incandescent lamps produced are ordinary 15 to 150 watt, 100 to 130 volt lamps. Less than 10 percent of the total nondecorative incandescents produced in 1994 were tungsten halogens, R lamps, or PAR lamps (**Figure** 4-4).

The Energy Policy Act of 1992 banned U.S. production of standard incandescent "R" and "PAR" floodlights and spotlights in October 1995. This action will continue to increase the market share of higher efficiency halogen and halogen-infrared incandescents.

4.4.2 Fluorescents

Figure 4-5 shows 1994 fluorescent lamp production by lamp type. Although census data only began to disaggregate linear T8s from other linear fluorescents over 30 watts (primarily T12s) in 1992, it shows that T8s are gaining an increasingly large share of the market. T8 production in 1994 was over three times what it was in 1992.

In 1991 EPRI surveyed lamp and ballast manufacturers and forecast the future shipments of T8s and T12s. They projected that by 1995, T8s would account for 20 percent of the 17 to 40 watt rapid-start lamps sold (**Figure** 4-6).[16] Census data from 1992, 1993, and 1994 shows T8 sales running significantly higher than EPRI had projected for those years.

T8 production in 1994 was over three times what it was in 1992.

Figure 4-4

U.S. incandescent lamp production (1994)

Most incandescents are ordinary tungsten-filament "A" lamps. Efficiency standards recently eliminated standard R and PAR lamps, which boosts the market for halogen and halogen-infrared lamps.

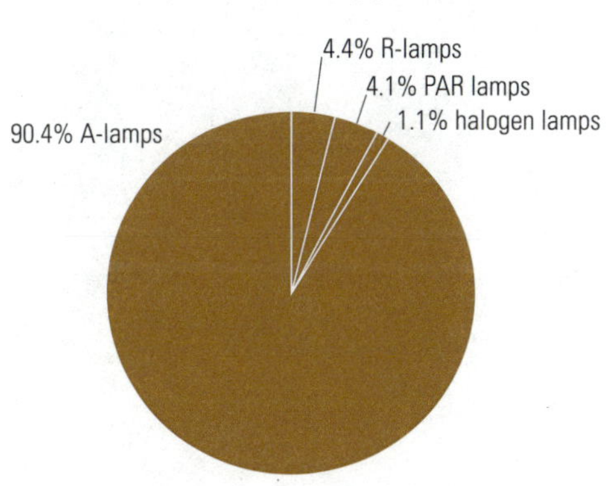

Source: U.S. Dept. of Commerce [15]

Figure 4-5

U.S. fluorescent lamp production (1994)

T8 lamps rose from five percent market share to 14 percent from 1992 to 1994 while the category containing linear T12s fell from 61 to 52 percent.

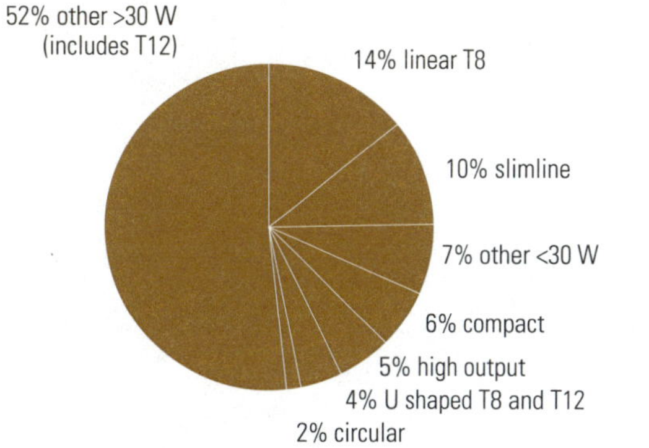

Source: U.S. Dept of Commerce [15]

Figure 4-6

T8 lamp shipments exceed predictions

EPRI's 1991 estimates of T8 and T12 U.S. lamp shipments are shown in the stacked bars. Census data for 1992–1994 show T8s exceeding the EPRI predictions by about 30 percent.

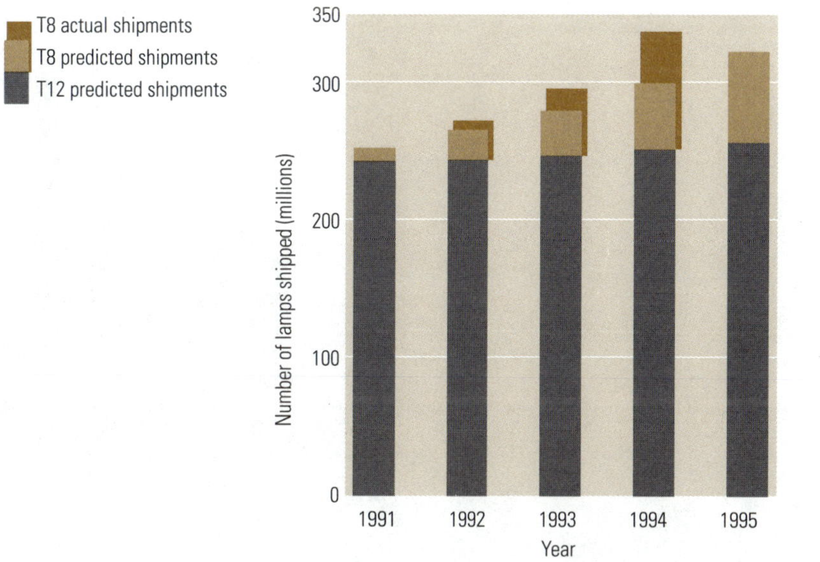

Source: EPRI [16], p. 5-4 and U.S. Department of Commerce [15]

Compact fluorescent lamp sales have risen dramatically in recent years (**Figure 4-7**), though not as rapidly as some analysts expected. EPRI predicted in a 1992 report that shipments of compact fluorescent lamps would increase to 47 million in 1993 and to almost 72 million lamps by 1995.[17] As Figure 4-7 shows, 1994 production was just over 35 million lamps.[18] Worldwide sales of CFLs reached an estimated 240 million units in 1995, more than five times the 1988 level.[19]

Figure 4-7

CFL sales on the rise

U.S. CFL sales rose from 200,000 units in 1982 to over 35 million units in 1994.

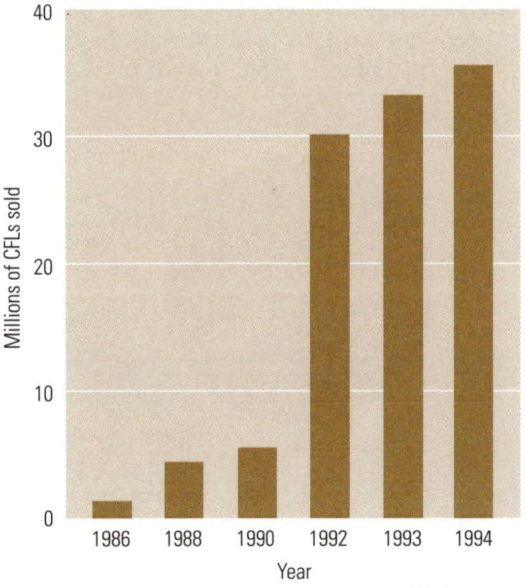

Source: EPRI [17] (1986, 1988 and 1990 data), and U.S. Dept. of Commerce [15] (1992, 1993 and 1994 data)

World CFL sales rose from 45 million units in 1988 to 240 million units in 1995.

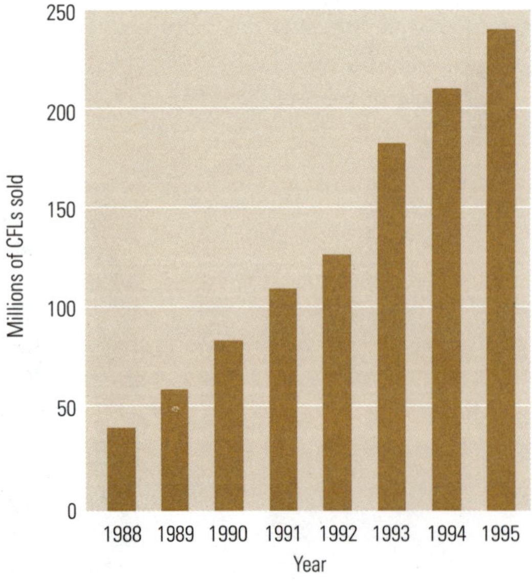

Source: Worldwatch Institute [19]

4.4.3 High-intensity discharge lamps

As shown in Figure 4-7, HIDs make up a very small share of total lamp production. HID production data (**Figure 4-8**) provide some insight into the mercury vapor, metal halide, and high-pressure sodium production trends.

Figure 4-8

U.S. HID lamp production trends

From 1992 to 1994, metal halide's share increased from 26 to 36 percent, while shares of high-pressure sodium and mercury vapor decreased.

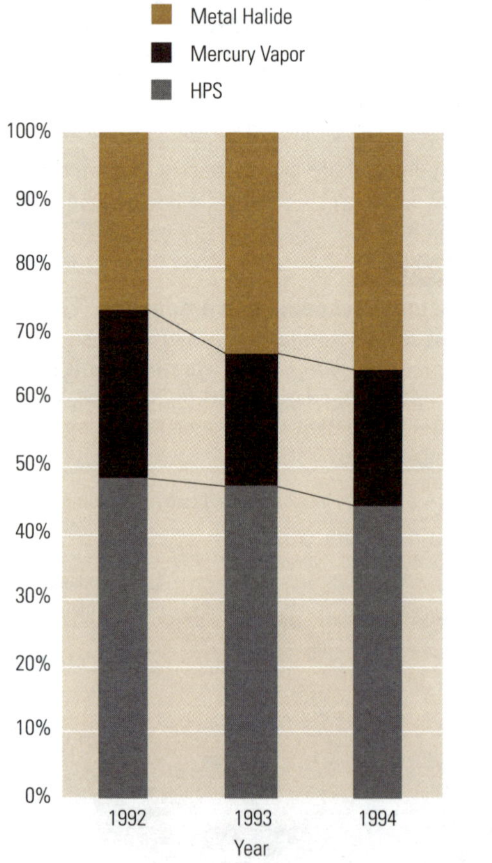

Source: U.S. Dept. of Commerce [15]

4.5 Trends in Ballast Shipments

The U.S. Bureau of the Census has recorded data on the U.S. manufacture and sale of fluorescent lamp ballasts since 1955.[20] Ballast shipment data from 1981 to 1996 show a clear trend of electronic ballasts' increasing share of production at the expense of corrected-power-factor magnetic ballasts (**Figure 4-9**). The shipments of electronic ballasts in 1995 was more than ten times that in 1990. Total shipments of electronic ballasts rose from about half a million units in 1986 to 32.9 million units in 1995 (**Figure 4-10**), but declined slightly in 1996 to 30.3 million units. Extrapolating data for the first two quarters of 1997 indicates shipments of about 35 million units for the year.

Electronic ballasts are ultimately expected to displace magnetic ballasts for most fluorescent lighting applications. This market transformation may come about even sooner if the U.S. Department of Energy succeeds in implementing new efficiency standards. In March 1994 DOE proposed a standard for fluorescent ballasts that could only be achieved with electronic ballasts with power factor of 0.9 or higher. After viewing public comments on the March 1994 standard, DOE is currently revising their analysis for the standard. DOE could issue a proposed rule as early as 1998, if the ballasts category remains a high priority for DOE rulemaking. If adopted, the new standards would take effect three years after publication of the *final* rule.[21]

While electronic ballasts will dominate the market in years to come, hundreds of millions of magnetic ballasts are currently in service and many of them will last another decade or more. Even if DOE bans magnetic ballasts in the upcoming standards, it will take several years for those rules to take effect, and vendors will be able to sell magnetic ballasts still in the distribution chain long after any production bans take effect. Even under the most aggressive efficiency scenario, then, it will be at least a decade before electronic ballasts outnumber magnetic units in the installed building stock.

While electronic ballasts will dominate the market in years to come, hundreds of millions of magnetic ballasts are currently in service.

Figure 4-9

U.S. fluorescent ballast shipments

Electronic ballasts are gaining market share relative to corrected-power-factor magnetic ballasts.

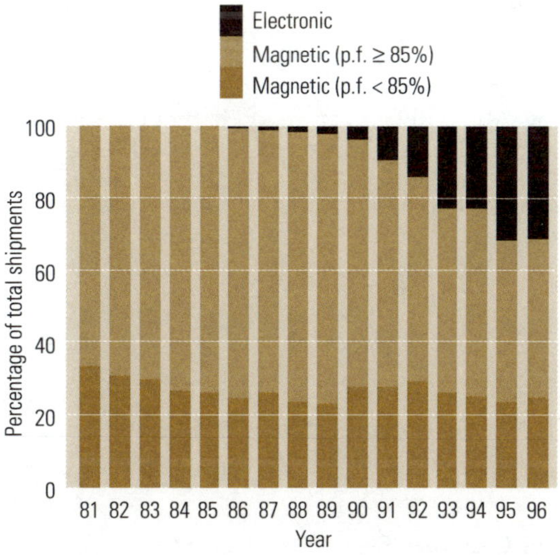

Source: U.S. Dept of Commerce [20]

Figure 4-10

U.S. shipments of electronic ballasts

Shipments of electronic ballasts have climbed rapidly through the 1990s. Although there was a slight dip in 1996, data for the first two quarters of 1997 indicate a rebound to about 35 million units for the year.

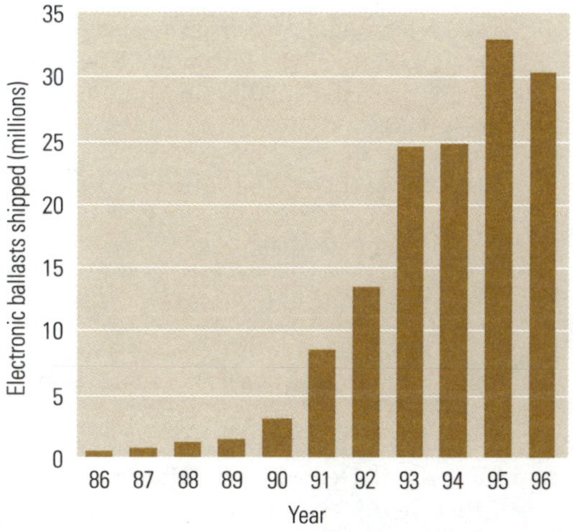

Source: Department of Commerce [20]

1 Total 1996 electricity use by sector (on the left-hand side of the graph) from EIA, *Monthly Energy Review*, DOE/EIA-0035 (March 1997), table 7.2.

Breakdowns of each sector's energy use (on the right-hand side of the graph) were obtained by applying percentage of energy for each to the sectoral energy use for 1996 from DOE/EIA-0035, table 7.2. The sources for the percentage breakdowns for each sector are:

• For residential: EIA, *Annual Energy Outlook 1997* (1997), DOE/EIA-0383(97), table A4, 1995 data year. Early release of this report was available from EIA's web site: http://www.eia.doe.gov.

• For commercial: EIA, *Annual Energy Outlook 1997* (1997), DOE/EIA-0383(97), table A5, 1995 data year.

• For industrial: EIA, *Manufacturing Consumption of Energy 1991* (December 1994), DOE/EIA-0512(91), table A36, 1991 data year.

• For other: EEI *Statistical Yearbook of the Electric Utility Industry 1995* (1997), table 41) to the 1996 "other" energy use from DOE/EIA-0035, table 7.2.

The "other" category as reported in table 7.2, includes public street and highway lighting, other sales to public authorities, sales to railroads and railways, and interdepartmental sales, but neglects most of the utilities' own use. Our "other" category includes an additional 8.6% of total electricity use to account for electricity generated by utilities for utilities' own use ("Company Use & Free Service") and line losses ("Lost and Unaccounted for"), from EEI *Statistical Yearbook of the Electric Utility Industry 1995* (1997), table 9.

2 J.W. Hanford, J.G. Koomey, L.E. Stewart, M.E. Lecar, R.E. Brown, F.X. Johnson, R.J. Hwang, L.K. Price, "Baseline Data for the Residential Sector and Development of a Residential Forecasting Database" (May 1994) LBL-33717, UC-1600, p. 106.

3 "Drivers of Electricity Growth and the Role of Utility Demand-Side Management," EPRI TR-102639 Final Report (August 1993), pp. 2-12, 3-15, 4-15, and 5-15.

4 EPRI [3], p. 4-1.

5 EIA, *Annual Energy Outlook 1997*, Table A5.

6 Barbara Atkinson, personal communication (February 24, 1997), Lawrence Berkeley National Laboratory, One Cyclotron Road, MS 90-4000, Berkeley, CA 94720, tel 510-486-7227, fax 510-486-6996, e-mail: BAAtkinson@lbl.gov. Previous projections were published in Barbara Atkinson, James McMahon, Evan Mills, Peter Chan, Terry Chan, Joseph Eto, Judith Jennings, Jonathan Koomey, Kenny Lo, Matthew Lecar, Lynn Price, Francis Rubinstein, Osman Sezgen, Tom Wenzel, "Analysis of Federal Policy Options for Improving U.S. Lighting Energy Efficiency: Commercial and Residential Buildings", LBL-31469, Lawrence Berkeley National Laboratory (December 1992).

7 Judith Jennings, Rich Brown, Mithra Moezzi, Evan Mills, Robert Sardinsky, Barbra Heckendorn, David Lerman, Lyle Tribwell, "Residential Lighting: The Data to Date" (March 1, 1996), draft report, presented at Illuminating Engineering Society Conference (1996), assumes 30 fixtures per household.

8 EIA, *Annual Energy Outlook 1997*, Table A4.

9 EPRI [3], pp. 3-1 and 3-17.

10 EIA, *Manufacturing Consumption of Energy 1991* [1].

11 John Freeman, et al. "Lighting Industrial Facilities: An Untapped Source of Energy Savings," *Proceedings of the ACEEE 1992 Summer Study on Energy Efficiency in Buildings*, v. 1 (Summer 1992), p. 1.68.

12 "Impact of Demand-Side Management on Future Customer Electricity Demand: An Update," EPRI CU-6953, Final Report (September 1990), p. 2-8.

13 EPRI Command Database, 1995 projections, Regional Economic Research, 12520 High Bluff Drive, Suite 220, San Diego, CA 92130-2014, tel 619-481-0081, fax 619-481-7550.

14 Sezgen et al., "Technology Data Characterizing Lighting in Commercial Buildings: Application to End-Use Forecasting with Commend 4.0," (May 1994) Final Report LBL-34243, Energy and Environment Division, LBL, One Cyclotron Road, 90-4000, Berkeley, CA 94720, p. 8; and DOE/EIA, "Lighting in Commercial Buildings," DOE/EIA-0555(92)/1, Energy Information Agency, Washington, DC.

15 U.S. Department of Commerce, Bureau of the Census, "Current Industrial Reports: Electric Lamps Summary for 1994," MQ36B(94)-5, (Superintendent of Documents, U.S. Government Printing Office, Washington, DC 20402). Available online at www.census.gov.

16 Electric Power Research Institute, "Survey and Forecast of Marketplace Supply and Demand for Energy-Efficient Lighting Products," EPRI TR-100288 (December 1992), p. S-4.

17 EPRI [16], p. S-3.

18 U.S. Department of Commerce [15].

19 Evan Mills, Lawrence Berkeley National Laboratory, and Nils Borg, National Board for Industrial and Technical Development, Sweden, as cited in Lester Brown, et al., *Vital Signs 1996*, Worldwatch Institute (New York: W.W. Norton Co., 1996), p. 61.

20 U.S. Department of Commerce, Bureau of the Census, "Current Industrial Reports: Fluorescent Lamp Ballasts Summary for 1995," MQ36C(95)-5, and "Current Industrial Reports: Fluorescent Lamp Ballasts–Third Quarter 1996," MQ36C(96)-3, Superintendent of Documents, U.S. Government Printing Office, Washington, DC 20402. Available online at www.census.gov.

21 Terry Loggi, personal communication (February 19, 1997), Mechanical Engineer, Office of Codes and Standards, U.S. Department of Energy, Mail Routing EE-43, Room 1J-018, Washington, DC 20585, tel 202-586-1689, fax 202-586-4617, e-mail TERRY.LOGGI@HQ.DOE.GOV

Technology

5 DAYLIGHTING

Virtually a lost art a few decades ago, daylighting is now a design strategy widely recognized for its energy-savings potential, as well as for its positive impacts on occupant satisfaction. Several spectrally selective glazing systems are currently available, and other daylight control devices are under development. Technologies such as prismatic light guides or fiber-optic distribution systems enhance the potential of bringing natural light deep into buildings. Daylight-sensing controls and dimming systems can yield large energy and demand savings by providing only the amount of electric light needed to supplement natural lighting.

Sunlight is free and uses no electricity. Human beings are designed to live in sunlight. Both economics and the less rigorous but more important imperatives of health and esthetics, therefore, favor the practical use of daylighting in buildings. Some electric utilities, mindful of potential benefits to themselves as well as to their customers, already offer rebates and design assistance with daylighting applications in new buildings.

Such assistance is important to avoid glare, excessive luminance ranges, and overheating from poorly executed daylighting. Much of that expertise, common among the architects of Victorian and earlier (back to classical) daylit buildings, was lost when several generations of architects were brought up on artificial light, and it must now be redeveloped. Simply adding lots of windows to a building to "let the sun shine in" can create so much glare, make other spaces look so dark by contrast, and admit so much unwanted heat gain that the space is virtually unusable. Poorly designed daylighting can be worse than none. Properly done, daylighting can make new and existing buildings both efficient and delightful. The tools to do it right exist, are available, and are being applied by an increasing body of talented design specialists.

Properly done, daylighting can make new and existing buildings both efficient and delightful. Poorly designed daylighting, however, can be worse than none.

HIGHLIGHTS

- Daylighting is both art and science. The key is to admit only as much light as necessary and to distribute it evenly and without glare. Poorly designed daylighting can be worse than none.

- Daylighting is applicable in both clear and cloudy climates. In fact, many designs work best under bright overcast skies.

- Daylight-sensitive dimming controls for electric lights can yield substantial energy and peak lighting demand savings.

- Since sunlight produces less heat per unit of light than typical electric lights, daylighting can reduce cooling load.

- Several glazing options for improved daylighting design now exist and more advanced products are under development. Spectrally selective tints or coatings—including advanced retrofit films—reduce heat gain but allow good visible transmissivity. The next decade may bring prismatic, refractive, and holographic glazings to better control incoming daylight.

5.1 Daylighting Design Approaches

Common daylighting design elements include ordinary façade fenestration, increased penetration into the plan via atria and reentrants (as in U- or E-shaped buildings that maximize exposure to the sun), clerestories, roof monitors (fenestration raised above and not parallel to the roof plane), skylights, and stepped-back building forms (which avoid the "urban canyon" effect). Much of the art of practical daylighting lies in the use of light shelves, wide windowsills, special reflectors, louvers, walls, blinds, and other ways to bounce light deep into the building's plan. These methods also avoid the glare and the strong modulation (intense light near the window, rapidly falling off with increasing depth) of direct-beam daylighting. Bi- or multilateral daylighting is much preferred over light from only one direction. Interior and exterior foliage can also be used to control light, temperature, humidity, views, privacy, and esthetics.

See the following publications for more information on daylighting:

Daylighting Design and Analysis, *by Claude Robbins, Van Nostrand Reinhold, New York (1986)*

Daylighting Performance and Design, *by Greg Ander (in press, 1995)*

The amount of glazing in a daylit building is far less important than where the glazing is located and how it is distributed. In fact, most daylighting concepts try to keep *direct* sunlight out of the building, since it can result in overheating, glare, and uncomfortable visual contrast, and can fade or bleach materials. **Figure 5-1** shows some simple daylighting techniques.[1] An example of exterior light shelves is shown in **Figure 5-2**. A good daylighting design will accommodate all types of skies—clear, partly cloudy, or cloudy.

Figure 5-1

Simple daylighting techniques

This schematic shows a mix of top- and side-lighting, light shelves, high-reflectance ceilings, and wall diffusion to provide fairly uniform deep-plan daylighting without the glare of direct sun.

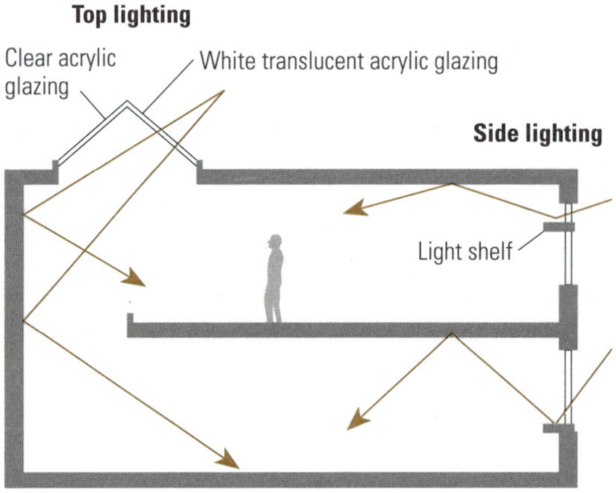

Source: Southern California Edison [1]

Figure 5-2

Exterior light shelves on the Bonderson Building in Sacramento, CA

5.2 Energy Savings and Demand Reduction

Designers, engineers, utility companies, and building energy standards are increasingly recognizing daylighting as a potential source for energy savings. Utilizing the free light from the sun and sky for interior illumination reduces the need for electric lighting in the daylit areas. Because up to 70 percent of the usable space and installed lighting capacity of many commercial buildings is in the perimeter zone, daylighting has a greater potential to reduce energy and peak demand than is commonly recognized, even in conventional buildings.[2] Depending on whether the photo-

CASE STUDY **DAYLIGHTING AT OCOSTA HIGH SCHOOL**

The Ocosta Junior/Senior High School (**Figure 5-3**), built in 1988, is a 67,000-square-foot building located in Westport, Washington, within five miles of the Pacific Ocean. Over 40 skylights bring light deep into the building's interior spaces, and 87 percent of the school has access to daylight. Electric lighting is installed at a relatively efficient density of 1.35 watts per square foot and photocells control 42 percent of the lights. During periods of bright daylight, 37.8 kW of connected lights are turned off by the photocell controls, while maintaining classroom illumination levels. During bright periods, the Ocosta school experiences a 99 percent reduction in lighting load connected to daylight sensors, and therefore a 42 percent reduction in whole building lighting load.[3]

Figure 5-3

High school uses daylighting

The top photo shows the exterior of Ocosta Jr/Sr High School; the gymnasium is on the right. The bottom photo shows the inside of the gym, with skylights supplemented by electric lighting.

Photos by David Melody
Source: Burr Lawrence Rising and Bates, Architects

sensitive controls step-dim the lamps (to, say, 50 or 30 percent of full output), continuously dim the lamps (down to 5 percent output), or turn them off completely when some threshold of daylight is available, the lighting energy savings of daylit spaces often range from 40 to 60 percent when averaged over the entire year (see **5.4**). Because daylight is readily available during hot summer afternoons, the peak demand benefits of daylighting controls are typically greater in percentage terms than the energy savings.[4]

Designing for daylighting can also result in a reduced building cooling load. All of the energy used for lighting eventually becomes heat, which adds significantly to the internal gains of a building. Since the luminous efficacy of daylight—from a clear or an overcast sky—is significantly higher than that of most common types of electric light, daylight contributes less heat to a space for a given amount of light (**Figure 5-4**).[5] Design expertise is required to ensure that the daylight is adequately and relatively evenly diffused throughout the space. Selecting a glazing which maximizes light transmission and minimizes total solar transmission can further reduce the heat gains resulting from daylight penetration, as we discuss next.

5.3 GLAZING SELECTION

The luminous efficacy of daylight is significantly higher than that of most common types of electric light.

Glazing should be selected with consideration for daylighting goals as well as the building's heating and cooling requirements. Most large commercial buildings are dominated by cooling loads, so glazing that transmits adequate light for the daylighting application and minimizes solar heat transmission is usually best. In buildings dominated by heating loads, glazing should be carefully chosen to minimize heat loss and, in some cases, configured to increase passive solar heat gain (see sections 5.1 and 5.5 of the 1996 *Space Heating*

Taken to the extreme, a focus on reducing cooling loads would lead one to abandon windows altogether in favor of opaque wall systems.

Technology Atlas), while maximizing daylighting. Horizontal skylights are particularly susceptible to heat loss as well as to direct solar gain, so they should be chosen discriminately for daylighting applications.

The spectral distribution of solar energy is shown in **Figure 5-5**. About half of the sun's energy is visible to the human eye, while the other half (longer-wavelength near-infrared or shorter-wavelength ultraviolet) is invisible, contributing only heat to the building interior. To reduce cooling load, the ideal window would transmit as much of the visible portion of the sun's energy as is desired for the particular application, while rejecting the rest of the solar radiation.

Glazing systems are rated using several parameters. The most common performance metric for rating total energy transmission is the shading coefficient (SC). The shading coefficient is the ratio of solar radiation (both visible and invisible) passing through a glazing system, to the amount that would pass through 1/8-inch clear glass. A similar metric that is becoming more common is the solar heat gain coefficient (SHGC), which is the fraction of the incident solar energy transmitted through the window.

Taken to the extreme, a focus on reducing cooling loads by using glazings with the lowest possible SC or SHGC values would lead one to abandon windows altogether in favor of opaque wall systems. Obviously, the reason for having windows is to transfer daylight from the outside to the inside (and vice versa, in the case of retail display windows). Therefore, reductions in SC and SHGC must always be considered in conjunction with the corresponding reduction in visible transmittance (T_{vis})—the fraction of visible light transmitted through the window.

Transmittance is important not only for visibility but also for using daylighting to offset interior electric lighting loads and associated cooling loads. Reflective glass products with T_{vis} values as low as 5 percent are common in many cooling-load-dominated regions. The dark, reflective building envelopes of many

Figure 5-4

Building heat gain from different sources of light

With proper glazing selection, daylight will contribute far less heat to a building per unit of light delivered to the interior than do electric lights.

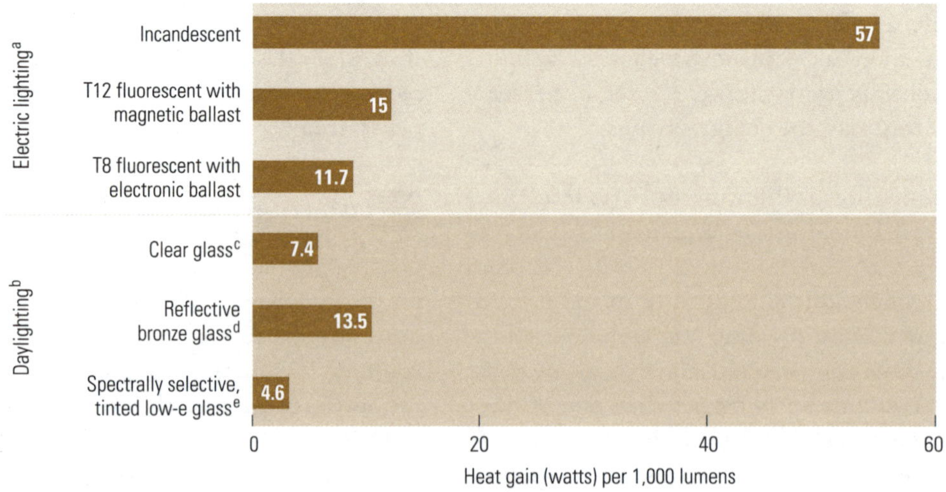

Notes

a: Assumes efficacies of 17.5 lm/w, 65 lm/w, and 85 lm/w, respectively, for the electric lighting options.

b: For daylighting 125 lm/w efficacy of daylight assumed (from Claude Robbins [5], p. 255). Glazing options data drawn from LBL's WINDOWS 4.0.

c: T_{vis} 0.82, SC 0.90, SHGC 0.78 and relative heat gain of 188 Btu/h/ft^2.

d: T_{vis} 0.20, SC 0.38, SHGC 0.33 and relative heat gain of 84 Btu/h/ft^2.

e: T_{vis} 0.40, SC 0.26, SHGC 0.23 and relative heat gain of 57 Btu/h/ft^2.

Source: E SOURCE calculations

sunbelt office buildings use glazings with such low T_{vis} that they create what solar architect Ed Mazria has dubbed, "all glass buildings with no windows." The windows are there for visual and economic reasons (glass is cheaper than masonry), but the highly tinted glass creates such dim interiors that they need almost as much electric lighting as if the walls were opaque. Ironically, the electric lights then typically release more waste heat year round than the sun would deliver through normal windows. The air conditioners thus must be sized to remove heat put there by lights that wouldn't be needed during the day if the right glazings were used.

However, some reduction in T_{vis} below that of clear glass is often desirable in commercial buildings for glare reduction, privacy, and esthetics. T_{vis} values in the range of 40 to 60 percent will generally meet these concerns while preserving adequate daylighting, since daylight is vastly more intense than indoor visual tasks require.

To indicate a window's relative performance in rejecting solar heat while transmitting visible daylight, a new performance parameter is gaining acceptance—the ratio of daylight transmittance to the shading coefficient. This dimensionless ratio, variously called the luminous efficacy constant (K_e) or lighting-and-cooling selectivity index (LCS), is beginning to appear in manufacturers' literature to permit direct comparison of the "coolness" of different glazings. Even where it is not published as a separate data point, it can be calculated as $K_e = T_{vis} \div SC$.

Clear glass, which lets in roughly equal amounts of visible light and solar near-infrared energy, has a K_e close to 1.0. A perfectly selective glazing, which would allow all visible light to pass through while blocking all of the invisible near-infrared and ultraviolet, would have a K_e of about 2.0, since about half of the solar radiation is in the visible spectrum.

Figure 5-5

The solar spectrum

The ideal window will transmit visible light and block ultraviolet and near-infrared radiation.

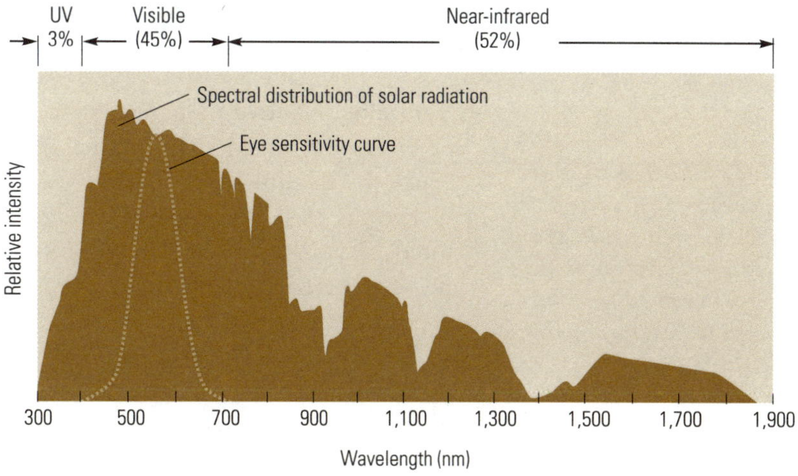

Using these numbers, glazing systems can be chosen for the desired balance of heat to light transmittance. **Figure 5-6** shows the shading coefficient and visible transmissivity of several types of windows.

OTHER BENEFITS OF DAYLIGHTING

Daylighting benefits go beyond energy savings and power reduction. Daylight—the light source that most closely matches human visual response—is considered the best source of light for color rendering and improves people's ability to perform visual tasks.[6] Not surprisingly, several studies have found that people prefer daylight over other sources of light, suggesting that daylighting positively affects worker satisfaction and perhaps productivity.[7] Other studies have found that access to windows, which is compatible with daylighting, is important for psychological well-being and physical health.[8]

5.3.1 Existing daylighting glazing technologies

There are three basic categories of spectrally selective glazings: those with spectrally selective tints, those with spectrally selective low-e coatings, and those that combine tints and low-e coatings.

Blue and green tints are better for daylighting since they naturally transmit a relatively high fraction of visible light compared to heat-producing infrared energy.

Spectrally selective tints. Energy-absorbing materials dispersed throughout glass will reduce the shading coefficient as well as give color to the glass. The most common colors of tinted glass—bronze and gray—block light and solar near-infrared heat in roughly equal proportions: they are not spectrally selective. Black-tinted glass (such as PPG's Graylite) is the worst choice for cooling load reduction, since it absorbs much more visible energy than near-infrared. Blue and green tints are better for daylighting since, unlike other colors, they naturally transmit a relatively high fraction of visible light compared to heat-producing infrared energy. These spectrally selective tints cost slightly more than ordinary tints, but that incremental cost may drop as market share grows.[9]

A number of spectrally selective tinted glass products are available. PPG Industries' blue-tinted Azurlite™ glass provides visible transmittance of 64 percent with an SC of 0.47, for a K_e of 1.36, while LOF's EverGreen™ glass offers T_{vis} of 59 percent with an SC of 0.44, for a K_e of 1.34. These values are for double-pane units with 1/8-inch glass.

Spectrally selective clear low-e coatings. Low-e coatings, originally designed for cold climates, have been modified to reject heat for cooling applications. While first generation low-e coatings were designed to be highly transparent throughout the solar spectrum (including near-infrared), the newer spectrally selective coatings reflect near-infrared energy, thereby reducing nonvisible transmittance. This option is perhaps the best glazing technology for daylighting and solar control for residential applications.[10] Low-e coatings are widely available at relatively low cost, and have become standard in most common high-end residential windows from Andersen and other major manufacturers.

One of the most exciting recent breakthroughs has come in the area of low-e retrofit films for application to existing windows. Southwall Corporation, the firm that pioneered Heat Mirror™ spectrally selective window films in new windows, released in early 1994 a retrofit product called Solis™.[11] Available in clear and grey-tinted versions, this product is unique in that other retrofit window films have little or no spectral selectivity; most reduce visible transmittance as much as they reduce heat gain. When applied to clear 1/4-inch glass, Solis cuts visible trans-

Figure 5-6

Total solar and visible light transmissions for selected glazing units

Glazing units with high visible light transmission and low shading coefficients (SC) are best for daylighting in buildings with cooling loads.

Glazing	Visible	Shading coefficient
Clear □	0.82	0.90
Low-e coating ■	0.77	0.69
Dark reflective bronze tint ○	0.20	0.38
Spectrally selective green tint ●	0.68	0.59
Spectrally selective low-e (clear) ■	0.70	0.49
Spectrally selective low-e with green tint ■	0.40	0.26

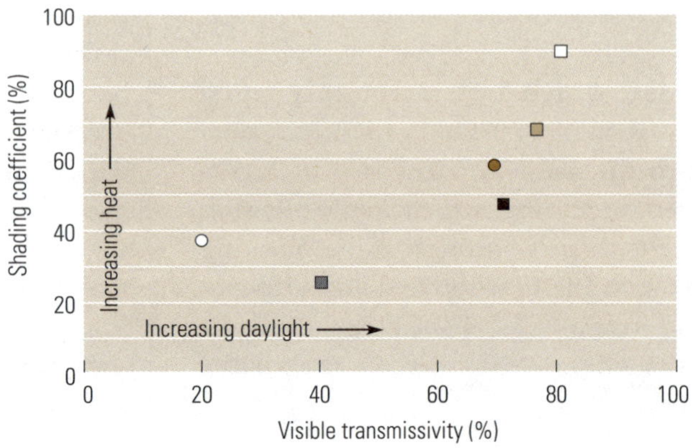

Source: E SOURCE calculations using WINDOWS 4.0

mittance from 89 percent to 70 percent but cuts solar heat transmission almost in half, from 83 to 45 percent, yielding a K_e of about 1.4. It blocks virtually all of the ultraviolet as well, which helps reduce fading of interior furnishings. **Table 5-1** summarizes the performance features of this window film, which could offer significant cooling load reduction without sacrificing daylight in existing commercial buildings. The cost is high, however—about $2.00 per square foot, wholesale, uninstalled.

Table 5-1

Performance of Solis spectrally selective retrofit window film

This window film cuts heat gain considerably more than it cuts visible transmittance.

Glass type	Visible light transmission (%)	Solar heat transmission (%)	Visible light reflection (%)	Ultraviolet transmission (%)	Shading coefficient
Clear 1/4-inch	89	83	8	67	0.96
Clear 1/4-inch with Solis™	70	45	9	2	0.50
Clear 1/4-inch with Solis™ Grey	40	38	7	2	0.44

Source: Southwall Technologies

Combination of spectrally selective tints and low-e coatings. By combining spectrally selective tints and spectrally selective low-e coatings, even higher values for K_e can be achieved. For example a commercially available double-pane glazing unit using green tinted outer glass, clear inner glass, and a spectrally selective low-e coating offers a T_{vis} of 0.4 and a SC of 0.6, for a K_e of 1.54.

5.3.2 Emerging daylighting glazing technologies

Several types of glazing materials are under development which use optical ingenuity to reject or redirect incident solar radiation for better daylighting control. A typical application of these panels is shown in **Figure 5-7**.

Prismatic panels. Made of optical prisms and insulating glass, prismatic panels incorporate principles of refraction and reflection to control daylight. Prismatic glazing systems typically have a cut-off angle range and a transmission angle range. The glazing system reflects direct sunlight striking the glass in the cut-off range, but changes the direction and intensity of sunlight striking the glass in the transmission range. Daylight striking the window in the transmission range is usually directed towards the ceiling. Prismatic glazings are currently available only from one European manufacturer (Siemens), but other types are in development throughout the world.[12] The manufacturer claims that prismatic panels increase daylight in a room by 30 percent under an average sky and 135 percent under clear skies, but LBNL suggests field studies are needed to corroborate these claims.[13]

Refractive daylighting device and light control panels. Researchers at the Queensland University of Technology in Australia are developing refractive daylighting devices and light control panels. The refractive device could be permanently mounted in a window to redirect sunlight towards the ceiling. Calculations indicate a potential payback on the order of two years for commercial applications. Light control panels consist of a laser-cut rotatable acrylic panel which can either reject light or deflect light deeply into the room. The device is still under development.[14]

Prismatic glazings are currently available only from one European manufacturer, but other types are in development throughout the world.

Figure 5-7

How light-bending panels can improve daylighting

Using refractive and reflective optics allows direct daylight to be distributed deeper into a building's core.

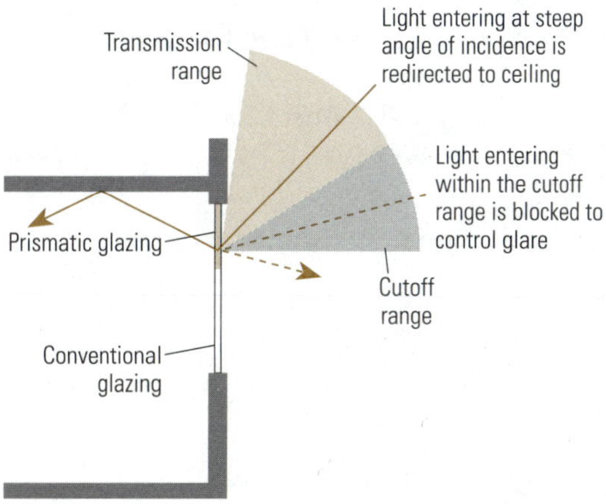

One analyst calculates that 15,000 MW of peak demand could be avoided by simply installing photo-sensitive controls in existing U.S. commercial spaces.

For more information about existing and emerging daylighting technologies

General information

Windows and Daylighting Group
Lawrence Berkeley National Laboratory
One Cyclotron Road, Building 90-3111
Berkeley, CA 94720
tel 510-486-6845
fax 510-486-4089

National Fenestration Rating Council
1300 Spring Street, Suite 120
Silver Spring, MD 20910
tel 301-589-6372
fax 301-588-0854

Prismatic glazing

Siemens AG
Lighting Division
Ohmstrasse 50
D-83301 Traunreut
Federal Republic of Germany

Refractive daylighting device R&D

Department of Physics
Attn: I.R. Cowling
Queensland University of Technology
GPO Box 2434
Brisbane Q 4001, Australia

Light control panels R&D

Department of Physics
Attn: Ian Edmonds
Queensland University of Technology
GPO Box 2434
Brisbane Q 4001, Australia
tel 61-7-864-22595
fax 61-7-864-15212
e-mail: i.edmonds@qut.edu.au

5.4 Electric Light Controls

Even spaces carefully designed for daylighting require electric lights for nighttime or very overcast sky conditions. Control of these electric lights is essential to save energy and reduce demand. Although conscientious building occupants can manually turn off electric lights when daylighting is adequate, only automatic control ensures savings.

One analyst calculates that 15,000 MW of peak demand could be avoided by simply installing photosensitive controls in existing U.S. commercial spaces—most of which incorporate no specific daylighting design strategies beyond ordinary windows.[15] The Florida Solar Energy Center found that T8 lamps with dimmable electronic ballasts controlled by daylight sensors consumed 40 percent less energy than a comparable system without daylight controls.[16] The importance of controls is also recognized by California's Title 24 building code and by ASHRAE Standard 90.1, which only offers daylighting credits if automatic controls are installed. Several utility programs offer rebates for daylight-sensing controls and design assistance for daylighting with automatic controls.

Case Study: The Importance of Controls

The 8,500-square-foot Herbert C. Templeman Education Center in Port Hueneme, California was built in 1987 to house the administrative offices for the local elementary school district. Its daylighting features include skylights, side windows with high-performance glazing, and low interior partitions to allow better access to daylight. The design includes a two-step on/off daylighting control scheme. Computer simulations show that the building uses 55 percent less electricity for lighting and 36 percent less electricity overall than it would have without controls (**Figure 5-8**). Peak demand is also reduced by 49 percent. Payback for the controls was less than one year.[17]

Defining appropriate control zones is essential to achieving energy savings. Since access to daylight changes throughout the day and seasonally, one area of the building may receive adequate illumination from daylight while another may require electric lighting. The size, location, and configuration of the daylight aperture, the sky conditions, the building location, and the cost of controls must all be considered in order to determine lighting control zones that will maximize energy savings from daylighting.

Figure 5-8

Savings from daylighting controls

With daylighting controls, the Templeman building uses 36% less electricity overall than it would have without controls. Controls also reduce peak demand by 49%, and payback is in less than a year.

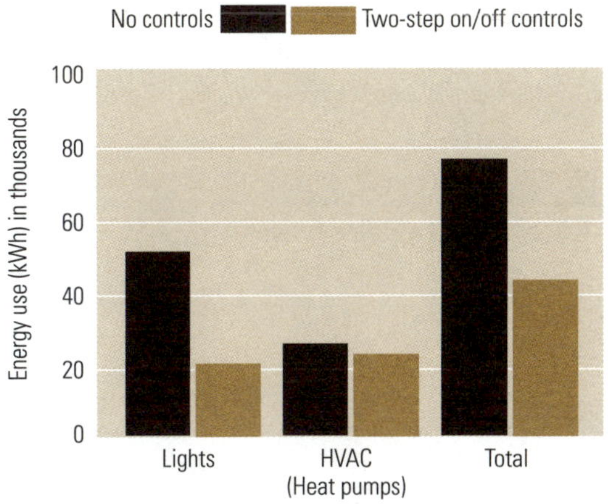

Source: Scott Ellinwood and Associates [17]

Automatic lighting controls for daylighting applications are readily available. The controls must include photosensors which detect the level of light in the space and respond accordingly to increase or decrease the electric light output. The lights can be operated for simple on/off control, multiple step control, or, for the maximum energy savings, they can operate in conjunction with dimming devices which raise or lower the illumination level depending on the available daylight (see **CHAPTER 11** for more information on electric light controls).

Entire packages of efficient lighting and controls measures will have more attractive economics than the controls evaluated alone.

In many cases, the payback for installing controls is five years or less. However, the payback depends on many factors, including the access to daylight, the controlled lighting density, the lighting hours of operation, the cost of electricity, and the capital cost of the equipment and installation. As discussed in section **3.1**, the entire package of efficient lighting and controls measures will have more attractive economics than the controls evaluated alone.

ASHRAE STANDARD 90.1 DAYLIGHTING CREDITS

ASHRAE Standard 90.1-1989 (*Energy Efficient Design of New Buildings Except Low-Rise Residential Buildings)* credits daylighting for its ability to reduce both the electric lighting demand and the building cooling load. The standard requires that the connected lighting power comply with the interior lighting power allowance (ILPA), which is calculated for each building. If the design calls for more connected power than the calculated ILPA, the designer can make use of lighting power control credits. Daylighting with controls can result in credits of up to 30 percent of the power of the lights connected to the controller.

Daylighting is also recognized in the standard's building envelope compliance requirements for its ability to reduce the building cooling load by reducing the internal gains due to lighting more than it increases heat gain through the daylighting aperture. To acknowledge this contribution of daylighting, a limited skylight area may be exempt from the thermal transmittance requirement for roofs, and additional vertical glazing may be permitted in certain climates. As with the lighting credits, building envelope credits are available only if automatic controls are installed. See **APPENDIX C** for more on the ASHRAE standard.

5.5 Light Transport Systems

As an alternative to complex tracking systems, simple self-contained light pipes are available at relatively low cost.

Light transport systems can introduce daylight deep into a building's core, although most of these systems are very costly, and cannot be justified on the grounds of energy savings alone. Transport systems typically consist of an outside light collection system, or heliostat, a light guide, and a luminaire which releases light into the interior space. The simplest form of a light guide is an empty shaft, along which a beam of light can travel. Other light guides include refractive ducts, fiber optics, lens guides, mirror ducts, acrylic rods, and fluid-filled tubes. **Figure 5-9** shows a variety of light guide systems.

As an alternative to complex tracking systems, simple self-contained light pipes are available at relatively low cost. The Sun Pipe, discussed on the next page, is an example of this type of product.

Refractive ducts or prismatic light guides. Prismatic light guides are sealed, empty cylindrical tubes or square-section ducts made of clear acrylic. Their inner surfaces are scored with precisely shaped prismatic grooves which cause total internal reflection and confine the light to the central air space. A light-emitting lens releases enough light to make the tube glow like a lamp.

Prismatic light guides can be used with purely artificial light, but lend themselves easily to daylighting or hybrid systems. In a hybrid system, daylight is fed into the "live end" of the light pipe network for distribution to fixtures in the illuminated space. A photocell at the "dead end" senses the amount of light flowing through the network and supplements it as needed with one or more large, highly efficient electric lamps.

TIR Systems in Burnaby, British Columbia is the main distributor of prismatic light guides in North America. 3M Corporation manufactures the prismatic inner film, but the guide is fabricated under license from TIR. TIR has installed daylighting light pipe systems (the Heliostat/Light Pipe System) in two buildings in Toronto, Ontario. One of the systems, which provides light to approximately 2,000 square feet of the core area of a commercial building, cost $200,000 plus installation costs in 1986 and has resulted in approximately 9,600 kWh savings per year. Although these applications have successfully demonstrated the technical feasibility of the system, the long payback time has made the system difficult to market. TIR Systems no longer actively markets the Heliostat/Light Pipe System for daylighting applications, although the firm might consider making it available on a custom basis.[18]

Mirror light guides. Mirror light guides use multiple specular reflectors to guide light from the input aperture to the output aperture. Since losses increase with the number of

Figure 5-9

Types of light guide systems

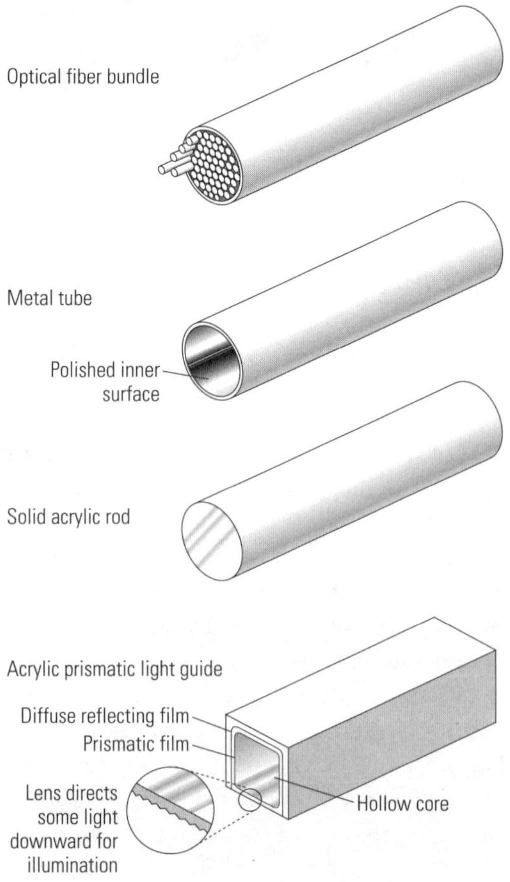

Source: Schuman et al. [9]

reflections, the efficiency of the light guide depends strongly on the ratio of length to effective diameter as well as the reflection coefficient of the mirror material used.

Lens light guides. The lens light guide is an arrangement of lenses in which each individual lens projects an image from the preceding lens to the following one. The details of the optical design of the lens system vary from design to design, but in each case a walled guide structure is required to keep the light source moving from source to target. This system requires a high-intensity point source of light.

Daylighting fiber-optic systems. The Himawari (Japanese for sunflower) system is a commercial daylighting fiber-optic system. The system tracks the sun, fresnel lenses filter out most of the ultraviolet and infrared radiation and concentrate the sunlight, and flexible quartz fiber bundles transmit the light to building spaces. The fiber-optic cables can be "wired" like electricity and fixtures can be plugged into fiber-optic "outlets."

Over 700 Himawari systems have been installed in Japan. Applications include private residences, an indoor swimming pool, hospitals, and rehabilitation centers, where daylight was desired but was not easily accessible.[19] The high capital cost of the Himawari has made it difficult to sell in the United States. The smallest system, which provides the equivalent light of three 150-watt incandescent bulbs, costs on the order of $10,000.[20]

Sun Pipe. The Sun Pipe, shown in **Figure 5-10**, is a low-tech solution for daylighting through ceilings. The product consists of two translucent dome-shaped diffusers connected by an internally reflective tube 2 to 8 feet in length. The two sizes of Sun Pipes are narrow enough (13 and 21 inches in diameter) to fit through most ceiling joists, and so they are easier to install than skylights, which they resemble in function. The diffusers provide a uniform glow, avoiding the direct glare of a skylight. E SOURCE knows of no independent evaluation of the technology or its efficiency. The Sun Pipe costs about $400, and can be used in residential, commercial, and industrial structures. The primary limitation of this type of daylighting device is its short reach; it can generally be used only on floors directly below a roof.[21] **Figure 5-11** shows an installation of several Sun Pipes in a small commercial office space.

Over 700 systems using active daylight tracking have been installed in Japan.

Figure 5-10

The Sun Pipe light pipe

This simple daylighting product resembles a skylight in function, but is easier to install and avoids direct glare.

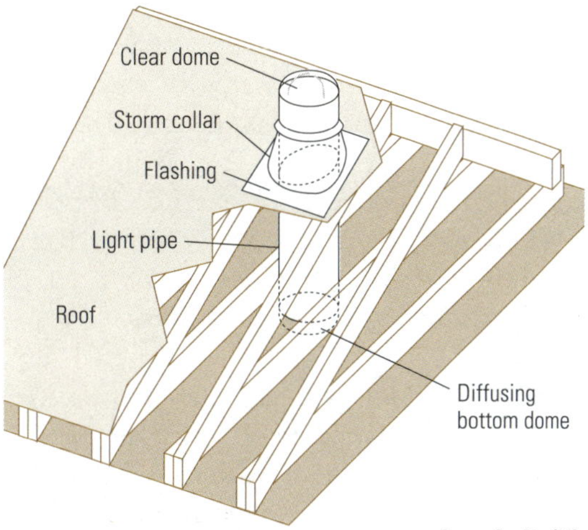

Source: Sun Pipe [21]

Figure 5-11

Sun Pipe installation

Workers in this space use 100% daylight when the sun is shining.

Source: Sun Pipe [21]

For more information on daylighting design tools, see:

Daylighting Design Tool Survey
(DA-170 summary, January 1992)
available from:
Energy and Environment Division
Lawrence Berkeley National Laboratory
Berkeley, CA 94720
tel 510-486-6845
fax 510-486-4089

For an additional listing of lighting design software, see **APPENDIX A**.

EVALUATION TOOLS

A number of software tools are available for predicting the energy savings that will result from daylighting design strategies.

ASHRAE Standard 90.1 method. ASHRAE Standard 90.1 offers a simple calculation procedure for determining the "daylighted area" of a building, and the lighting power credit resulting from installing automatic daylight-sensing lighting controls in those daylighted areas.[22]

Daylighting nomographs. Daylighting nomographs assist designers in determining potential daylighting benefits and costs, and provide a checking strategy for energy conservation and load management. These are available free from the Windows & Daylighting Group, Lawrence Berkeley National Laboratory, 1 Cyclotron Road, Building 90-3111, Berkeley, CA 94720, tel 510-486-6845, fax 510-486-4089.

BEEM. BEEM is a menu-driven simplified calculation methodology which permits an economic evaluation of various lighting controls as well as the optimization of many fenestration parameters for sidelighting applications. The BEEM software outputs include daylighting illumination levels, and annual and peak impacts of daylighting on lighting, cooling, and heating costs. BEEM is compatible with EPRI's Lightcad, an autoCAD-based electric lighting evaluation tool. BEEM is available for $50 from R.A. Rundquist & Associates, 56 Ward Avenue, Northampton, MA 01060, tel 413-586-7743, fax 413-585-5912, or free to EPRI member utilities from the EPRI Lighting Information Office at 510-444-8707.

Controlite. Controlite is a menu-driven program that performs a point-by-point daylight analysis, and results in a lifecycle cost analysis of daylighting options. It is available free from Francis Rubinstein, Lawrence Berkeley National Laboratory, 1 Cyclotron Road, Building 90-3111, Berkeley, CA 94720, 510-486-4096, fax 510-486-4089.

Lumen Micro 6.0. Lumen Micro is a menu-driven program that calculates daylight and electric light levels. It is available for $595 from Lighting Technologies, 2540 Frontier, Suite 107, Boulder, CO 80301, 303-449-5791, fax 303-449-5864.

Superlite. Superlite is a sophisticated computer tool that calculates interior light levels, including internally reflected daylight, by analyzing detailed information about the geometrical and optical properties of building components. A graphics package is optional. Version 2.20 is available at no cost from the Windows and Daylighting Group at Lawrence Berkeley National Laboratory, contact information above.

1 Southern California Edison's guidebook, *Daylighting: Performance and Design,* is now out of print, but a book by the same name written by Gregg Ander, SCE's chief architect, will be published by Van Nostrand Reinhold in late 1994.

2 Robert A. Rundquist, "Daylighting Controls: Orphan of the HVAC Design," *ASHRAE Journal,* v. 33, no. 11 (November 1991), p. 31.

3 Case study provided by the Lighting Design Lab, 400 East Pine Street, Suite 100, Seattle, WA 98122, tel 206-325-9764, fax 206-329-9532.

4 Rundquist [2]. See also Danny S. Parker, Ted C. Stedman, and Jeff K. Sonne, "Side-by-Side Testing of Commercial Office Lighting Systems," Florida Solar Energy Center FSEC-PF-265-93, 300 State Road 401, Cape Canaveral, FL 32920, tel 407-783-0300, fax 407-783-2571, which measured a 40% power savings from daylight dimming controls.

5 Claude L. Robbins, Daylighting Design and Analysis (New York: Van Nostrand Reinhold Company, 1986), table 13-4, p. 255.

6 Claude L. Robbins [5], p. 4.

7 Claude L. Robbins [5], p. 4.

8 S.B. Hartleb Puleo and R.P. Leslie, "Some Effects of the Sequential Experience of Windows on Human Response," *Journal of the Illuminating Engineering Society,* v. 20, no. 1 (1991), p. 91.

9 Jennifer Schuman et al., "Technology Reviews: Glazing Systems," LBL-33204, Energy and Environment Division, Lawrence Berkeley National Laboratory, Berkeley, CA 94720, tel 501-486-5001; LBNL, "Demand-Side Efficiency Technologies Summaries," Draft LBL-33887, Energy and Environment Division, LBNL, Berkeley, CA 94720, tel 510-486-5001, fax 510-486-5454, pp. 86–90.

10 LBNL, "Demand-Side Efficiency Technologies Summaries" [9], p. 97.

11 Southwall Technologies, 1029 Corporation Way, Palo Alto, CA 94303, tel 800-786-2468 or 415-962-9111, fax 415-967-8713.

12 LBNL, "Demand-Side Efficiency Technologies Summaries" [9], p. 75.

13 Jennifer Schuman et al., "Daylighting Optical Systems," LBL-33203, Energy and Environment Division, Lawrence Berkeley National Laboratory, Berkeley, CA 94720, tel 510-486-5001, fax 510-486-5454.

14 Schuman et al. [13], pp. 35–37.

15 Rundquist [2], p. 31.

16 Danny S. Parker, Ted C. Stedman, and Jeff K. Sonne, "Side-by-Side Testing of Commercial Office Lighting Systems," Florida Solar Energy Center, 300 State Road 401, Cape Canaveral, FL 32920 FSEC-PF-265-93, tel 407-783-0300, fax 407-783-2571.

17 Scott Ellinwood & Associates, 790 East Santa Clara Street, Suite 101, Ventura, CA 93001, tel 805-643-7301, fax 805-643-9174.

18 Chris Pekar, TIR Systems, Ltd, 3935 2nd Avenue, Burnaby, BC, Canada, V5C 3W9, tel 604-294-8477, fax 604-294-3733.

19 Kenneth E. Eben, "Let the Sun Shine In," *Lighting Design and Application*, v. 23, no. 5 (May 1993), p. 44.

20 Kenneth Eben, personal communication (February 7, 1994), Manager of Optical Fiber Products, Textile Division, Mitsubishi International Corp., 520 Madison Avenue, New York, NY 10022-2423, tel 212-605-2392, fax 212-605-1746.

21 Sun Pipe, P.O. Box 2223, Northbrook IL 60065, tel 708-272-6977, fax 708-272-6972.

22 ASHRAE/IES Standard 90.1-1989 User's Guide, pp. 8–37.

6 INCANDESCENT LIGHTING

6 INCANDESCENT LIGHTING

For several years, incandescent lamps have been challenged by alternative sources with lower lifecycle costs. For some applications, however, incandescents are still the best choice. They have low initial cost, provide easy dimming, and still come in a wider array of sizes, shapes, wattages, and distribution patterns than any other light source. Refinements in incandescent technology continue to boost their performance, allowing for their selective use in energy efficient designs.

HIGHLIGHTS

- While better incandescents can provide slight efficiency gains, the biggest savings come from replacing incandescents with more efficient fluorescent or high-intensity discharge (HID) sources. Improving incandescent fixtures without changing sources provides only marginal savings.

- The U.S. Energy Policy Act of 1992 banned production and domestic distribution of most standard incandescent floodlights and spotlights—100 million of which are sold in the U.S. each year—and will pave the way for expanded use of halogen and halogen infrared lamps, as well as HID and fluorescent sources.

- When switching to more-efficient incandescents, avoiding "snapback" to low-efficiency sources is a critical concern. Because standard Edison bases can accommodate just about any screw-in lamp—from the most efficient halogen-infrared flood to the least efficient standard A-lamp—savings from an efficiency upgrade can be easily lost when the lamp is replaced.

- Utility incentives for high-efficacy incandescents are rare. The most effective utility stance on incandescents is to be aware of the best models available for the few applications that demand them, and to allow their restrained use to enhance lighting esthetics while maintaining an efficient overall lighting design.

The incandescent lamp was the first technology to separate light from fire. While others tried unsuccessfully to harness electricity to make light, in 1879 Thomas Edison produced the first commercially viable incandescent lamp, using a carbon filament supplying about one lumen per watt. Tungsten filaments were introduced in 1910, and the basic incandescent lamp changed little over the next 60 years.

6.1 INCANDESCENT LAMP TECHNOLOGY

An incandescent lamp consists of a glass-enclosed wire filament that glows when a voltage is applied across it, producing visible light (**Figure 6-1**). Unfortunately, 70 to 90 percent of the power consumed by such "incandescence" is emitted as infrared (heat) radiation. While inefficient from an energy standpoint, the luminous filament can be made

Figure 6-1

Incandescent lamps

Incandescent lamps come in a wide variety of shapes, sizes, and base types.

Source: GE Lighting

quite small, offering excellent opportunities for beam control in a very small package.

EPACT IMPACTS

The National Energy Policy Act of 1992 (EPACT)[1] created minimum efficacy standards for incandescent reflector lamps having these characteristics: 40 to 205 watts, medium base, 115 to 130 volt, and diameter over 2.75 inches. This standard took effect on November 1, 1995.

To meet EPACT requirements, lamps manufactured in or imported to the U.S. must have the minimum efficacies shown in **Table 6-1**.

Table 6-1

EPACT minimum efficacies

Under the National Energy Policy Act of 1992, incandescent reflector lamps sold in the U.S. must meet or exceed these values. Many lamp types are exempted, however.

Nominal lamp wattage	EPACT minimum required efficacy (lumens/watt)	Typical efficacy of current models (lumens/watt)
40–50	10.5	10.5
51–66	11.0	10.4
67–85	12.0	10.2–12.0
86–115	14.0	12.0
116–155	14.0	11.4–13.3
156–205	15.0	11.3

Source: EPACT [1]

While most of the 100 million standard and "energy saving" R and PAR lamps previously sold in the U.S. each year did not meet EPACT standards, many lamps are exempted, including those less than 2.75 inches in diameter, low-voltage lamps, ER- and BR-shaped lamps, rough-duty lamps, and colored lamps. Manufacturers have dropped those lamps that do not meet the EPACT standard and filled the gap in two ways: by expanding their offerings of improved-efficiency lamps, and by substituting similar but EPACT-exempt lamps, most notably the BR (bulge-reflector) shaped lamps. EPACT allows existing lamp stocks to be sold at any future time.

Screw-in incandescent lamps come in a wide variety of shapes, sizes, and base types. **Figure 6-2** shows the most common lamp shapes and their functions. In this text, we use the term "reflector lamp" to refer to any lamp with directional output—a group that includes R, PAR, ER, BR, AR, and MR type lamps. While the National Energy Policy Act of 1992 has removed many of the least efficient versions from the market (primarily old-style reflector lamps), most others will still be available (see box on EPACT impacts).

INCANDESCENT TERMINOLOGY

beam spread: Angle measured between two points at which intensity is 50 percent of center beam intensity.

bulb: Glass enclosure surrounding filament or halogen capsule.

center beam candle power (CBMP): For directional reflector-type lamps (R, BR, ER, PAR, etc.), the intensity of the center of the light distribution pattern in candelas.

dichroic coating: Interior surface coating of bulb that separates infrared radiation from visible light, reflecting one and transmitting the other.

field angle: Angle measured between two points at which intensity is 10 percent of center beam intensity.

filament: Tungsten wire, usually coiled, that glows when current passes through it.

halogen lamp: An incandescent lamp with a tungsten filament and a filling of halogen gas at greater than atmospheric pressure. Also called quartz lamps, or tungsten-halogen lamps.

lamp base: Part of the lamp through which electric power is passed to the filament.

quartz lamp: See halogen lamp.

Figure 6-2

Principal incandescent lamp shapes and photometric distribution

A-lamp

The A-lamp is the standard household incandescent. It provides roughly uniform lighting intensity in all directions.

R-lamp

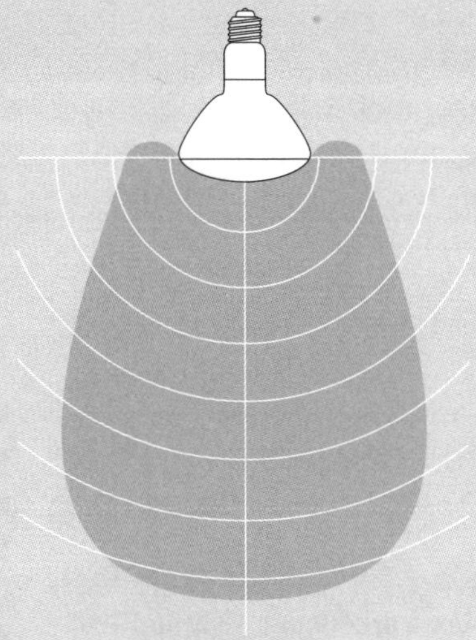

The R-lamp projects light preferentially in one direction via an internal reflector. Most household "floodlights" are R-lamps.

BR-lamp

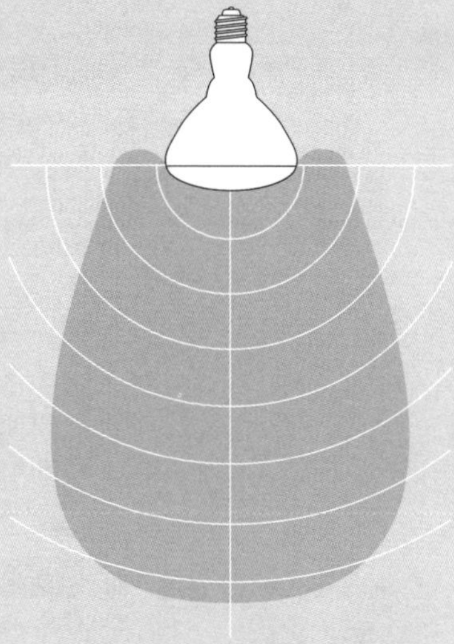

The BR-lamp has a bulge in the neck of the reflector. It provides similar performance to the R-lamp, but gathers more light from the filament and focuses it into the main beam.

PAR-lamp

The PAR-lamp uses a parabolic reflector for a tighter beam. PAR lamps (and their smaller MR relatives) are the primary incandescent lamp types that use halogen and infrared-reflector technology.

MR-lamp

The MR-lamp runs at low voltage (12V) and comes in two sizes, both less than 2 inches in diameter. These lamps offer excellent optical control but are often overused.

ER-lamp

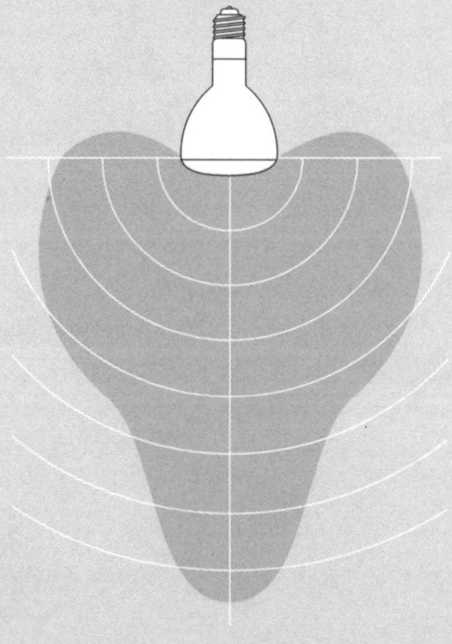

The ER-lamp uses an ellipsoidal reflector that focuses light a few inches in front of the bulb. The original intent of the ER-lamp was to reduce internal light loss in "black-baffle" ceiling cans.

Source: Osram Sylvania

Incandescents also come in a double-ended tubular form for use in specially designed dedicated fixtures. These "T-lamps" include the most efficacious high-wattage incandescent sources available.

As **Figure 6-3** shows, the vast majority of incandescent lamps produced are inexpensive and inefficient A-lamps. Halogen and PAR lamps are slowly increasing their market share relative to A-lamps, however.

Lawrence Berkeley Laboratory estimates that there are three billion incandescent lamp sockets in the U.S. residential sector, and that only about one-third of those are candidates for cost-effective replacement with compact fluorescents.

Figure 6-3

U.S. incandescent lamp production 1980 and 1994

Although the vast majority of incandescent lamps produced in the U.S. are standard A, R, and PAR models, the EPACT lamp efficiency standards will increase production of higher-efficiency halogen models.

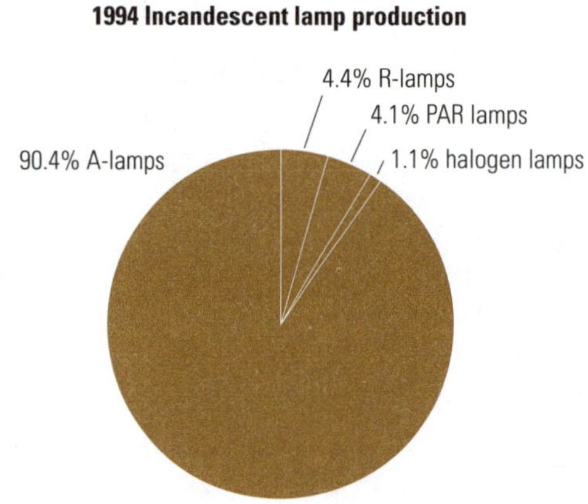

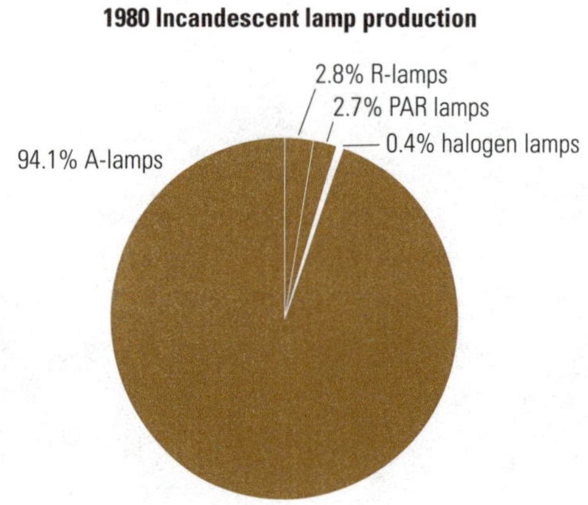

Source: U.S. Dept. of Commerce

6.2 Incandescent improvements

Incandescent *lamp* efficacy (the efficiency of converting electrical energy into light) ranges from under 10 to over 35 lumens per watt, depending primarily on filament temperature. *Photometric* efficiency (the efficiency of directing light produced by a lamp) also varies widely where reflective surfaces and shapes are involved.

Several developments have improved the incandescent lamp, including:

- halogen encapsulation,
- longer lasting filaments,
- different mixtures of fill gases,
- reshaping and coating of bulbs to focus light output,
- selective coatings to control heat output, and
- powering lamps at lower voltages.

While each of these innovations individually contributed to improving incandescent lamps, together they yield significant synergisms for improving both lamp life and efficacy. **Figure 6-4** shows the range of efficacies of incandescent sources. Incandescent lamps come closest to their theoretical maximum efficacy with the halogen infrared-reflecting lamp (see **6.2.5**). That maximum (53 lumens per watt) could be achieved only if the tungsten filament became molten.[2]

6.2.1 Halogen encapsulation

One of the most significant improvements to incandescent lamps is to surround the filament with a separate capsule containing halogen gases (usually bromine or iodine) that capture vaporized tungsten and redeposit it on the filament, extending its life while preventing condensed tungsten from darkening the bulb. **Figure 6-5** shows one type of halogen lamp, and **Figure 6-6** shows how the relative output and efficacy of halogen lamps compares with conventional tungsten lamps over their operating lifetimes.

Figure 6-4

Relative incandescent lamp efficacies

Incandescent sources have seen an order-of-magnitude improvement in efficacy since their invention. Most now range from 10 to 20 lumens per watt.

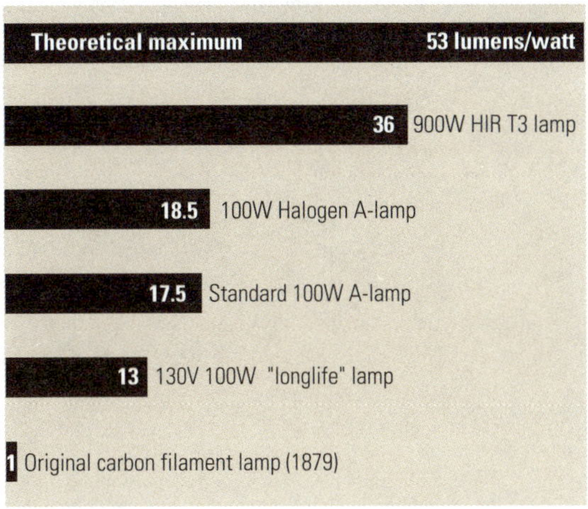

Source: IES, GE, Osram Sylvania

The halogen cycle allows the filament to operate at a higher temperature, creating more light per unit of input electricity than standard incandescents. (Halogen lamp efficacies are shown in Figure 6-9.) To deal with the high filament temperatures, manufacturers strengthen the capsule by using quartz in the glass mix, which is why halogens are sometimes referred to as "quartz lamps." Color temperature also increases, providing a generally better quality of light.

Dimming halogen lamps for long periods may cause significant tungsten condensation on the bulb walls, greatly shortening lamp life. Manufacturers recommend that lamps be burned at full output for at least an hour once a week to re-vaporize tungsten off of the halogen capsule wall so it can be redeposited onto the filament to maintain lamp life.

The high temperatures and pressures created in halogen lamps mandate some caution. Lamps in close contact with combustible materials (such as wooden moldings), or prone to physical contact (such as in a track light in a low ceiling), create a burn hazard unless appropriate ventilation, enclosure, training, and/or mounting precautions are taken. One particular application of halogen lamps—the halogen torchiere—deserves special caution because of its potential fire danger (see **6.3**, and E SOURCE *Tech Update TU-96-10* "Halogen Torchieres: Cold Facts and Hot Ceilings" (September 1996). Severe thermal stresses, such as splashing the quartz glass with water or a dirty spot on the glass absorbing too much heat, can also cause halogen capsules to explode. Many halogen lamps use a thick glass outer envelope to contain exploding fragments.

Due to the extreme intensity of the center beam of halogen spot lamps, the level of ultraviolet radiation (UV) emitted with visible light may contribute to sunburn or even skin cancer. To create a hazard, however, requires very high light levels and prolonged exposure, as might occur in fine jewelry manufacture or by a visually impaired person requiring such intensity. Manufacturers of optical equipment, such as Bausch & Lomb, produce glass UV filters for fixtures using MR-16 and other lamps to reduce this hazard to acceptable levels. In addition, many manufacturers include a UV-filtering coating in the quartz glass envelope.

Figure 6-6

Tungsten and halogen performance over rated life

The halogen cycle improves both efficacy and lumen maintenance.

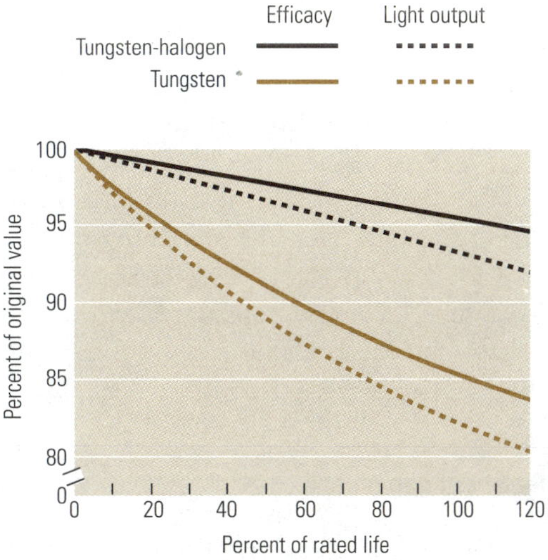

Source: IES

Figure 6-5

Halogen incandescent spotlight

This particular halogen lamp—the MR-16—is very popular for retail stores.

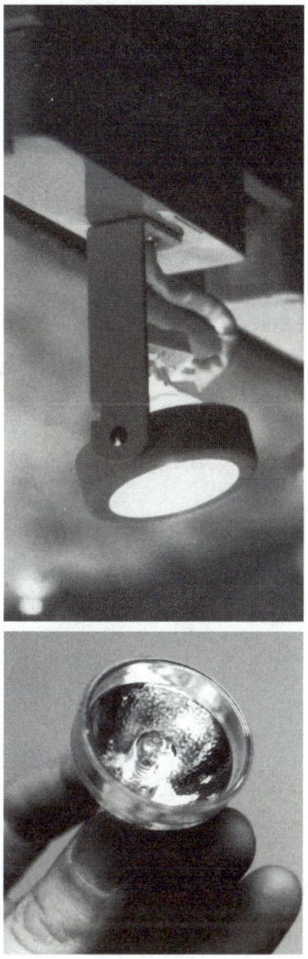

The halogen cycle allows the filament to operate at a higher temperature, creating more light per unit of input electricity than standard incandescents.

6.2.2 Filament modifications

Because the filament is an electromagnetic radiator, the spectral distribution of an incandescent lamp's output is a function of filament temperature. At higher temperatures, the lamp's radiation shifts toward violet, placing more of the lamp's output in the visible part of the spectrum, and improving efficacy. Most efforts to improve incandescent efficacy thus focus on increasing filament temperature, using methods such as halogen encapsulation, infrared-reflecting coatings, and "overdriving" the filament.

By operating at a reduced voltage, lamp life can be extended with a relatively small reduction in light output, although efficacy is reduced.

Incandescent lamp lifetime is limited by the strength of the tungsten filament. While light is being emitted, the tungsten is gradually boiling off. Vaporized tungsten condenses on the inside of the lamp's glass bulb, darkening it over time. This darkening is responsible for much of the lamp's lumen depreciation. The lamp dies when the filament breaks at its thinnest point. To avoid this problem, extended-life or "rough-service" lamps typically have extra-thick filaments. Due to their greater lifetime, however, even more tungsten condensation occurs, causing such lamps to lose over a third of their output before burning out.

A similar life extension technique is to run 130-volt rated lamps at the standard 120 volts. This "underdriving" reduces filament temperature, wattage, and efficacy, cutting light output by 20 to 25 percent but nearly tripling lamp lifetime (**Figure 6-7**). Color temperature also drops slightly, creating a more yellow light. This particular trick is commonly used in large institutions to minimize relamping labor by stretching lamp life (though often at the expense of esthetics and light level). Careful consideration of energy costs versus maintenance costs will indicate if this practice is cost-effective.

Figure 6-7

Lifetime, output, and efficacy vs voltage

Operating 130-volt lamps at 120 volts to extend lifetime is a common technique. Unfortunately, lamp efficacy suffers.

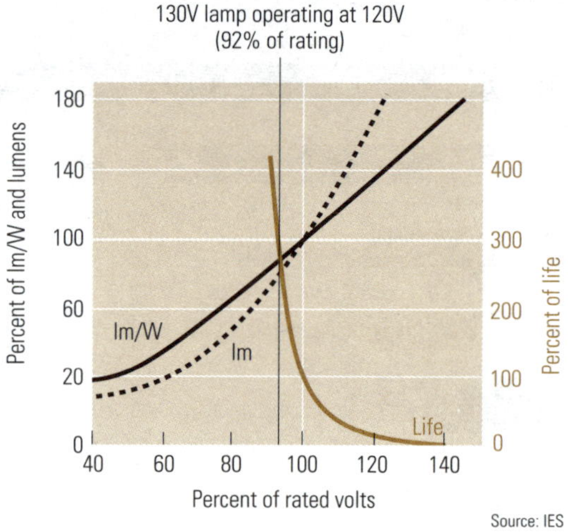

Source: IES

6.2.3 Fill gas

Adding krypton gas to the usual inert argon and nitrogen fill gases retards tungsten vaporization, resulting in an increase in lifetime of 50 percent or more. The heavier krypton is also a better insulator than argon, so it allows a higher temperature inside the bulb, which in turn improves efficacy. The Illuminating Engineering Society notes that conversion to krypton fill may increase efficacy by seven to 20 percent.[3]

6.2.4 Beam control

A critical part of lighting design is to get the light only where it's needed. Narrow-beam sources can provide dramatic highlighting, sometimes with relatively low lighting power. The ability to focus light is related to the ratio of a luminous source's size to that of its surrounding reflector. This is one area where incandescent lamps excel, because their small filaments allow good beam control within a relatively small reflective glass bulb.

The tiny filaments used in low-voltage reflector lamps offer the greatest optical control of incandescent light. These lamps and their applications are discussed in more detail below.

The advantage of beam control makes R, PAR, and MR lamps very popular in retail lighting, where they can make merchandise stand out from its surroundings. **Figure 6-8** illustrates the range of beam spreads; some low-voltage MR and AR lamps have beam spreads as narrow as three degrees.[4] In comparison, the narrowest comparable beamspread available with a compact fluorescent source is about 30 degrees; fixtures for compact HID PAR sources are now available with beamspreads as narrow as 10 degrees.

6.2.5 Halogen infrared-reflecting (HIR) coatings

This recent advance in incandescent efficacy exploits the fact that hotter filaments produce a higher proportion of visible light. Halogen infrared-reflecting (HIR) lamps feature a microns-thin spectrally selective optical coating on the outside of the lamp's quartz filament capsule. The coating transmits visible light but reflects infrared (heat) radiation back onto the filament, achieving even higher filament temperatures than standard halogen capsules and correspondingly higher efficacy. Larger lamps exhibit higher efficacy. **Figure 6-9** shows the variation in efficacy with size for halogen and halogen-IR lamps; see **6.4** for an example of the energy savings from HIR lamps.

GE Lighting is the only North American manufacturer of halogen lamps with the IR coating; this line currently includes PAR lamps and double-ended tubular lamps. These lamps are the most efficient incandescent sources available today, generating 20 to 40 percent more lumens per watt than standard halogens. There are no patents to keep other manufacturers from producing such products, but GE has the technical advantage—it's mastered the complex process of applying a spectrally selective film to a curved surface at reasonable cost.

Philips and Osram-Sylvania are both reportedly working on developing HIR technology.[5] Meanwhile, GE is working on improved processes which could boost HIR lamp efficacy an additional seven percent.

It is critical for HIR performance that the filament be precisely positioned in the center of the filament tube, or most of the gains from the reflective coating will be lost. This is relatively easily accomplished in the linear, double-ended halogen-IR lamps, because additional filament supports can be readily added. PAR lamps do not contain any filament supports, however, presenting a greater design challenge. One result of this issue is that the filaments in HIR PAR lamps are quite fragile, and can be seriously damaged if the lamps are "hot-relamped"—screwed into energized sockets. These lamps should always be installed in non-energized sockets, or they may burn out well before their rated life.

6.2.6 Reflector coating materials

Despite recent efficiency improvements, most of the output from incandescent lamps is radiant heat. Two types of reflector lamp coatings offer different ways of dealing with this heat.

Dichroic coatings, originally developed for projection equipment, transmit about two-thirds of the lamp's heat through the *back* of the reflector. This effect is useful when subjects in the illumination path—artwork, merchan-

Figure 6-9

Efficacy vs size for halogen and halogen infrared (HIR) lamps

Using halogen or HIR lamps does not guarantee high efficacy. Because hotter filaments have higher efficacy, only the largest HIR lamps exceed 25 lm/W.

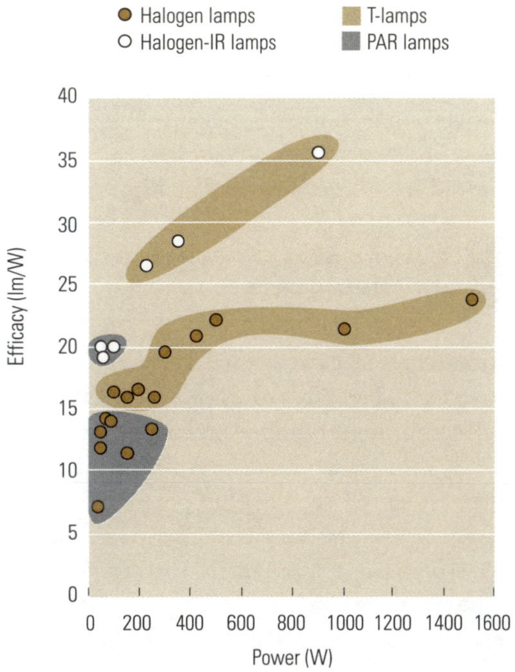

Source: GE

Figure 6-8

Typical incandescent beam spreads

The beam spread (sometimes called beam angle) of a lamp's light distribution pattern is defined as the angle between two beams of light which are each 50 percent of the maximum-intensity beam.

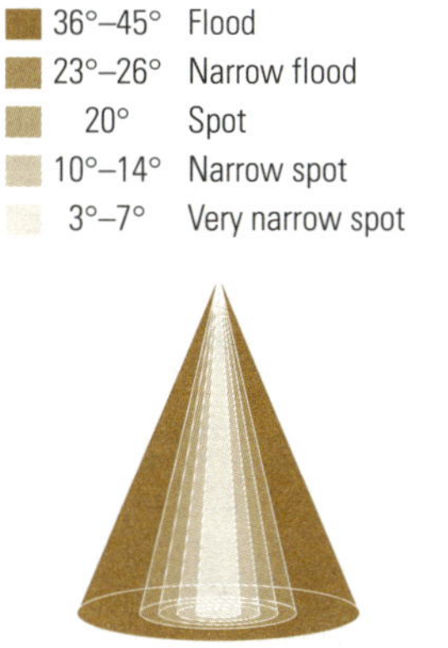

Source: Osram Sylvania

dise, or museum artifacts, for example—are sensitive to heat. Dichroic coatings are widely used in low-voltage MR lamps, and are available in a limited variety of line-voltage PAR lamps as well.

Dichroic reflector coatings limit the radiant heat in the illumination path.

One problem with dichroic coatings is that they can degrade severely over time as high temperatures and chemical oxidation take their toll. This effect, which can reduce light output by 40 to 70 percent, is being addressed by new dichroic formulations such as the titanium oxide outer coating in Osram Sylvania's True Aim Titan lamps.[6] GE Lighting uses a titanium/silica coating in their ConstantColor Precise lamps for the same purpose.[7] These improved designs are claimed to maintain 90 to 100 percent of the initial lamp light output.

Aluminum coatings do the opposite: they reflect most of the lamp's heat out the front of the reflector and directly into the illumination path. This can be beneficial if the lamp fixture itself is more sensitive to heat than the illumination target. Fixtures such as MR-16 track heads with built-in electronic transformers are candidates for aluminum reflector lamps—as long as the object being illuminated can handle the heat.

These two types of reflector coatings have very different functions, but the lamps themselves look quite similar. In facilities that use both types of lamps, stocking and replacement procedures must be followed carefully to avoid installing the wrong lamp.

FUTURE DEVELOPMENTS IN INCANDESCENTS

Researchers at Lawrence Berkeley Laboratory have looked at ways to improve the efficiency of the huge number of incandescent A-lamps that are (and will probably remain) in the residential sector. One possibility is to apply low-emissivity coatings to the bulb wall of halogen A-lamps to reduce heat transfer out of the lamp and run the filament hotter, which in turn improves efficacy.

6.2.7 Low-voltage lamps and transformers

Low-voltage lighting systems offer two key advantages to lighting designers: the lamps use small, thick filaments that provide superior optical control, and the wiring is generally safe enough to run bare wires out in the open.

Although many shapes and sizes of low-voltage light sources are available, the most common is the MR16 (Figure 6-5). Because the MR16 is compact (two-inch diameter reflector) and available in many beam spreads and wattages, some designers feel they are ideal for practically all applications, especially in the retail environment. This has resulted in some installations being "MR16'ed to death."An appropriate role for MR16 lamps (and other halogen sources, for that matter) is to provide accent lighting to supplement higher-efficiency ambient light sources.

Low-voltage lamps require a step-down transformer to convert 120- or 277-volt AC (line voltage) to 12- or 24-volt AC (**Table 6-2**). Depending on the application and luminaire type, the transformer may be an integral part of the fixture or mounted remotely. Integral transformers are commonly used with recessed downlights, track fixtures, and single-lamp can fixtures, while remote transformers are typically used with strip-type lights and live conductors such as cable or rail systems. There are three types of transformers for low-voltage incandescent lighting:

- *Standard magnetic core and coil transformers.* These units, the most common type used for remote applications in the U.S., are relatively inexpensive, but are large, heavy, noisy, and have low efficiency (typically, 80 percent or less) and relatively poor voltage regulation. (Proper voltage regulation is important because of the effects described in Figure 6-7.)

- *Electronic transformers.* These are smaller, lighter, and quieter than standard magnetic units, and have efficiencies as high as 95 percent. They are not as reliable, however, require special dimmers, generate some harmonic distortion, and have limited output options over 150 watts. Because of their small size, they are widely used in track fixture assemblies and task lights.

■ *Toroidal transformers* (**Figure 6-10**). Toroidal magnetic transformers provide excellent performance through a combination of high-grade materials and careful design and construction. The toroid is a doughnut-shaped transformer in which copper primary and secondary windings encase a steel core. The best products are 95 to 97 percent efficient and operate very quietly due to epoxy impregnation of the core and coil. Toroidal transformers are used widely in Europe.

An important characteristic of transformers: their internal losses are roughly constant and independent of the load placed on them. Thus a transformer with 20 percent losses at full load will suffer 40 percent losses at half load.

Figure 6-10

Toroidal transformer

Toroidal transformers for low-voltage lamps offer reliability and low internal losses. This system from Q-Tran runs cool enough to be UL-listed for installation in insulated wall cavities—even with zero clearance to combustible materials.

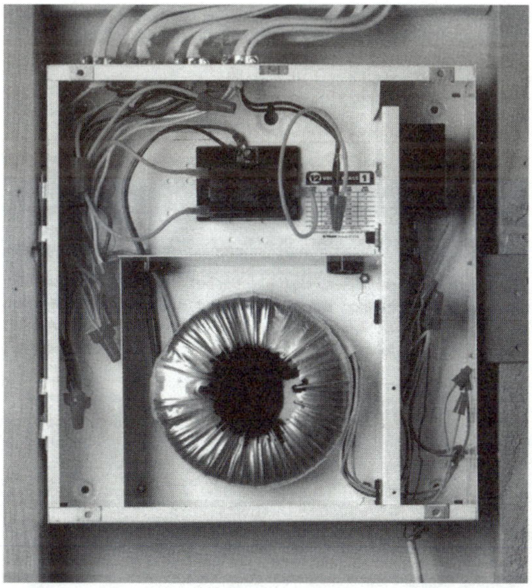

Source: Q-Tran

Table 6-2

Low-voltage transformers

There are three types of low-voltage lighting transformers: magnetic, electronic, and toroidal. Toroidal transformers offer high performance and are a relatively new entrant to the North American market.

Manufacturer	Model	Maximum output (W)	Approximate transformer losses (W)	Transformer losses (% of full load output)
Magnetic				
Toki America Technologies, Inc.	C2-24/55	55	11	20
1278 E. Katella Avenue	C1-100	100	20	20
Anaheim, CA 92805	C1-150	150	30	20
tel 714-937-5410				
fax 714-937-8995				
National Specialty Lighting	TR-12/24-150	150	30	20
333 Exchange Drive	TR-12/24-250	250	45	18
Arlington, TX 76011				
tel 800-527-2923				
fax 817-860-9055				
Electronic				
B plus L Technologies	CV 10/75-12	75	<4	5
3675 Rue Grand-Allee	CV 10/150-12	150	<8	5
Boisbriand, Quebec				
Canada J7H 1H5				
tel 514-434-7884				
fax 514-434-7638				
Transcraft Canada	OF50126	50	<3	6
7140 Alexandra	OF75126	75	<4	5
Montreal, Quebec	IR15012CH	150	<8	5
Canada H2S 3J4				
tel 514-278-1675				
fax 514-278-1677				
Toroidal				
Q-Tran Inc.	QT-0150MV	150	7.5	5
114-B Washington Street	QT-0300MV	300	15.0	5
Norwalk, CT 06854	QT-0450MV	450	22.5	5
tel 203-866-8111	QT-0600MV	600	30.0	5
fax 203-866-8823	QT-0750MV	750	22.5	3
	QT-0900MV	900	27.0	3
	QT-1050MV	1050	31.5	3

GETTING THE MOST OUT OF LOW-VOLTAGE HALOGEN LIGHTING

Use low voltage in these places:

- Where a very tight and/or intense beam is necessary, such as in merchandising or accenting a sculpture.
- When no other source is small enough: miniature T lamps will fit into coves and moldings, while two-inch MR lamps can be easily concealed.
- Where standard voltage would create a possible hazard, such as in wet locations (for example, kitchens or outdoors) or areas where human contact is more likely (for example, children's rooms).
- Where standard wiring costs would be very high (most codes allow low-voltage wiring to be run in plastic sheathing or wiremold, instead of metal conduit).

Don't use low voltage in these places:

- Where other sources, such as compact fluorescent, could create a similar effect.
- In coves over two inches wide, where a small fluorescent source, such as subminiature compact, or linear fluorescent lamps, would easily fit.
- In a quiet space, such as a library or bedroom, where a humming magnetic transformer would be distracting, or where the distance from the transformer to the lamp will cause a large voltage drop.
- Where an electronic transformer would be located very near the lamp—such as an enclosed downlight in an unventilated ceiling—and get overheated.

CASE STUDY: INEFFICIENT SOURCE MAKES AN EFFICIENT SYSTEM AT TUCSON MEDICAL CENTER

When the Tucson Medical Center (TMC) shifted to a "patient-focused care" process, paperwork tasks moved from a central office to small fold-down tables outside each patient's door, creating a lighting challenge for facility engineers. The indirect cove lighting in the hallway was never designed to accommodate detailed taskwork, so Steve Bragg, a member of the TMC facilities team, worked with vendors, engineers, and contractors to devise this elegant solution. Pulling the table down from the wall activates a single 20-watt MR11 spotlight mounted on the wall above, providing excellent task illumination **(Figure 6-11)**. Since each table is self-retracting and is used for only a few minutes per shift, the energy consumption of this lighting system is extremely low. The hospital staff has been delighted with the results—even the doctors' handwriting is nearly legible!

This example highlights the importance of design philosophy in energy-efficient lighting. A conventional solution to this problem would have been to use higher ambient lighting levels in the hallways or group-switched downlights over the doorways. Even using efficient fluorescent sources, this approach would have had a significant energy penalty compared to Bragg's task-based design that uses a much less efficient incandescent source.

Figure 6-11

Halogen MR-11 perfect for low-duty application

A single MR-11 spotlite illuminates fold-down work tables in this Tucson hospital.

Source: Tucson Medical Center

6.3 Applications

The most appropriate applications of incandescent lighting are:

- retail, accent, and case lighting (such as jewelry cases) to highlight small items,
- store windows where ambient outdoor light level is already very high,]
- in restaurants and areas where dimming is prevalent to allow varying scenes and moods,
- in theatrical presentations, and
- for traditional decoration, such as Christmas trees.

Table 6-3 summarizes the best remaining incandescent applications, and what may replace them in the future.

Accent and display lighting is used to make a subject stand out from its surroundings by making it brighter. Doing so requires it to be at least three times (and preferably five to ten times) more luminous than any adjacent surface. For residential applications and spaces with very short burn hours, incandescent lamps may still be chosen to maintain use of existing fixtures that cannot accommodate fluorescent retrofits. (Even chandeliers now have high-efficiency alternatives: fiber-optic distribution of a central light source.)

Similarly, spaces equipped with occupancy sensors may be better served with incandescent lamps if occupied periods are very short (a supply closet, for instance). Other sources, such as fluorescents, suffer from severely shortened lives when cycled frequently, while incandescent lifetime is primarily a function of total burn hours.

The most *in*appropriate applications of standard incandescents are for general area lighting, uplighting, cove lighting, downlighting, and industrial applications (except where very high levels are needed for very localized tasks). Most incandescent task lighting, such as desk lamps and spotlights, can be retrofitted or replaced with compact or linear fluorescent light sources.

One incandescent application that has attracted attention recently—and a flurry of replacement product development—is the halogen torchiere.

One incandescent application that has attracted attention recently—and a flurry of replacement product development—is the halogen torchiere. Approximately 40 million of these inexpensive and popular units are currently in use in the United States. Numerous incidents of death, injury, and property damage from fires caused by halogen torchieres have been documented,[8] and the Consumer Product Safety Commission (CPSC) is discussing actions to minimize the risk of operating torchieres, including possible recall of some products.[9] Underwriter's Laboratories (UL) has recently instituted a stringent fire-test standard for these units to limit the danger of exposure to the high-temperature halogen source used in torchieres.

Table 6-3

Remaining incandescent applications (and what may replace them)

While the need for incandescent lamps is rapidly dwindling, some appropriate applications remain.

Application	Potential replacement technology for higher efficiency
Retail, accent, and case lighting (such as jewelry cases) to highlight small items	Fiber-optic display lighting with HID sources
Store windows where ambient outdoor light level is already very high	Small HID sources
General ambient lighting where dimming is required	Dimming electronic ballasts for CFLs and linear fluorescents
In theatrical presentations	None
For traditional decoration, such as "flame" lamps and chandeliers	Manufacturers are developing fiber-optic distribution systems for many decorative applications. Several manufacturers produce fluorescent-lit decorative fixtures.
Short-burn-hour locations (such as closets)	None
For track lighting	Small HID and compact fluorescent fixtures for track lighting are now available with good color rendering and high centerbeam candlepower
In small coves	Subminiature T2 fluorescent lamps will fit many small installations

Halogen torchieres use 300- to 600-watt double-ended T3 lamps in an upturned bowl. The temperature of the protective glass cover that sits above the lamp can reach 1,000°F or more, hot enough to ignite many common materials.[10] The top-heavy design of the torchiere and its upturned bowl that can collect dust, insects, and other debris add to the fire risks of these fixtures. In addition, the CPSC has documented 160 cases of 500-watt halogen torchiere lamps shattering, in some cases sending hot pieces of quartz glass flying out of the fixture and starting fires.[11]

Halogen torchieres carry a heavy energy penalty in addition to their fire risk. A typical $15 300-watt torchiere operated over a 12,000-operating-hour lifetime will consume $288 in energy costs (and $30 in replacement lamp costs).[12] To address these safety and energy cost concerns, alternative products are being developed by manufacturers. Possible designs may use 2D fluorescent, PL or quad-lamp fluorescent, or metal halide lamps—all with major energy savings. For example, a 38-watt 2D fluorescent torchiere could provide three lighting levels and nearly the same light output as a 300-watt halogen torchiere at full power. At a projected cost of $60, such a product would consume only $37 in energy costs over a 12,000-operating-hour lifetime, and a lifecycle cost savings of $236. The extra $45 to buy the lamp would have a simple payback of 1.3 years.

For more detail and analysis of the issues surrounding halogen torchieres, see E SOURCE's *Tech Update TU-96-10* "Halogen Torchieres: Cold Facts and Hot Ceilings" (September 1996). For

CASE STUDY — INCANDESCENT ACCENT LIGHTING IN A DEPARTMENT STORE

Incandescent lamps dominate lighting plans in specialty and high-end retail stores, where making products look attractive takes a higher priority than energy savings. If incandescents are carefully blended with other source types, however, the design can be efficient *and* visually appealing.

In the Atlanta department store shown in **Figures 6-12** and 6-13, ceiling-mounted incandescent accent lights provide dramatic highlighting, while fluorescent cove lights, wall washers, and downlights take on the bulk of the area lighting. Just over one-third of the lighting wattage is used by the strategically placed halogen spotlights. Using this same equipment, a new department store in Salem, New Hampshire has a total lighting power density of 1.9 watts per square foot, including all showcase lighting. Designs that rely more heavily on incandescent sources typically use three to five watts per square foot for high-end retail stores.

Lighting design by David Apfel Lighting Design (New York).

Figure 6-12
Lighting design for shoe display

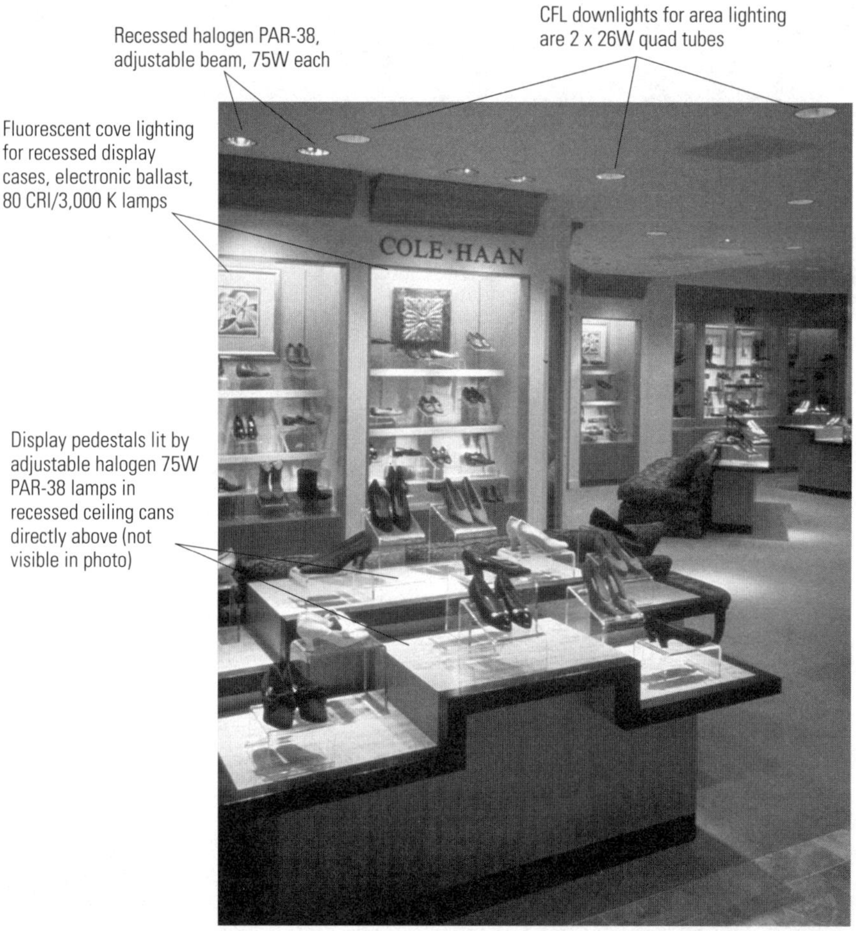

Source: David Apfel Lighting Design

an update on the progress of new torchiere designs, see E SOURCE *Tech Update TU-97-8*, "Energy Saving Torchieres: Coming to a Store Near You" (July 1997).

6.4 Economics

Incandescent lamps typically consume ten times their purchase cost in energy consumption over their life. Generally, the most favorable economics will be created when switching from an incandescent source to a non-incandescent source with much higher efficacy. The economics of retrofitting with fluorescent, HID, and other sources are discussed in the respective sections on those sources.

Upgrades within the incandescent family can also prove to be cost-effective, however. One of the best examples is switching from a standard R or PAR lamp to a lower-wattage HIR with similar or better photometrics. **Table 6-4** shows the paybacks from three possible retrofits of this type.

Table 6-4

Economics of upgrades from standard incandescents and halogens to HIRs

Significant wattage reductions are possible with halogen infrared-reflecting lamps that maintain high luminous intensity.

	Application		
	Floodlight	Spotlight	High-intensity spotlight
Original lamp	BR40 Standard flood	PAR30 Halogen narrow spot	PAR38 Halogen spot
Power (watts)	120	75	150
Centerbeam candlepower (lm)	1,450	15,000	23,000
Lamp cost ($)	5.5	8.1	8.5
Lifetime (hours)	2000	2000	3000
Annual lamp & labor cost ($/yr)	10.5	13.1	9
Annual energy cost ($/yr)	19.2	12	24
Replacement lamp	PAR38 HIR wide flood	PAR30 HIR narrow spot	PAR38 HIR spot
Power (watts)	60	50	100
Centerbeam candlepower (lm)	1,250	17,000	29,000
Lamp cost ($)	11	11.8	13.3
Lifetime (hours)	3,000	3,000	3,000
Annual lamp & labor cost ($/yr)	10.67	11.2	12.2
Annual energy cost ($/yr)	9.6	8	16
Total annual savings ($/yr)	9.43	5.90	4.80

Assumptions: Duty is 2,000 hours per year
Electricity cost is $0.08 per kWh
Relamp labor is $5.00 per lamp

Source: Lamp costs from GE Lighting

Figure 6-13

Lighting design for ladies' wear display

High brightness from recessed halogens makes models and displays stand out.

Source: David Apfel Lighting Design

6.5 MODIFICATIONS TO STANDARD INCANDESCENT LIGHTING

Diodes. The use of diodes in incandescent lighting is a controversial attempt to improve lamp characteristics such as lifetime and beam control. Placing a diode in series with the lamp filament creates a half-wave rectifier that reduces electrical consumption by clipping every other AC voltage wave, effectively reducing the RMS voltage seen by the lamp. Diodes have been sold as add-on devices in the form of "energy-saving buttons" and have also been directly built into lamps. The principal stated benefit of diode rectifiers is that they dramatically increase the lifetime of incandescent lamps. Any life extension, however, comes at a significant penalty in efficacy and degradation of power quality. Claims of energy savings from diode-based products are insupportable, given the results of tests and studies.[13] The use of diode rectifiers is receding; GE, Philips and Osram Sylvania have eliminated diodes completely from their products.

Electronic control. Developments in microelectronics now allow some interesting control functions to be incorporated into incandescent lamps at relatively low cost.

The Bulb Boss® (**Figure 6-14**) is a small disc with a microchip that attaches with an adhesive pad to the base of any ordinary incandescent light bulb. Four different functions are available: step dimmer, emergency flasher, automatic off (preset at 10 or 30 minutes), and night light. Beacon Light Products, manufacturer of the Bulb Boss, claims the Bulb Boss also makes bulbs last twice as long because the electronics limit inrush current, reducing filament shock. The suggested retail price of the Bulb Boss is $9.99; promotional prices have been reported as low as $7.88. Beacon reports that the most popular model is the 30-minute auto-off unit, which is "best used in rooms without chairs—including porcelain chairs."[14]

Figure 6-14
Bulb Boss

Source: Beacon Light Products, Inc.

A product test of the Bulb Boss was conducted by the Imperial Irrigation District in California. Installations of five of the automatic shut-off devices per household in 100 homes showed that users, who pay 7.5¢/kWh, saved an estimated $12 per unit in electric utility costs per year. The greatest energy savings were achieved in rooms where limited time is spent, such as garages, children's rooms, and bathrooms.[15]

Through a licensing agreement, ABCO has acquired Bulb Boss technology and integrated it into a line of incandescent lamps. (Philips Lighting had formerly manufactured the IQ Bulbs with Bulb Boss technology built into incandescent lamp bases. This line was discontinued in late 1996 due to lower-than-expected sales.)[16] ABCO's five "Smart Bulb" products include the four Bulb Boss functions (auto-off, dimmer, nightlight, and flasher) plus a timer model that cycles lamps on and off on a fixed schedule. These bulbs are available through mass merchandise retail outlets, and cost about $5 to $8. The rated lifetime of the Smart Bulbs is 2,000 hours.[17]

6.6 UTILITY CONSIDERATIONS

Few North American utility programs provide incentives for high-efficiency incandescent lamps. The experience of Consolidated Edison demonstrates one of the biggest reasons why. After announcing incentive payments for retrofits that reduced wattage in standard light sockets (for example, switching a 75 W for a 100 W bulb, reducing both illumination level and energy consumption), many commercial customers stepped up for rebate payments. After the initial inspection, however, there was nothing to keep the customers from switching back to the original wattage as soon as the new lamp burned out (or even sooner). Con Ed eliminated incentives of this type in 1991.

In certain cases, however, significant wattage reductions can be achieved by selectively using high-efficacy incandescents for those applications that cannot use fluorescent, HID, or other light sources. By acknowledging this reality and informing customers of the potential benefits of high-efficacy incandescents, utility staff can help avoid the commonly held perception that efficiently lit spaces must be limited to fluorescent or HID lighting. Although they are the least efficient family of light sources, the considerable benefits of incandescent lamps—low first cost, high color rendering ability, easy full-range dimming, light weight, instant-on/instant-restrike operation, and wide variety of source shapes, distributions, and intensities—ensure that they will continue to fulfill a number of important roles in well-lit spaces.

Ensuring Persistence When Replacing Low-Efficacy Incandescent Sources

When replacing standard incandescent lamps with higher-efficacy sources, there are techniques to block or limit opportunities for these fixtures to revert to low-efficacy incandescents:

- Avoid replacing one medium-based incandescent with another; instead, alter the fixture to accept *only* pin-based compact fluorescents or other efficient lamp types.
- Keep *only* the right lamps in stock at all times to eliminate any excuse or opportunity for using another lamp that fits the same socket.
- Consider installing a baffle that screws into the housing of the fixture (such as a downlight) to secure the lamp from thieves.
- If possible, retrofit with a theft-resistant compact fluorescent device requiring special tools for removal; restrict access to the tools.
- Try to eliminate incandescent fixtures completely; while first cost may be higher, savings will be greater with almost any other type of light source.

1 Energy Policy Act of 1992, Public Law 102-486.

2 Mark S. Rea, ed., *Lighting Handbook: Reference & Application,* 8th Edition (New York: Illuminating Engineering Society of North America, 1993), p. 179.

3 Rea [2], Chapter 6, p. 184

4 Osram Sylvania 35AR111 provides 45,000 center-beam candlepower with a 3° beamspread.

5 Bill Middlebrook, personal communication (February 1997), Halogen Product Manager, Philips Lighting, 200 Franklin Square Drive, P.O. Box 6800, Somerset, NJ 08875-6800, tel 908-563-3724, fax 908-563-3525.

6 Tim Lamberth, personal communication (February 1997), Commercial Engineer, Osram Sylvania, 1401 17th Street, Suite 440, Denver, CO 80202, tel 303-295-3994 x221, fax 303-295-3755.

7 Bill Gregory, personal communication (February 7, 1997), West Coast Specification Area Manager, GE Lighting, 1442 Paloma Avenue, Burlingame, CA 94010, tel 650-342-6197, fax 650-342-4279, e-mail gregorywr@msn.com.

8 Karen Fernau, "Fire Danger of Halogen Bulbs Highlighted in 3 Tempe Blazes," *The Arizona Republic* (January 23, 1996), p. B1.

9 Renae Rauschwalbe, personal communication (August 12, 1996), Compliance Office, Division of Corrective Action, Consumer Products Safety Commission, tel 310-504-0608 ext. 1363, fax 301-504-0359.

10 Erik Page, Senior Research Associate, Lawrence Berkeley National Laboratory, 1 Cyclotron Road, MS 46-125, Berkeley, CA 94720, tel 510-486-6435, fax 510-481-6940, e-mail erpage@lbl.gov.

11 Rauschwalbe [9].

12 Assumptions: energy cost $0.08/kWh, lamp life $2,000 hours, replacement lamp cost $6.

13 Alexander Emanuel, "Light-Bulb Saver Performance," *IEEE Transactions on Industry Applications,* v. IA-22, no. 3 (May/June 1986); and T. Murnane, "Bulb-Life Extenders Save No Energy, DOE Study Says," *Energy User News* (September 15, 1980).

14 Contact: Darlene Blakeslee, Sales and Marketing, Beacon Lighting Products, 723 West Taylor Avenue, Meridian, ID 83642, tel 208-888-5905, fax 208-888-7433, web www.beacon-light.com.

15 Kenneth N. Johnson, "DSM Product Test Report" (January 21, 1992), Supervisor of Energy Conservation, Imperial Irrigation District, Box 937, Imperial, CA 92251, 619-339-9024.

16 Mark Rausch, personal communication (February 1997), Director of Application Marketing, Philips Lighting, 200 Franklin Square Drive, P.O. Box 6800, Somerset, NJ 08875-6800, tel 908-563-3518, fax 908-563-3740.

17 Shellie McSloy, personal communication (February 1997), Product Specialist, Angelo Brothers Co. (ABCO), 12401 McNulty Road, Philadelphia, PA 19154, tel 215-671-2000, fax 215-464-4115.

7 FULL-SIZED FLUORESCENT LIGHTING

7 FULL-SIZED FLUORESCENT LIGHTING

Fluorescent lighting dominates the commercial sector, making an examination of its efficiency—and the potential for improvements—particularly compelling. The installation of improved fluorescent lamps, ballasts, fixtures, and controls has been the bread and butter of both utility- and end-user-based efficiency programs for several years. Exciting developments continue to mount in this important category of lighting equipment.

Fluorescent lamps are the primary technology for general area lighting. They combine high efficacy (**Figure** 7-1) and long life with light quality ranging from acceptable to excellent. Fluorescent sources generally pose no safety hazards and have few operational limitations. Some of those limitations—such as low beam power, poor operation at low temperatures, lack of tiny light packages, and expensive dimming controls—are being erased by ever-improving technology.

This chapter covers the basics of fluorescent technology and looks at the performance of noncompact products. Because of their wide variety and special applications, compact fluorescents—including circular lamps and single-ended T5 lamps—are discussed separately in **CHAPTER 8**. We discuss fluorescent lighting systems in three sections: lamps, ballasts, and fixtures, the latter including specular reflectors and lenses. Fluorescent lighting controls—including a complete discussion of dimming products—are treated in **CHAPTER 11**.

HIGHLIGHTS

- Modern rare-earth (tristimulus) phosphors with high efficacy and excellent color rendering are now available in nearly all types and sizes of fluorescent lamps.

- The lamp-ballast efficacies of advanced fluorescent lighting systems now exceed 80 lumens per watt, with the very best reaching around 100 lumens per watt.

- The reliability and power quality of electronic ballasts have improved, and they continue to be the most energy efficient way to drive fluorescent lamps.

- There are now several dozen models of continuously dimming ballasts for full-size fluorescent lamps on the U.S. market, and prices are slowly dropping. Dimming systems are discussed in more detail in **CHAPTER 11**.

- Reflectors for retrofit and new installations have improved, but they still must be applied carefully.

Figure 7-1

Fluorescent lamp efficacy range

Modern full-sized fluorescent lamps combine efficiency with good light quality, and have few limitations.

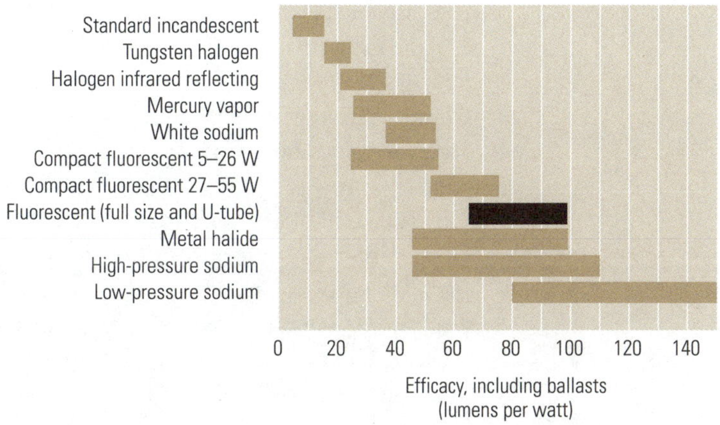

Full-sized fluorescent systems are most appropriate for general lighting in commercial, institutional, and industrial spaces with low to medium ceiling height. Those with high ceilings (over 15 feet or so), are generally better served by HID light sources (**CHAPTER 9**), but high-performance fluorescent fixtures incorporating specular reflectors, high ballast factor ballasts, and multiple lamps are increasingly being used in such applications. Residential applications are generally limited to kitchens, workshops, bathrooms, laundry rooms, garages, and other utilitarian areas that require high levels of illumination. The 1992 National Environmental Policy Act (EPACT) legislation has significant impacts in the U.S. on fluorescent systems (see box).

EPACT'S IMPACTS ON FLUORESCENT LIGHTING

EPACT affects about half of the estimated two billion full-size fluorescents (two-foot U-lamps and four-foot and eight-foot straight lamps) currently installed in the U.S. Lamps covered by the Act must meet the specific efficacy and CRI standards shown in **Table 7-1**. The following lamps are exempt from EPACT requirements:

- four-foot Slimline and high output (HO) lamps
- both four- and eight-foot very high output (VHO) lamps
- 18- to 24-inch T12 and T8 lamps
- 2-, 3-, 5-, and 6-foot T12 and T8 lamps
- any U-lamp shorter than two feet (from socket to bend)
- all Circline lamps
- specialty types, such as those used in plant growth.

The primary impact of these rules is to replace standard full wattage lamps with energy-saver reduced wattage ones and/ or to increase the color rendering index (CRI) of standard and high-output cool- and warm-white lamps by switching to new rare earth phosphor lamps. Most existing lamps meet the efficacy standards, although many "deluxe" varieties do not. "Energy saver" T12 lamps meet both the efficacy and CRI standards, but only because the CRI standard for these reduced wattage lamps is so lax.

EPACT became effective in October 1995 for two-foot U-lamps and four-foot standard lamps, and in April 1994 for eight-foot Slimlines and HO lamps. No lamps manufactured after those dates can be sold in the U.S. unless they meet the standards. For many types of lamps not meeting EPACT specifications, merely upgrading to rare-earth phosphors will bring them into compliance.

It should be noted that the lamp efficacy standard is calculated without ballast losses. While the 1987 National Appliance Energy Conservation Act (NAECA) ballast efficacy standard remains in effect, it is now outdated by the industry's complete shift to "energy efficient" ballasts. The second round of standard-setting for ballasts is now underway, and could have a significant impact on ballast efficacy.

Table 7-1

EPACT lamp standards

Lamp group	Wattage	Minimum efficacy (lm/W)	Minimum CRI
4-foot medium bipin	>35	75	69
	≤ 35	75	45
2-foot U-shaped	>35	68	69
	≤ 35	64	45
8-foot slimline	65	80	69
	≤ 65	80	45
8-foot HO	>100	80	69
	≤100	80	45

7.1 FLUORESCENT LAMPS

The basic fluorescent lamp contains low-pressure mercury vapor and inert gases in a partially evacuated glass tube (**Figure** 7-2). The mercury atoms' electrons are energized by an electric arc, elevating them to a higher energy level. As they descend back to the lower energy state, the mercury atoms emit ultraviolet photons, which then strike phosphors, specially formulated compounds that line the tube. Another round of energization and discharge causes the phosphors to emit visible photons, or "fluoresce." Altering the phosphors produces different qualities of white light.

Figure 7-2

Fluorescent lamp operation

Fluorescent lamps maintain an electric arc through gas, unlike the continuous metal filaments used in incandescent lamps.

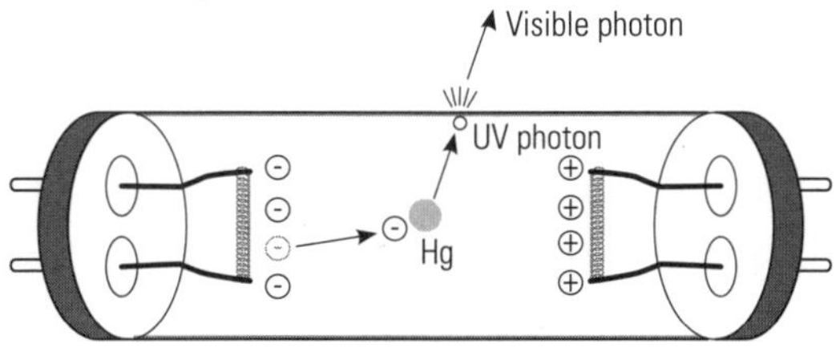

Figure 7-3

Fluorescent lamp cathode construction

Cathode erosion eventually causes lamp failure.

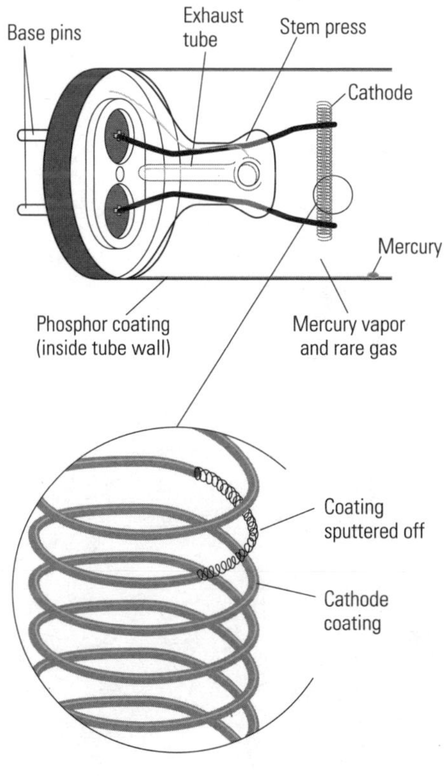

While several mechanisms exist to strike the arc, all fluorescent lamps use a power conditioning device called a *ballast* which amplifies line voltage to start the lamp, and limits current to maintain its arc. Ballasts also ensure control and safety in a variety of failure modes. Ballasts and starting methods are discussed in **7.2**.

Ballast-conditioned power is supplied through contacts in metallic end caps (called "bases"), which feed cathodes at each end of the lamp. The cathodes are conductive wires coated with an electron-emissive material such as barium oxide. When heated, this material showers electrons into the tube to create and maintain the arc. The primary reason for fluorescent lamp failure is the eventual erosion of the cathodes (**Figure** 7-3). This erosion, which occurs mostly during lamp ignition, also sputters cathode material onto the lamp walls, creating the familiar lamp-end blackening.

Lamp bases contain contacts designed to control compatibility with different starting methods and power loadings. In general, two pins or recessed contacts on each end implies rapid starting or preheat starting, while a single electrical contact indicates instant starting. An important exception: all manufacturers offer T8 lamps with two-pin bases that can be rapid-started *or* instant-started (GE's Trimline T8™, Osram Sylvania's Octron™, and Philips' TL70/TL80). For operation with instant-start ballasts, the socket contacts at each end of these lamps are simply wired together to act as one.

Fluorescent lamps differ primarily in their size, type of phosphor coating, fill gas, and base/cathode design. By varying these four things, manufacturers create lamps that span a wide range of wattage, light output, color, and lifetime. Diameters range from 0.25 to 2.125 inches and lengths from 6 to 96 inches. Nominal power consumption (without ballasts) ranges from under 20 watts to over 200 watts, and light output also spans an order of magnitude, from under 1,500 lumens to over 15,000. Lamp-only efficacy covers a smaller range: from 60 to 100 lumens per watt. As shown in **Figure 7-4**, lamps are designated primarily by their wattage and tube diameter.

7.1.1 Lamp families

The large selection of fluorescent lamps available has a sometimes confusing array of names and features. The majority of full-sized fluorescent fixtures use rapid-start 40-watt or 34-watt F40T12 lamps. **Figure 7-5** shows 1994 and 1996 sales figures for four-foot lamps broken down by lamps diameter and phosphor type. Two trends are evident: a large number of users are now buying T8 lamps, which offer the highest combination of energy efficiency and lighting quality, while another class of users is more interested in low purchase cost and is specifying the commodity-grade 34-watt and 40-watt lamps that are either compliant with or exempt from EPACT. **Figure 7-6** shows the fluorescent lamp family tree, and the sections below discuss the pros and cons of each of the family members.

Figure 7-5

1994 and 1996 sales of 4-foot fluorescent lamps

T8 and Energy Saving T12 halophosphor lamps are increasing their shares of the fluorescent market.

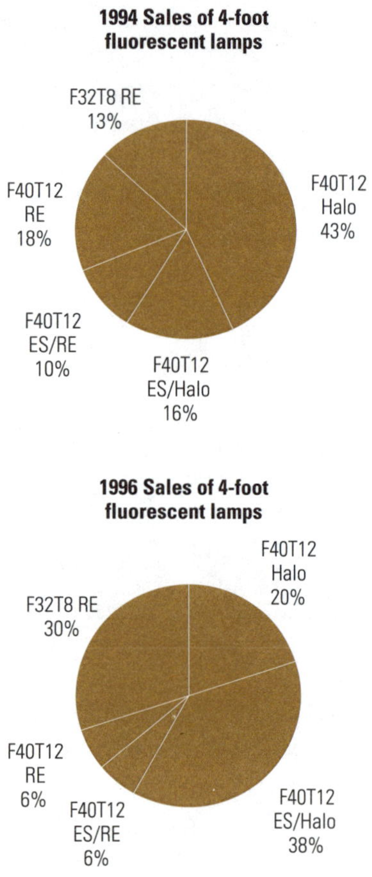

Notes and abbreviations:
RE=Rare Earth phosphors (high CRI)
Halo=Halophosphors (i.e. Cool White)
ES="Energy Saving" lamps (argon fill)
1996 Sales figures are estimates

Source: U.S. Census

Figure 7-4

Lamp designations

The first number in a lamp's designation usually—but not always—indicates its nominal wattage.

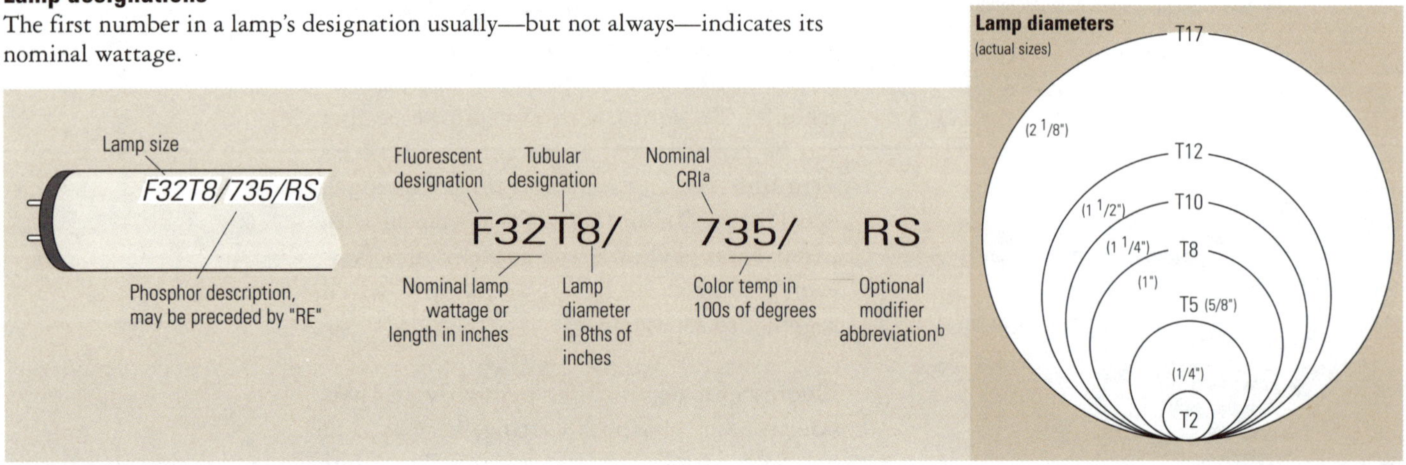

Notes: a: Nominal CRI range (7 = 70–79, 8 = 80–89, 9 = 90–99)
b: Optional modifier abbreviations (RS = rapid start, ES = energy saving, HO = high output, etc)

Figure 7-6

Fluorescent lamp families

This schematic chart shows the fluorescent lamp families and their relationships to the three starting methods (circle sizes do not represent market share). Most lamps are only compatible with one starting method, except the popular 265-mA T8s, which can be rapid- or instant-started, and some rapid-start lamps that can be preheat-started. Most of the special types of lamps (high output, "premium", etc.) are rapid-start only. The so-called "energy-saving" krypton-filled lamps are available in nearly all categories.

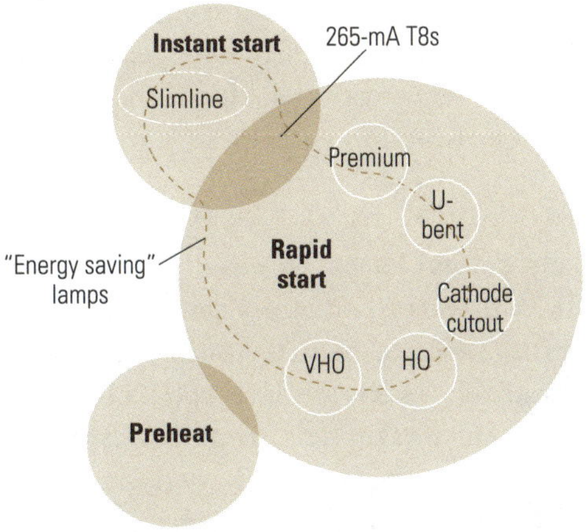

Preheat lamps. Preheat lamps are generally small tubes (6 to 36 inches in length) for use in applications such as drafting lights, undercounter lighting, and appliances. They have relatively low efficacy and short lifetimes, and are limited to magnetic ballasting. New versions with good color rendering are available. Most magnetically ballasted compact fluorescent products (**CHAPTER 8**) are preheat lamps. Users seeking to maximize energy efficiency should avoid preheat lamps when possible.

Standard rapid-start lamps. Standard rapid-start lamps are the most common fluorescents in use today, and come in the widest variety of sizes and features. (Rapid starting itself is a ballast function described in **7.2** below.) Rapid-start lamps are reasonably efficient, and include most of the "family member" subgroups available (see Figure 7-6).

Most double-pin T8 rapid-start lamps can also be instant-started, an option that allows flexibility for facilities that want to only stock one lamp but employ both starting types (rapid start for use in shorter duty cycle applications and instant start where the lights will be left on for longer periods). Table 7-2 lists brand names for these lamps, which are available in lengths from 24 to 96 inches. Since the popular four-foot versions of these lamps are rated to run at a current of 265 mA, this category is sometimes referred to as "265-mA T8s." Paired with high-frequency electronic ballasts, these lamps are the foundation of most energy efficient fluorescent systems. They should not be confused with T8 instant start or preheat lamps which use the same envelope but have different current loadings and usually have poor-color-rendering cool white phosphors.

Table 7-2

T8 rapid/instant start lamps

Company	Lamp name
General Electric	Trimline T8™
Philips	TL70, TL80
Osram Sylvania	Octron®

"Premium" lamps. Some manufacturers offer premium-grade fluorescent lamps that combine thick coats of rare earth phosphors with heavier-duty electrode construction (Table 7-3). This combination yields high efficacy, color rendering, and lumen maintenance as well as extended lamp life, but at a much higher price than conventional lamps. Premium lamps are avail-

Table 7-3

Premium fluorescent lamps

Company	Lamp name	Catalog I.D.	Comments
Duro-Test	Ultra 9	Classic-42	Twisted T10 tube, 30W (on a T8 ballast), 2,750 initial lumens, 22,000 to 26,000 hours rated life **based on 10–12 hours per start**
General Electric	Staybright™ XL	F40 SXL/SPX	T12 tube, 40 W, 3,400 initial lumens, 24,000 hour rated life
General Electric	Staybright™ XL	F32T8SXL/SPX	T8 tube, 32W, 2,950 initial lumens, 24,000 hour rated life
General Electric	Staybright™ XL	F96T8SXL/SPX	T8 tube, 59W, 5,950 initial lumens, 18,000 hour rated life
General Electric	Staybright™ XL	F96T12SXL/SP/WM	T12 tube, 60W, 5,900 initial lumens, 15,000 hour rated life
General Electric	Staybright™ XL	F96T12SXL/SPX	T12 tube, 75 W, 6,900 initial lumens, 15,000 hour rated life
Philips	Advantage X	F40 AX35	T10 tube, 40 W, 3,700 initial lumens, 24,000 + hour rated life

Source: manufacturers' data

"Energy saving" lamps have a higher rate of lumen depreciation than standard lamps, are very sensitive to operating current, and cannot be deeply dimmed or run on low-ballast-factor ballasts.

able in a limited selection of four-foot and eight-foot lengths, and may cost two to three times as much as comparable CRI T8s (see Table 7-6).

One way these lamps may be useful and cost-effective is in a delamping retrofit of existing fixtures. Because of the 5 to 24 percent higher output of these lamps, it may be feasible to remove some of a fixture's lamps and use the same fixture, reducing the cost of the retrofit. This option should be weighed against another method of extracting high light output: reballasting with high-ballast-factor ballasts.

"Energy saving" lamps. In the 1970s, lamp manufacturers found that adding krypton gas to the standard argon gas fill suppressed both energy consumption and light output. Thus was born the "energy saving" (ES) lamp—but the term can be misleading. Although lamp-only efficacies for ES lamps are relatively high, these lamps are no more efficient than standard magnetically ballasted T12s, and electronically ballasted T8s can beat them in most cases. Since EPACT took effect, ES lamps account for over 40 percent of all new four-foot fluorescent sales and approximately 70 percent of all new eight-foot Slimline and HO sales.[1] ES lamps come in three-foot to eight-foot lengths, with all common phosphor options. Nearly all "energy saving" lamps are T12s, however, and they have several operational limitations. Brand names for this lamp category are listed in **Table** 7-4.

Table 7-4

"Energy saving" lamps

Company	Brand name
Duro-Test	Watt-Saver®
General Electric	Watt-Miser®
Osram Sylvania	SuperSaver®
Philips	Econ-o-Watt

The primary attraction of ES lamps is that they offer a simple energy-saving relamping option for the multitude of existing T12 applications, especially those that are overlit. Some facility managers prefer ES "retrofits" since no ballast change is required, eliminating the need for any real electrical work.

Unfortunately, ES lamps have a higher rate of lumen depreciation than standard lamps, are very sensitive to operating current, and cannot be deeply dimmed or run on low-ballast-factor ballasts. ES lamps also perform poorly in cold temperatures and should not be run with cathode-cutout ballasts. Perhaps the greatest drawback to ES lamps is that their use may either impede the path to future upgrades (such as T8 conversions) or convey the impression that because "energy saving" lamps are in place, maximum efficiency has been achieved. When investigating ES lamps, compare their rated "mean lumens" with "initial lumens" to assess their long-term performance, and remember that the primary means of wattage reduction with these lamps is their lower light output.

Cathode cut-out lamps. In rapid start systems, about two watts per lamp can be saved by stopping cathode heating once a lamp has been started. There are two ways of accomplishing this: cathode cutout ballasts (discussed in **7.2** below) and cathode-cutout lamps. Only one or the other should be contemplated; they cannot be used together. Cathode-cutout lamps incorporate a bimetal switch that eliminates resistive cathode heating above a certain temperature. These specialty products are only offered in four-foot 32-watt energy-saver models (**Table** 7-5).

There are several reasons to be wary of cathode-cutout lamps. They can only be used with magnetic rapid-start ballasts, eliminating the upgrade path to more efficient electronic ballasts. Because the bimetal switch must cool before cathode heating can resume, a restrike time of up to one minute is required when the lamp is turned on immediately after being turned off. In addition, these lamps are variations of the "energy saving" krypton

Table 7-5

Cathode cutout lamps

Manufacturer	Trade name	Catalog number	Power consumption (W)	Light output (initial lumens)	Lifetime (hours)
General Electric	Watt-Miser®Plus	F40/SP35/RS/WMP	32	2,650	15,000+
Sylvania	Super-Saver Plus®	F40/D35/SSP	32	2,750	20,000

Source: manufacturers' data

lamps described above, and are subject to the temperature limitations, higher relative ballast losses, and limited dimming capability inherent to this lamp type.

HO/VHO/twist/groove lamps. Part of the rapid-start family includes lamps designed to be driven at higher current loadings by specially designed ballasts. Getting more light out of the same size of lamp requires heavier cathode construction, a thicker phosphor coating, and a ballast that supplies higher current to the lamp ends. While conventional four-foot T12 lamps run at a current of 430 mA, high-output (HO) lamps run at 800 mA, and very high output (VHO) lamps run at a whopping 1,500 mA. All HO and VHO lamps use the T12 envelope (except for a few that use the larger T17 lamp size). HO and VHO lamps use recessed double-contact bases that distinguish them from double-pin rapid-start and single-pin instant-start lamps.

Lamps in the HO/VHO categories have relatively low efficacy, and even with electronic ballasting top out at about 80 lumens per watt. Because of their very high luminous intensity, they find applications in outdoor signage, especially signs that are backlit through colored materials and/or operate during the day, such as convenience stores, billboards, and roadway signs. High-bay spaces also use HO/VHO lamps for general illumination, but HID sources are more appropriate in these applications because of their ability to focus and direct light output, and because of their longer lifetimes.

Figure 7-7 shows one of the main problems with high-output lamps: significantly increased lumen depreciation because of the electrical battering inflicted on the phosphors. Light output can decline by nearly 40 percent over 10,000 hours. In addition, the rated lifetime of HO/VHO lamps is only 9,000 to 12,000 hours. Another problem with this technology is the large size of the ballasts: they are nearly a foot long, and are larger in all dimensions and much heavier than those for normal-output fluorescent units. Limited phosphor availability has been recently overcome; high-output lamps can now be purchased in nearly the same range of color temperatures and CRIs as normal fluorescents.

Figure 7-7

Lamp lumen depreciation as a function of current loading

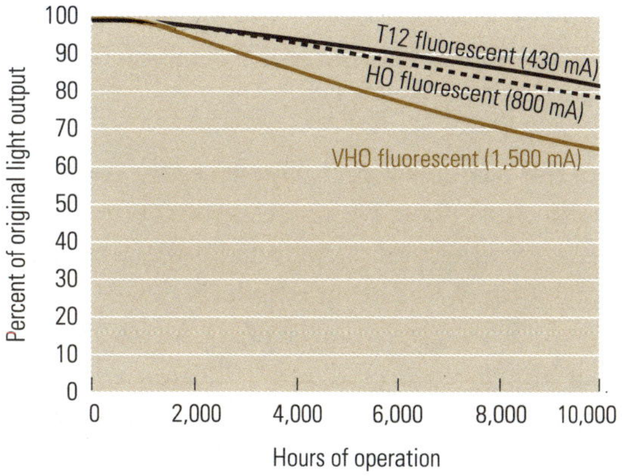

Source: National Lighting Bureau

Lamps with twisted, grooved, or dimpled tubes are variations on the VHO lamp category that wring even more light out of a fixed-size tube by providing a larger surface area for UV photons to impact phosphors. GE's F96PG17/WW, one of the few remaining examples of twisted lamps, puts out 15,700 initial lumens—12 percent more than GE's comparable VHO lamp, and 175 percent more than standard eight-foot lamps. These costly lamps are relatively inefficient and short-lived and are applicable only for special situations that call for the highest light output available from a fixed-volume tube.

One of the main problems with high-output lamps is significantly increased lumen depreciation because of the electrical battering inflicted on the phosphors.

Slimline (instant-start) lamps. Widely used by supermarkets and mass-merchandisers, these instant-start-only, 425-mA lamps are readily recognizable by their single-pin ends, and are also called Slimline lamps for opaque historical reasons. These lamps have some interesting pros and cons. Because they use no power for cathode heating, they effectively save two to three watts per lamp. But the starting method—a short, high-voltage surge that literally shocks the electrons right off the cathodes—is tough on the lamp and reduces lifetime. Typical instant-start lifetimes range from 7,500 hours to 15,000 hours, while

An emerging generation of linear T5 lamps offers improved efficacy and optical control compared to T8 and T12 lamps.

standard 430-mA rapid-start lamps are usually rated at 20,000 hours. (All ratings are for three hours of operation per start.)

The best application for instant-start lamps is in locations where they remain on for long periods. If the average runtime per start exceeds three hours, lifetime is extended significantly (see Figure 7-12). Instant-start lamps do not make sense for spaces controlled by occupancy sensors.

Several manufacturers have recently come out with eight-foot T8 instant-start lamps that have excellent performance, especially when used with the electronic ballasts developed for them. With rated power of 59 watts and initial light output of about 5,900 lumens, these lamps operate at or near 100 lumens per watt, and because of their rare earth phosphor blends, they provide excellent color rendering. These lamps should not be confused with the original eight-foot Slimline (200-mA) T8 lamps that are still available, which operate at lower efficacies with cool-white phosphors. To get the better lamps, make sure they carry the rare earth phosphor designations as shown in Figure 7-4, and have 70+ CRI.

Subminiature lamps. Two recent additions to the fluorescent family are the narrow T5 lamps and the even narrower T2 lamps. Both developments represent significant breakthroughs in fluorescent lighting, offering high light output from small diameter tubes with excellent color rendering and lumen maintenance.

T2 lamps. T2 lamps are about as thick as a pencil, and are currently produced only by Osram Sylvania. The lamps are available in lengths ranging from 8.6 inches to 20.6 inches with power consumption ranging from 6 to 13 watts. Applications for these lamps include undershelf fixtures, task lights, display case and store fixture lighting, valence and cove lighting, and appliance lighting. These lamps have rare earth phosphors and specially-designed instant start electronic ballasts that produce 50 to 60+ lumens/watt, similar to the efficiency of compact fluorescents. Lamp life is rated at 10,000 hours.

The T2 lamp and other small diameter, high power density lamp types (including some T4 and T5 compact fluorescent and new T5 linear fluorescent lamps) present some safety issues because the cathodes and lamp wall occupy close quarters. Unless the ballast uses end-of-life shutoff circuitry, a filament that breaks or touches the lamp wall can overheat the socket or lamp, possibly fracturing the lamp. Even when the cathode is dead, an arc can be struck across the cathode supports and cause overheating problems. Make sure that ballasts for small-diameter fluorescent lamps incorporate end-of-lamp-life shutoff circuitry.

T5 lamps. An emerging generation of linear T5 lamps offers improved efficacy and optical control compared to T8 and T12 lamps. T5 systems have established a significant market share in Europe and are becoming available in the U.S.—Philips now offers a complete line of T5 lamps, and GE and Osram Sylvania plan to introduce competing lines sometime in 1997. Philips and Energy Savings Inc. currently offer electronic ballasts for T5 lamps, and fixture manufacturers are introducing products designed for the new lamp sizes. T5 lamps offer:

- up to 18 percent higher efficiency compared to T8 systems;
- lumen maintenance as high as 97 percent;
- greater optical control and fixture design flexibility because of the smaller lamp diameter; and
- a high optimal operating temperature that improves performance in enclosed fixtures.

Because of their European genesis, T5 lamps are manufactured in metric lengths that are about two inches shorter than comparable U.S. lamps. Four sizes are currently available, ranging from 550 mm to 1,450 mm, These lamps are just a bit shorter than conventional 2-, 3-, 4-, and 5-foot fluorescent lamps, with power consumption of 14 to 35 watts. (These lamps should not be confused with older preheat T5 lamps. Although both new and old T5s share the same 5/8-inch tube diameter, preheat T5s have low efficacy, short lifetimes, and come only in lengths up to 21 inches.)

Philips' 1,150-mm lamp, for example, is nearly four feet long, generates 2,900 initial lumens (2,750 design lumens) and has a lifetime of 16,000 hours. With 28 watts of input power, the system efficacy is over 100 lumens per watt. The lamp has a CRI of 85 and is available in color temperatures of 3,000, 3,500, and 4,000 K.

T5 applications include low-profile cove and soffit lighting and pendants and recessed downlights for general area lighting. The light output of T5s per unit length is about the same as that for T8s, so their surface brightness is quite high—careful fixture design is required to prevent glare. As with T8 systems, ballast manufacturers plan to develop products with varying ballast factors so designers can fine tune lighting levels.

Curved. U-shaped fluorescent lamps have been available for many years, but have enjoyed only limited application. While circular (also known as "Circline") lamps have seen widespread use in residential ceiling and wall sconce fixtures, U-shaped two-foot T12 and T8 lamps are used primarily in square ceiling fixtures (typically 2 x 2 configurations). A market also exists for the smaller T8 U-tubes (available in 16- and 24-watt models) in decorative fixtures originally designed for incandescent lamps, but few fixture manufacturers offer any such options. Most applications for these lamps exist only in smaller, square, recessed fixtures similar to those using compact fluorescent lamps.

7.1.2 Efficacy

The efficacy of fluorescent lamps is closely linked to the type of ballast used to drive them. (High-frequency operation typically improves the efficacy of the lamp by 10 to 12 percent, and electronic circuitry reduces the parasitic losses of the ballast itself. See **7.2** for more on ballasts.) It can still be useful to examine the lamp-only efficacy of different parts of the fluorescent family, however. A few general relationships hold:

- Longer lamps are generally more efficient than shorter lamps, because the fixed losses of the cathodes become a smaller fraction of the total lamp energy consumption (**Figure** 7-8).
- Some parts of the fluorescent family are more efficient than others, but differences within groups are greater than differences between groups (**Figure** 7-9).

Figure 7-8

Lamp-only efficacy vs length for fluorescent lamps

Longer lamps are generally more efficient because they suffer proportionately smaller cathode losses.

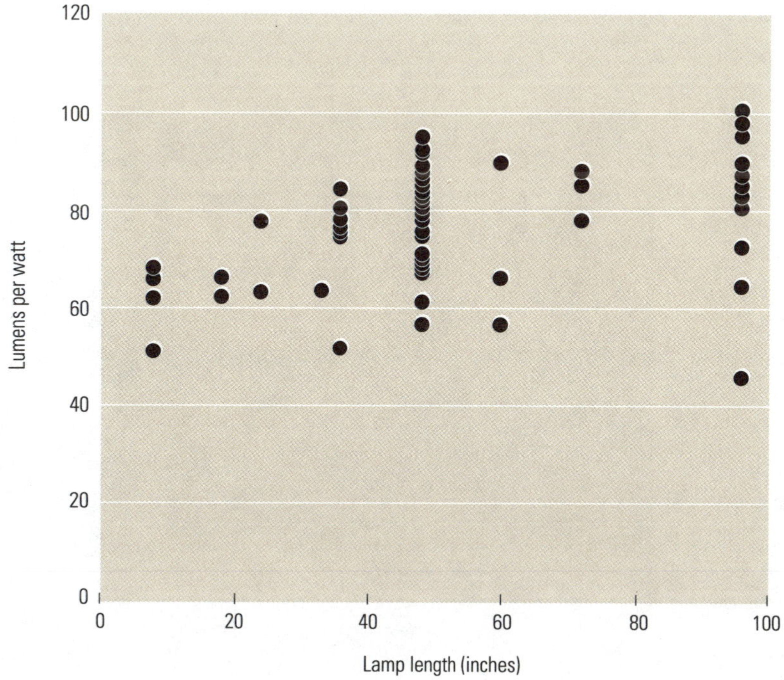

Source: manufacturers' data

Figure 7-9

Lamp-only efficacy of fluorescent lamp groups

Some parts of the fluorescent family are more efficient than others, but differences within groups are greater than differences between groups.

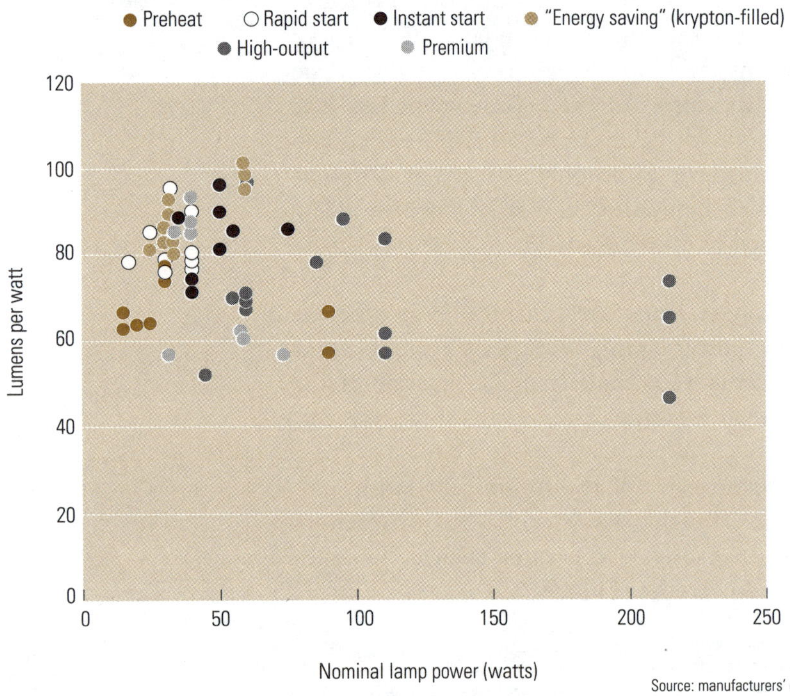

Figure 7-10

Lamp-only efficacy vs diameter for fluorescent lamps

T8 lamps are generally—but not always—more efficient than comparable larger-diameter T12s and T17s.

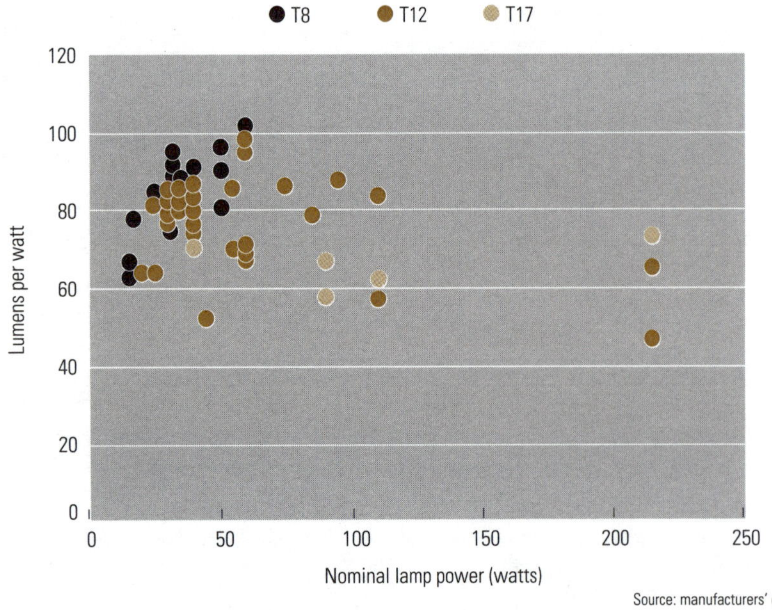

■ For a given current loading, phosphor blend, and operating temperature, an optimal lamp diameter exists that will maximize luminous efficacy. But because these variables can change significantly from lamp to lamp, it is not accurate to say that T8 lamps are intrinsically "more efficient" than T12s (**Figure** 7-10).

■ In the past, there was an inverse relationship between color rendering and efficacy. Modern phosphor blends have negated this relationship, and now some of the highest-efficacy lamps also provide the best CRI (**Figure** 7-11).

7.1.3 Lifetimes

Like other lamp types, fluorescent lamps follow a mortality curve; their "lifetime" is the statistical point at which 50 percent of the lamps in a large batch have failed. The primary issues in fluorescent longevity are the method and frequency of starting. Fluorescent lamps fail most commonly when their cathodes, weakened by erosion through starting and maintaining the electric arc, physically break (see Figure 7-3). Fluorescent lamp life is rated at three hours of continuous operation per start, unlike HID lamps, which are rated at 10 hours per start. Incandescent lamp life is relatively insensitive to the number of starts.

A few manufacturers rate their fluorescent products at conditions other than three hours per start. When evaluating claims of "extended life" products, check the rating conditions. If lamps are rated at more than three hours per start, their lifetimes will be artificially inflated. Another way to look at this situation: lamps will survive about 5,000 instant starts or 7,000 rapid starts. Use them wisely.

Figure 7-12 shows the effect of reducing the number of lamp starts: it increases expected lifetime by as much as 50 to 75 percent. Continuous-burn applications, such as interior spaces that operate 24 hours per day, can thus take advantage of this phenomenon by extending the time between relampings.

Because of the impact of starting method on lamp life, the different fluorescent groups exhibit varying lifetimes. **Figure 7-13** shows these ranges for a three-hour duty cycle.

The other lamp lifetime consideration is lumen maintenance, which for fluorescent lamps is mostly a function of phosphor type and current loading. **Figure 7-14** shows typical lumen depreciation curves for several members of the fluorescent family. Because of the drop in light output over time, group relamping is usually recommended to keep lighting levels consistent in large spaces. **CHAPTER 12** discusses relamping strategies in more detail.

7.1.4 Phosphors and color rendering

The phosphor coating inside a fluorescent lamp's walls is the primary determinant of its light quality. Over the past decade, careful chemistry in phosphor blending has significantly boosted the quality of fluorescent light output. Color rendering index (CRI) and color temperature (in kelvin), discussed in **CHAPTER 2**, are the basic metrics of light quality. Figure 7-11 above shows that lamps no longer must use more energy to achieve a high CRI. Color temperature has little or no direct effect on lamp efficacy, but the availability of different color temperatures is important for matching light appearance with

Figure 7-11

Lamp-only efficacy vs CRI for fluorescent lamps

Some of the highest efficacy lamps also provide the best color rendering.

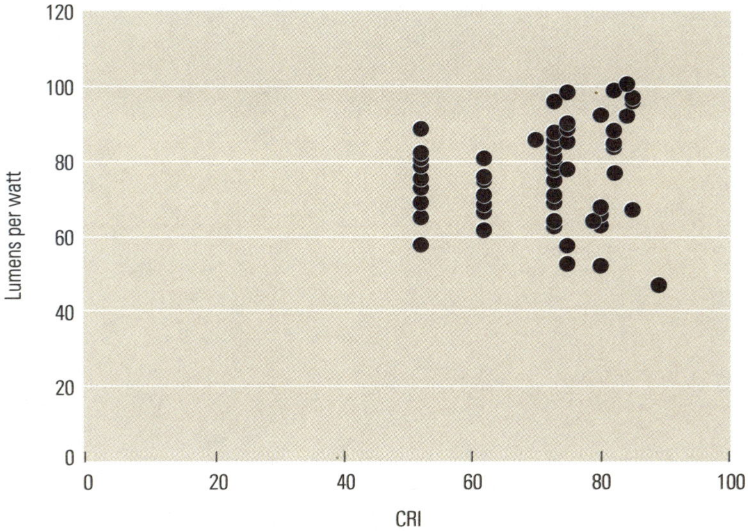

Source: manufacturers' data

Figure 7-12

Lamp-life vs run time per start

Fluorescent lamp life is very sensitive to the number of times the lamp is started.

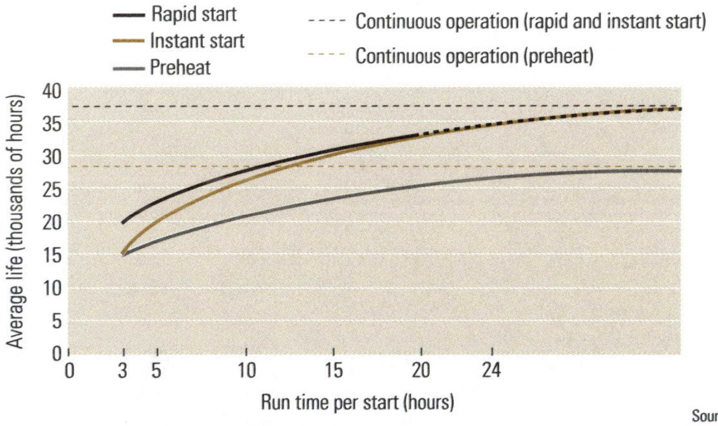

Source: Sylvania

Figure 7-13

Rated lifetimes for fluorescent lamp types

These lifetimes are based on three hours of operation per start.

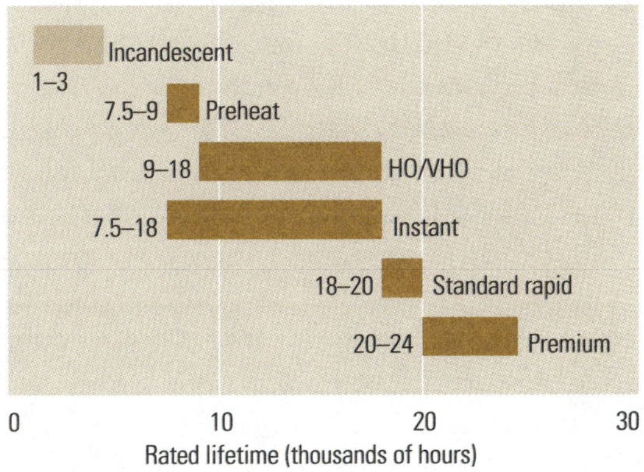

Source: manufacturers' data

Figure 7-14

Lamp lumen depreciation of fluorescent lamps

The light output of fluorescent lamps declines over time, and is primarily a function of phosphor type and current loading.

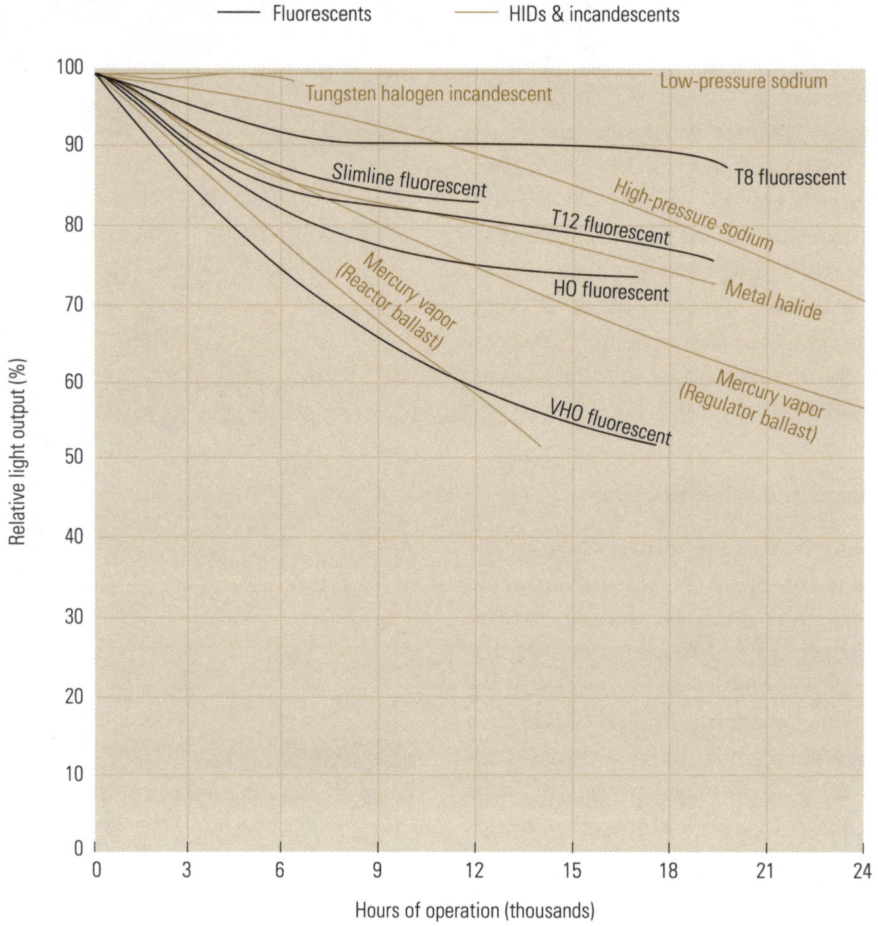

Source: National Lighting Bureau

other types of lamps operating in the same area or with daylighting. Modern fluorescent lamps can provide color temperatures from under 3,000 K to over 6,000 K and CRIs from 52 to 91 in most common lamp sizes.

Lamps with high-CRI phosphors can be identified in several ways. Figure 7-4 above shows the designation codes now used by most manufacturers to identify CRI in their lamp catalog numbers. The better phosphors are also referred to as "rare earth phosphors"—in reference to their places in the periodic table—or "tristimulus phosphors," indicating the common practice of selecting three primary materials, each with a specific color of light output, that combine to form high-quality white light. The chemical compositions of phosphors used in fluorescent lamps are generally available in the literature.[2]

As the name rare earth implies, the newer materials are relatively costly—approximately 20 times more than standard halophosphors. Economy is achieved by using the high-cost materials in smaller-diameter lamps, hence the popularity of the T8 envelope. Most lamps that use rare earth phosphors have a thin coat of them applied over a base layer of halophosphors. One problem with the new phosphors: manufacturers developed their own proprietary blends with similar—but not exactly equal—spectral output. Thus a 75-CRI lamp from Philips might have a slightly different color than a 75-CRI lamp from General Electric, hampering the ability of facility managers to shop between brands.

Manufacturers have adopted an x-y coordinate color standard so that their lamps will be more readily interchangeable. However, the exact color of lamps can vary even between batches of a specific lamp type from a single manufacturer. This fact, combined with the slight variations between manufacturers, provide another compelling reason to pursue group (rather than spot) relamping (see **12.5**).

The newer phosphors also have better lumen maintenance than standard cool-white and warm-white halophosphors. (The 34-watt "energy-saving" halophosphor lamps in particular have poor lumen maintenance.) Standard halophosphor lamps lose about 20 percent of their output during their rated lives, while equivalent rare earth phosphor lamps lose 10 to 12 percent. Enhanced rare earth phosphor blends and reflective coatings, such as GE's new Starcoat™ technology, promise even greater maintained light output with end of life losses at around eight to nine percent. The energy impact: a reduction in initial installed wattage of as much as 10 percent can be tolerated when switching from a halophosphor lamp to one with rare earth phosphors, while maintaining the same *average* light output over the life of the lamp. The example in **3.1** illustrates this principle in a typical design.

7.1.5 Temperature effects

The operation of fluorescent lamps is sensitive to ambient temperature, since temperature affects the vapor pressure of the enclosed mercury. Lamp type, ballast type, and fixture characteristics are all pieces of the temperature puzzle. As **Figure 7-15** shows, electronically ballasted T8 lamps have a higher optimal temperature than magnetically ballasted T12 lamps, and so will perform better in a closed fixture. Illuminance and efficacy excursions of 15 percent or more can occur across the temperature bands commonly found in installed fixtures. Compact fluorescent lamps are especially sensitive to operating temperature (see **8.4.5**).

Cold-temperature operation generally precludes systems that compromise cathode heating, such as instant starting, cathode-cutout lamps or ballasts, deep dimming, and "energy-saving" lamps.

Although HO and VHO lamp systems are commonly specified for cold applications, T8, T5 (compact fluorescent), and T12 (slimline) systems are available which offer equal or superior performance at low temperatures. Electronic ballasts are available which will reliably start lamps down to 0°F—check with both lamp and ballast manufacturers to con-

Figure 7-15

Temperature vs light output, power consumption, and efficacy for four-foot fluorescent lamps

The maximum efficacy for electronically ballasted T8 lamps occurs at a higher temperature than that of magnetically ballasted T12s, which gives the T8 an added performance advantage in hot, enclosed fixtures.

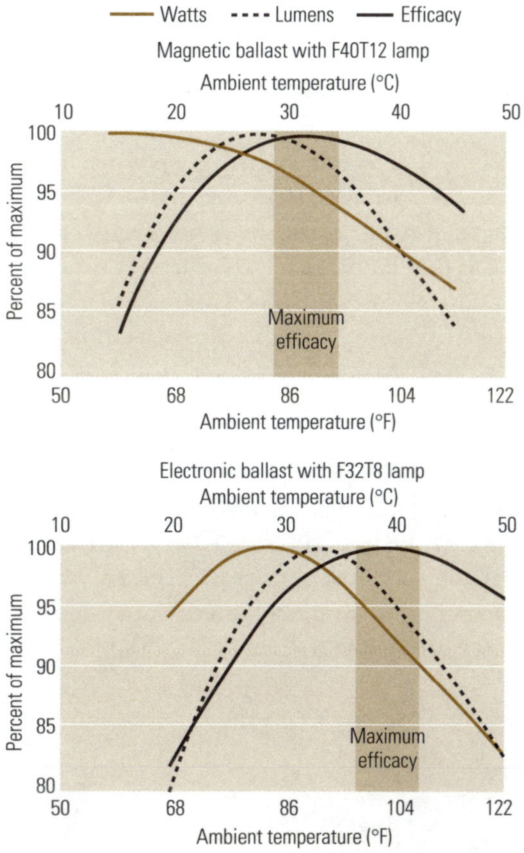

Source: Bleeker and Veenstra [7]

A reduction in initial installed wattage of as much as 10 percent can be tolerated when switching from a halophosphor lamp to one with rare earth phosphors, while maintaining the same average light output over the life of the lamp.

firm the design starting temperatures for their products. In addition, the following steps can help insure reliable fluorescent operation at low temperatures:

- full line voltage
- ballasts close to lamps (no remote or tandem wiring)
- good electrical connections
- properly sized conductors
- enclosed fixtures

7.1.6 Lamp economics

Retrofits or installations that rely on expensive lamp types for higher output or longer lifetimes always carry the potential of being thwarted by future unauthorized or incorrect lamp replacements. For this reason, if higher output is required from a fluorescent system, it is generally safer to use a conventional lamp driven by a high-output ballast.

With fluorescent lighting systems, lamps usually cost less than the labor to install them, and much less than the energy to run them (see Figure 12-1). **Table 7-6** lists typical prices for four-foot fluorescent lamps.[3] Most fluorescent lamps cost between one and ten dollars, although specialties such as eight-foot very-high-output (VHO) lamps can cost twenty dollars or more. In general, prices climb for lamps with 80+ CRI or those in unusual sizes. For example, a four-foot 70+ CRI T8 lamp costs $3.45, while a two-foot version of the same lamp costs $5.96. The economics of complete fluorescent systems are described in **3.1**.

Table 7-6

Fluorescent lamp prices (1997)

These list prices are for four-foot lamps, based on case orders for GE lamps. Discounts below these prices are usually available.

Lamp type	Typical wholesale price per lamp ($)
Energy saving 34W cool white T12	2.19
Energy saving high output cool white T12	9.17
Energy saving instant start cool white T12	10.27
70+ CRI T8	3.45
70+ CRI T12	4.78
80+ CRI T8	4.61
80+ CRI T12	9.74
Cathode cut-out T12	4.80

Source: GE Lighting [3]

7.2 BALLASTS AND POWER SUPPLIES

The primary functions of a ballast are to initiate the lamp arc with high voltage and/or cathode heating, then to stabilize the arc by limiting the electrical current to the lamp. Secondary functions include power quality correction and control features such as lamp dimming or compensating for lumen depreciation. Fluorescent lamp ballasts typically cost between $5 and $50, last for 40,000 to 100,000 hours, and consume parasitic power of a few to a dozen or so watts.

As the central connection point for fluorescent lighting systems, ballasts are key to achieving synergistic energy savings through control of light output. Ultimately, the most promising avenue for this control is the fully dimmable fluorescent ballast, which allows users to take advantage of strategies such as automatic daylight dimming, lumen maintenance, occupancy-controlled and scheduled dimming, and manual task dimming. As described in **CHAPTER 11**, the current state-of-the-shelf array of dimming fluorescent products does not yet make these features available at low cost; developments in this area may well represent the last frontier in the improvement of ballast technology.

Observers vary in their optimism about the coming of a reliable, inexpensive (say, under $25 for a two-lamp F32 or F40 unit) dimming fluorescent ballast—some say never, but others claim that well-designed dimming ballasts can be built to sell for no more than nondimming products. While new products such as Electronic Lighting Inc.'s (ELI) dimming ballasts are expanding immediate possibilities, promises of products have failed to materialize more often than not. In any case, the arrival of inexpensive dimming equipment will dramatically increase the potential energy savings in fluorescent lighting systems.

Typical ballasts and their internal components are shown in **Figure 7-16** (next page). Like lamps, ballasts also come in several varieties that form a family (**Figure 7-17**). While most existing ballasts still use magnetic technology, electronic (high-frequency) ballasts offer the greatest potential for energy efficiency. After discussing starting methods, the bulk of this section therefore focuses on electronic ballasts.

Figure 7-17

Fluorescent ballast types

This schematic ballast family diagram shows that VHO and preheat ballasts are only available as magnetic (low-frequency) units, while all integrated-circuit dimming ballasts are electronic units.

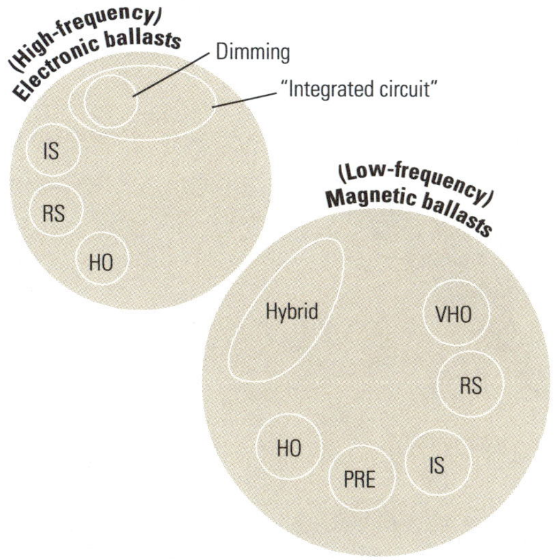

7.2.1 Starting methods

Three different methods can be used to start fluorescent discharge in lamps. With a few exceptions, the intended starting method of the lamp and the ballast must match. (Mismatches either don't start, or prematurely burn out the lamp, the ballast, or both.) The electrical characteristics of these starting methods are shown in **Figure 7-18** (page 137).

During *preheat* starting, a lamp's cathodes are heated for several seconds at low power before an arc is initiated. An external "starter" (basically a capacitor in a small can) provides the necessary pulse to initiate the lamp's arc. Cathodes remain heated to continuously supply free electrons from their emissive coatings. Lamps are wired in series, so failure of one causes any others on the same ballast to shut off. Preheat starting is used extensively with small lamps, and is not particularly effective or efficient (instant start electronic ballasts are beginning to be used to drive these lamps).

Rapid starting also begins by warming the cathodes, but at higher power for much less time (about half a second). Full voltage across the lamp then strikes the discharge arc. After starting, continuous electric heating of cathodes supplies electrons, consuming two to three watts per four-foot lamp. Most rapid-start ballasts wire lamps in series, so failure of one causes the other to shut off. In late 1993, several ballast manufacturers (such as Toshiba and ETTA) introduced parallel-wired rapid-start electronic ballasts that avoid this problem. A "hybrid" version is also available which rapid-starts lamps but then discontinues cathode heating to save those two to three watts.

bal•last *n* : 1: something that gives stability

Instant starting ignores cathode heating, and simply jolts the lamp with a high-voltage (400 to 1,000 V)[4] pulse that starts the lamp in under a tenth of a second. Because the cathodes are not actively heated, instant start systems save two to three watts per lamp compared to rapid start systems. The primary disadvantage of this technique is that it accelerates erosion of the cathode's emissive coatings, reducing lamp life by about 25 percent relative to rapid starting (at three hours per start). In multiple-lamp instant-start systems, the lamps are wired in parallel to supply the same high voltage to all lamp ends. Failure of one lamp allows the others to continue burning.

Figure 7-16

What's in a ballast?

Most ballasts in service today are magnetic units that consist of little more than a transformer, a capacitor, and a thermal cutoff switch to protect the ballast from overheating. Hybrid ballasts use electronic circuitry to control power to the lamp's cathodes, but magnetic components still drive the main arc. In high-frequency electronic ballasts, magnetic devices are required for current and voltage control, surge/spike suppression, and other protective functions.

Standard magnetic ballast

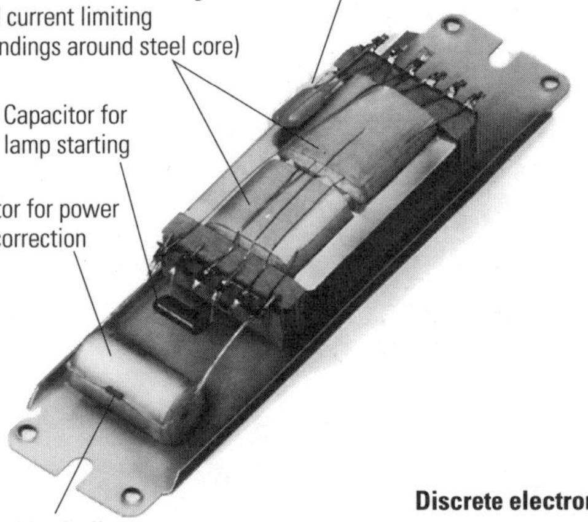

"Hybrid" magnetic ballast
(cathode cut-out)

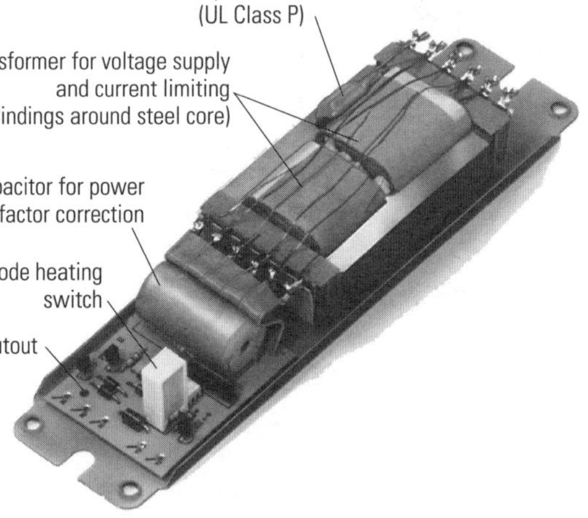

Discrete electronic (high-frequency) ballast

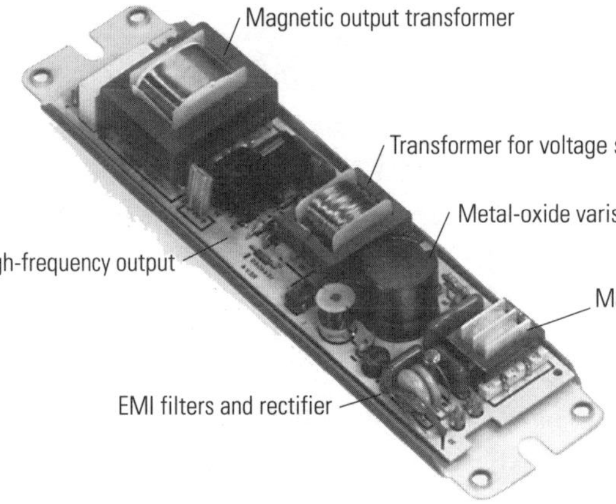

Integrated circuit electronic ballast

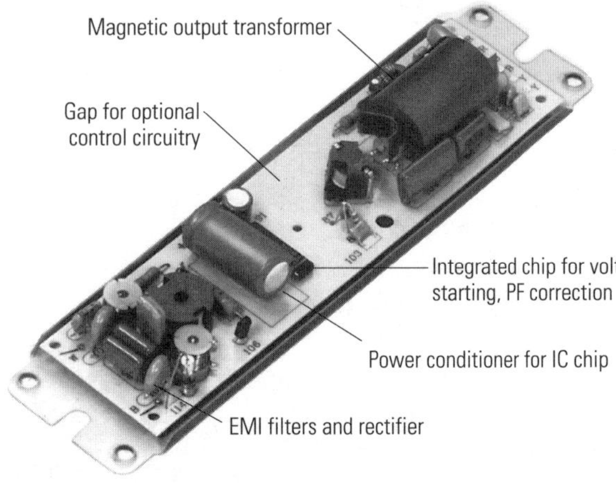

Dimming electronic ballast

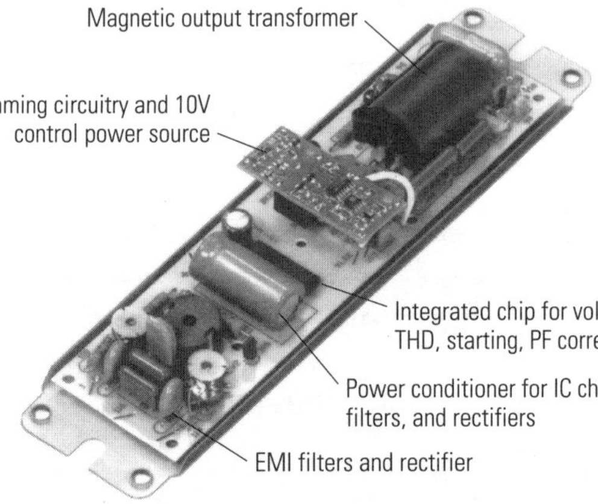

Source: Advance Transformer

7.2.2 Magnetic ballasts

Since most existing ballasts use magnetic transformers to adjust line voltage to cathode voltage, they are described as "magnetic." Defining magnetic ballasts more accurately can be difficult, however, since even "electronic" ballasts use magnetic equipment for some functions. One good way to make the distinction: any ballast that provides output power to the lamp at line frequency (60 Hz in North America, 50 Hz elsewhere) is a magnetic ballast.

Improvements to materials and design have reduced the internal electrical resistance of magnetic ballasts, making those manufactured after 1978 about 10 percent more energy efficient than those before that date. Present U.S. federal ballast regulations mandate that only these "energy efficient" magnetic ballasts be used for the majority of applications in the U.S., except for some small residential and appliance lamps that continue to be produced with very low-power-factor ballasts.

With EPACT driving the increased sales of "energy-saving" lamps, ballast manufacturers are developing magnetic ballasts optimized for their operation. For example, Advance Transformer's E-PAK34 magnetic ballast operates 34-watt energy-saving lamps six percent more efficiently than its standard Mark III ballast, saving four watts per two-lamp system.

Hybrid ballasts. Ballasts that cut out cathode heating after starting the lamp are known as hybrid ballasts, since they use magnetic technology to power the lamp and electronic technology to control power to the cathodes. (Some people also use the word "hybrid" to describe discrete electronic ballasts (see page 142). In this book, we equate "hybrid" ballasts with cathode-cutout operation at 60 Hz.) These 60-Hz devices have been overshadowed by electronic ballasts, which offer greater efficiency gains for little extra cost. Light output and efficacy are about five to seven percent lower than that of an equivalent electronic ballast. Hybrids are limited to two-lamp varieties, and lamps are wired in series. Most sources agree that eliminating cathode heating does not reduce lamp life, as the normal operation of the lamp keeps the cathode warm enough to ensure a reliable supply of electrons.[5] Combining low ballast factors with

Figure 7-18

Electrical activities during fluorescent lamp starting and operation

There are three primary methods of initiating and maintaining a fluorescent discharge arc. These graphs show arc and cathode voltage for the two most important methods: rapid and instant starting. Preheat starting (not shown) is similar in nature to rapid starting, but with a longer delay.

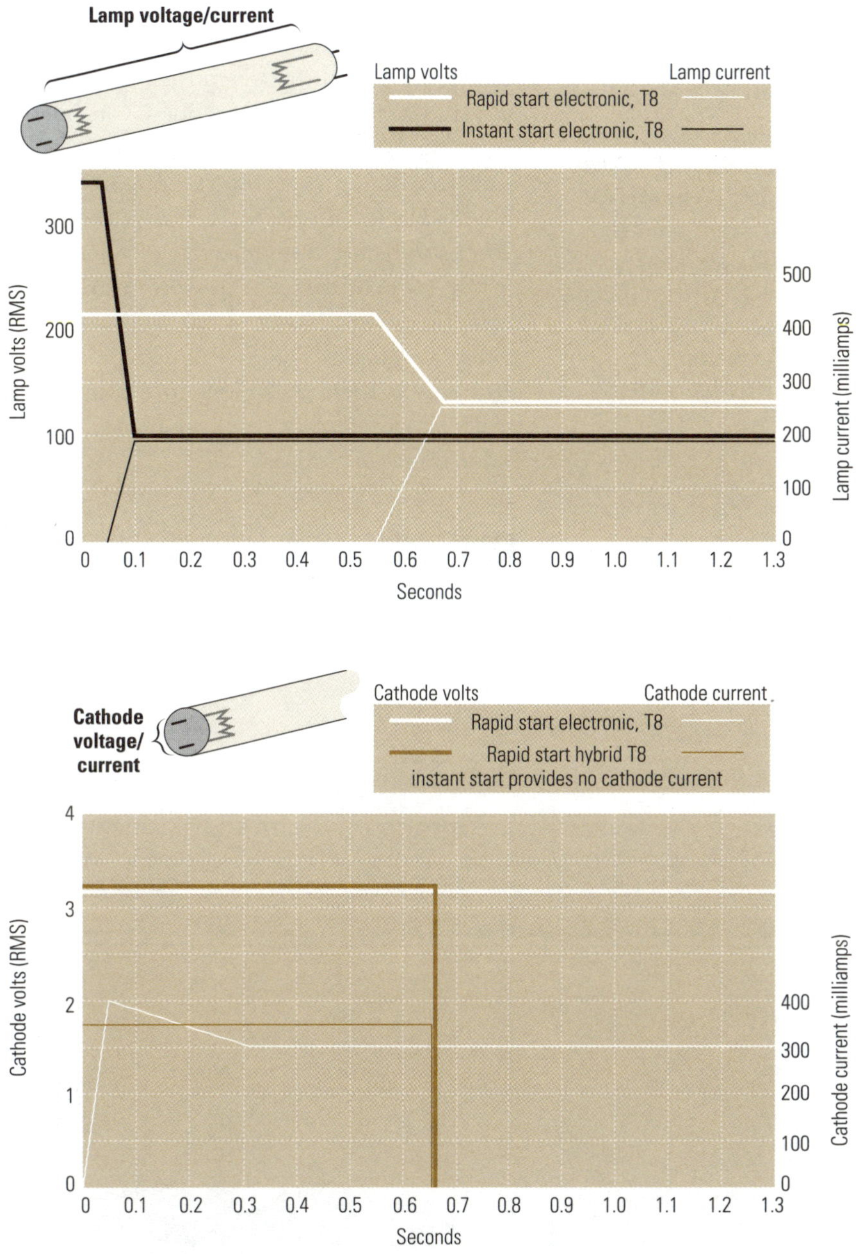

Source: Advance Transformer

There are several mechanisms for energy savings with electronic ballasts, including their intrinsic efficiency, reduced design margins, control savings from continuous dimming, and other effects.

cathode cutout may reduce lamp life, however (just as with instant-start ballasts) because lamp current and attendant cathode heating are significantly reduced. Hybrids also should not be used in cold locations.

7.2.3 Electronic ballasts

Fluorescent lighting technology took a major leap forward with the advent of the electronic ballast, which uses semiconductors to rectify incoming 60 Hz to direct current (DC), and a high-frequency inverter to convert DC to 20,000-Hz (or higher) current to the lamp. Other solid-state components control or filter power to minimize harmonic distortion, maintain a high power factor, and shape the power waveform. The growing popularity of electronic ballasts is evident in **Figure 7-19**, which shows steadily increasing sales volume and market penetration.

Figure 7-19

U.S. shipments of electronic ballasts

By the end of this decade, the conventional magnetic ballast may be virtually extinct.

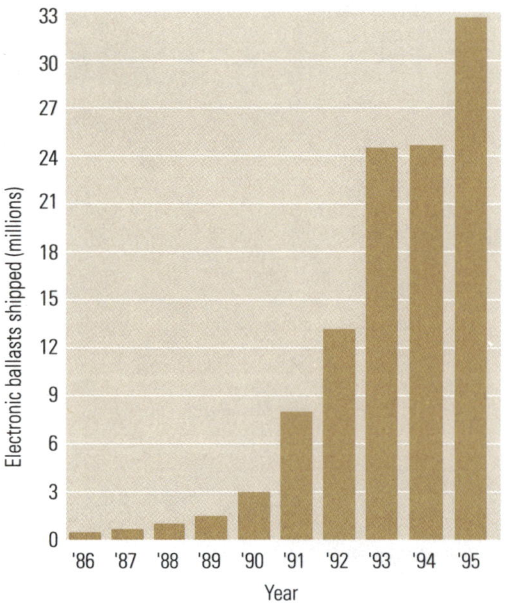

Source: U.S. Census Bureau, Osram/Sylvania

There are several mechanisms for energy savings with electronic ballasts, including their intrinsic efficiency, reduced design margins, control savings from continuous dimming, and other effects. Section **3.1.1** presents an example of how these improvements combine to create very large energy savings—up to 90 percent—in fluorescent systems.

Intrinsic efficiency

- Electronic ballasts waste less power internally than magnetic ones, reducing direct ballast losses by about three to eight watts per ballast, depending on the number of lamps being fired by a single ballast.

- Lamps operate about 10 percent more efficiently at the electronic ballasts' high frequency.[6]

- Since less heat is dissipated within the ballast, more lamps' worth of ballast circuitry can be fitted into a single case, eliminating some duplicate circuitry and saving about another one watt per lamp.

Reduced design margins

- Lighting systems are sometimes overdesigned to ensure adequate illuminance with abnormally low supply voltage or nonoptimal lampwall temperature. Because electronic ballasts are less sensitive than magnetic ballasts to these variables, they require smaller design margins, potentially saving up to an eighth of the energy normally required.

Control savings related to continuous dimming

- With the aid of light-level sensors, dimming electronic ballasts can provide just enough power for the fluorescent lamps to supplement whatever natural light is available at the time, maintaining the design illuminance but no more. This "daylight dimming" can save upwards of half the lighting energy in daylit perimeter zones (see **CHAPTER 5**).

Continued on page 141

Electronic Ballast Terminology

Electronic ballast. A fully *electronic ballast* uses solid-state circuitry to provide current limitation, voltage modification, and high-frequency output. In contrast, *magnetic ballasts* use induction chokes and transformers to limit current and modify voltage, and don't alter frequency at all. *Hybrid ballasts* are magnetic units with a cathode-cutout feature.

Ballast factor (BF). Ballasts can underdrive or overdrive a given lamp for lower or higher light output. Ballast factor is the ratio of a lamp's light output with a given ballast to the same lamp's light output with a standard reference ballast, tested according to ANSI standard procedures on an open-air bench, and is between 0.86 and 0.92 for most "normal" ballast factor ballasts. "Low" ballast factor ballasts are available down to 0.75 and "high" ones up to 1.28.

Ballast efficacy factor (BEF). BEF is used when comparing two ballasts to determine which supplies more light for a given wattage. It is calculated by dividing the BF (as a percentage) by the input wattage to the ballast. BEF comparisons between ballasts must be made using the same quantity and type of lamp. For example, while a BEF greater than 1.3 for a single *two-lamp* T8 ballast is not uncommon, the BEF for each of a pair of *one-lamp* ballasts could be 2.8, while providing no more light. Caution is also needed when making this comparison because a ballast with a higher BEF may have greater efficiency, but lower light output.

Current crest factor (CCF). CCF is the ratio of peak voltage to RMS voltage, and indicates the sharpness of the waveform sent by the ballast to the lamp. A perfect sine wave has a CCF of 1.414, and lamp manufacturers caution against exceeding 1.7; many ballasts have CCFs between 1.5 and 1.7. Standard lamp life ratings are based on current crest factors of 1.7 or less.

Power factor (PF). There are two types of power factor: phase power factor and total or true power factor. When alternating current drives a purely resistive load (for example, an incandescent light bulb), the voltage and current are in phase—they reach their maximum, minimum, and zero values at the same times. More complex loads have elements that store magnetic and electric energy—known as inductance and capacitance, respectively—that force voltage and current to slip out of phase with respect to each other. Phase power factor is a measure of the "out of phase-ness" of a device's circuit, and ranges from 1.0 (in phase) to 0 (90° out of phase). Most quality ballasts have PF exceeding 0.95. Low-power-factor circuits require the electric utility to supply additional "reactive power" that is not recorded by the meter. Reactive power incurs resistive losses in the distribution system and requires oversized distribution lines and transformers.

Total power factor reflects the fact that electrical power has a shape component, indicating waveform distortions such as harmonics which, like out-of-phase current, dissipate energy against any resistance but perform no work. See **3.2.2** for more on power factor.

Total harmonic distortion (THD). Total harmonic distortion describes the amount of distortion in the power waveform that is being caused by a ballast. Utilities often call for 20 percent or less, but the U.S. electric manufacturing industry is examining a standard that would accept up to 32 percent. Many existing magnetic ballasts create 20 to 30 percent, so such limits are designed to avoid any worsening of line distortion when ballasts are changed. The most popular electronic ballasts are less than 20 percent THD, while a number of products are available with THD below 10 percent. Lower is not necessarily better, however, since low-THD

Continued on next page

Electronic Ballast Terminology

Continued from previous page

ballasts may cause problems from high inrush currents. High THD (common from electronic devices such as personal computers) can disrupt powerline carrier controls and create unacceptably high currents on the neutral wires of three-phase systems.

Flicker index (FI) and flicker factor. Flicker index and flicker factor (the latter is sometimes called *percent flicker*) are two ways to indicate the variation in light output over a single cycle of alternating-current (AC) lamp operation. The mathematical definitions of these quantities—one cannot be directly converted to the other—are shown in **Figure 7-20**. Flicker index is replacing flicker factor as the metric of choice, because its area-based calculation provides a better indication of the eye's flicker perception than does the peak-based flicker factor. Flicker index can range from zero to one, but most values are below 0.20, while flicker factor can range from zero to one hundred percent, with most values below 33 percent. Flicker values are typically specified for ballasts rather than lamps, although they actually describe a characteristic of the lamp-ballast system. With low frequency operation, flicker is largely a function of phosphor composition—some combinations of phosphors "smooth out" the discharge arc's sine wave better than others. With high-frequency operation, the driving characteristics of the ballast dominate flicker effects.

Relying purely on flicker index to describe the constancy of light output misses two key effects. First, the frequency of operation has a strong impact on phosphor behavior and the visual perception of flicker. At a typical electronic ballast frequency of 30 kHz, each cycle only lasts 1/30,000th of a second; variations in light output over that cycle are evened out by the time lag of phosphor decay and are difficult or impossible for the eye to perceive.

Secondly, with high-frequency operation, variations between cycles can be much more important than variations within a given cycle. In high-frequency ballasts without proper line frequency filtration, for example, a 60-Hz cycle can be imprinted on top of the 30-kHz cycle, causing visible flicker that will not show up in measurements of "flicker index."

Since minimizing flicker improves overall lighting quality and is believed to reduce eye fatigue, a flicker index of less than 0.01 or a flicker factor of less than 5 percent is recommended. An uncommon but potentially serious consequence of flicker is the triggering of seizures among epileptic or photosensitive individuals (see **3.4.2**).

Audible sound rating. Audible sound from a ballast is rated by a letter indicating its intensity relative to a quiet background. Most ballasts are "A" rated, meaning they meet the most stringent level of sound emissions. "B" rated ballasts are acceptable in noisy or industrial locations.

Radio interference. The Federal Communications Commission sets an allowable level of radio interference. Nearly all ballasts meet FCC standards, though magnetic ballasts often have radio-frequency emissions *below* those of electronic models.

IEEE transient protection. Ballasts certified as having IEEE transient protection can withstand a defined voltage spike without being damaged. Most electronic ballasts meet this standard, and nearly all magnetics exceed it.

Figure 7-20

Determining flicker index and flicker factor

Flicker index is replacing flicker factor as the metric of choice, because its area-based calculation provides a better indication of the eye's flicker perception than does the peak-based flicker factor.

$$\text{Flicker index} = 100\left(\frac{\text{Area 1}}{\text{Area 1} + \text{Area 2}}\right)$$

$$\text{Flicker factor} = 100\left(\frac{A - B}{A + B}\right)$$

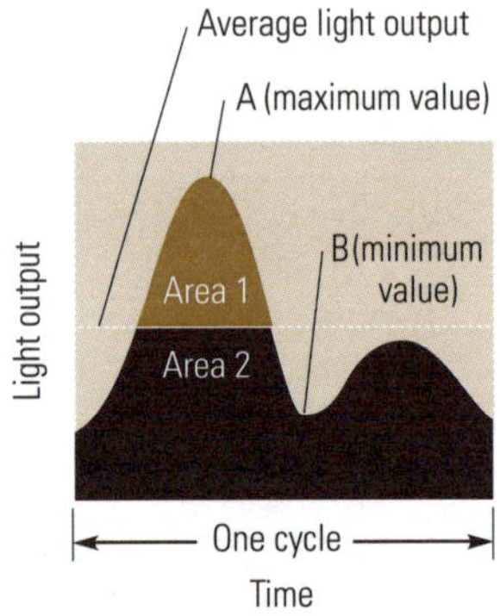

Source: IES

Breaking the Rules with Lamp-Ballast Combinations

As shown in **Table 7-7**, some electronic linear fluorescent ballasts can be used to drive multiple circular, U-shaped, or compact fluorescent lamps in ways that provide greater light output with less wattage than the magnetic ballasts normally specified for such lamps. These arrangements "break the rules" because they couple lamps and ballasts that are not specifically designed for each other. While many of these combinations use instant-start ballasts, in facilities where normal on-off cycling is 8 hours or more (or lamps run continuously), the lifetime of such lamps is minimally affected by such retrofits. Users should consult with lamp manufacturers and conduct small-scale trials before employing nonstandard combinations.

Table 7-7

Selected nonstandard electronic ballast/lamp combinations

In some cases, ballasts and lamps not designed for each other can be successfully mated. Combinations such as the ones shown here can expand the range of applications for high-frequency electronic ballasts.

Ballast make	and model	Lamp no./type	Input watts	Power factor	THD (%)
Advance Transformer	REL-2532-LW-RH-TP	Three FBO16[a]	54	0.96	22
	REL-IP32-RH-TP	One FC12T9[b]	21	0.80	35
	REL-2532-LW-RH-TP	FC12T9+FC8T9	35	0.95	26
	REL-2P32-RH-TP	FC16T9[d]+FC12T9	48	0.97	23
	REL-2532-LW-RH-TP	FC16T9+FC12T9	51	0.98	18
Osram Sylvania	QT 4X32/120 IS	Four 26 W CFL	80	0.95	20
Magnetek	B332I120HP	Three FBO16	48	0.89	15

Notes:
a: FBO16 = 16 W T-8 U-tube.
b: FC12T9 = 32 W Circline.
c: FC8T9 = 22 W Circline.
d: FC16T9 = 40 W Circline.
Circline combinations have BF approx. 0.75–0.80. CFL and U-tube combinations have BF of about 0.8–0.9+.

Continued from page 138

■ Because lighting systems are typically designed to provide the required light levels at the end of lamp life, they often provide excess light when first installed. The dimming feature can automatically brighten lamps as they dim with age and dirt, so they needn't be unnecessarily bright when they're young, fresh, and clean. This automatic compensation for lumen depreciation—which also requires the use of light-level sensors—can save about a seventh of the energy over a group-relamping cycle.

■ The ability to adjust the light from each ballast's lamps allows the user to customize and modulate the distribution of light across a space. This "task tuning" to provide more light on the desk and less in the corner or hallway can save about 10 to 20 percent of office lighting energy.

■ Dimming the lamps better maintains their output and their lifetime (as long as cathode current is maintained), since lumen depreciation from causes within the lamp (and from some outside, such as dirt deposition by convective air movement) is roughly proportional to the product of current times operating hours.

■ Occupancy-controlled *dimming* can provide savings nearly as large as on-off occupancy control without the attendant reduction in lamp life.

Additional effects

■ Since the eye handles a 10^{13}-fold brightness range, from starlight to sunlight, dimming lamps very gradually by about 20 to 30 percent is usually imperceptible. An energy management system can do this during peak-demand hours, saving both direct lighting loads and the resulting space cooling and air handling loads, hence markedly reducing utility demand charges.

■ High-frequency operation raises the efficacy-maximizing lampwall temperature by about 5.4°F (3°C), making it easier to achieve this optimum in enclosed fixtures.[7]

■ Electronic ballasts are generally designed to de-energize failed lamps, saving the several watts of excitation energy that a magnetic lamp would otherwise use pending relamping.

Compared to conventional, magnetically ballasted fluorescent lighting systems, these effects can collectively reduce watts of electrical input per usefully delivered footcandle by 30 to 90 percent, depending on whether the application is in a core or perimeter zone, and whether the ballast can dim to accommodate daylighting (see example in **3.1**).

Some high-frequency ballasts are more "electronic" than others.

Electronic ballast variations. Some high-frequency ballasts are more "electronic" than others. Discrete-component electronic ballasts—the original type introduced in the mid-1980s—use separate electrical components soldered onto a circuit board, and have relatively large magnetic coils. These ballasts have high component counts, require significant assembly effort, and, because of their magnetic coils, have relatively high internal losses. They still provide most of the benefits of high-frequency electronic operation, however. These ballasts should not be confused with "hybrid" electronic-magnetic ballasts that operate at line frequency (described above).

Discrete electronic ballasts are slowly giving way to integrated-circuit (IC) electronic ballasts, which combine the functions of dozens or even hundreds of separate resistors, transistors, and capacitors into single chips. While requiring more complex R&D, these units reduce the number of components in a ballast and avoid manual soldering, improving reliability. IC ballasts generally provide advanced performance features as well, such as automatic compensation for voltage reductions.

7.2.4 Power quality

The power quality characteristics of ballasts are measured by two quantities: power factor and harmonic distortion. Both harmonic distortion and power factor have been improved in the newer models from most electronic ballast manufacturers, to levels better than those of magnetic ballasts. **CHAPTER 3** includes a more complete discussion of power quality. Recommended power quality specifications are shown in Table 7-9.

If the concern with power factor is overheating the neutral wire with reactive current, it is important to remember that neutral current is the product of two quantities: the deviation of the power factor from unity and the total current draw of the lighting system. Efficient electronically ballasted systems significantly reduce the latter, and so may allow a larger power factor excursion without exceeding specified neutral-wire loading levels.

CASE STUDY — FIELD MEASUREMENTS OF THD

Utility engineers measured harmonic distortion before and after a retrofit with electronic ballasts at a college in South Dakota. The original system used 2,500 1960s-vintage magnetic ballasts to drive 5,000 fluorescent lamps, and had measured THD of 2.5 percent. After a retrofit to 1,250 four-lamp electronic ballasts in 1990, the measured THD dropped to 1.5 percent.[8] The measurements were made by the local utility, Northern States Power, at the building's main distribution panel.

Unfortunately, many of these ballasts failed (see Table 7-8). The entire group of ballasts was replaced by the manufacturer with a reduced-harmonic design, but the THD of those ballasts was not measured.

Total harmonic distortion (THD) levels are typically given for a ballast when operating a four-foot lamp. THD may increase significantly, however, when an electronic ballast is used to power shorter lamps, whether straight, U-shaped, or circular. If THD is a concern (because of either an existing high building level or a utility rebate requirement), the ballast's level (when operating such lamps) should be determined prior to installation. The advent of low (under 10 percent) THD electronic ballasts has, however, created new options for powering two- and three-foot lamps, while standard (up to 20 percent) THD electronic ballasts can often provide the same flexibility with circular or short U-lamps—all without creating power quality problems or violating utility requirements.

In the effort to reduce harmonic distortion, another problem has been created: inrush current. Electronic ballasts that use integrated circuits instead of inductors (choke coils) for power factor correction and THD reduction make the ballast appear to the power source as a capacitor. Upon startup, the capacitor gulps power to charge itself up. Without an inductor to act as a brake on the circuit, the inrush current to charge the capacitor can be extremely high.

High inrush currents create problems at lighting panel and relay switches when large groups of lamps are turned on at the same time. An analysis by Johnson Controls—inspired by high failure rates of lighting contactor switches—shows that a particular fluorescent lighting circuit (37 low-THD ballasts driving 148 T8 lamps) that draws 16 amps at steady-state conditions can draw inrush currents as high as 600 amps. These currents are far higher than typical switch ratings (typically around 150 amps) and can damage the face of the contactors and eventually destroy them by welding them together.[9]

There are several possible solutions for the inrush current problem, including the following:

- Use ballasts that have higher THD specifications (20 percent or even 33 percent maximum) rather than the more ambitious goal of 10 percent maximum. These higher-THD ballasts typically have enough inductance to control the inrush current, but still have low enough THD to control neutral-wire heating.

- Specify maximum inrush current for ballasts, although this figure is not typically supplied by manufacturers.

- Use zero-crossing switches, which sidestep the problem by closing the contactor at the zero point in the 60-Hz cycle.

- Use heavy-duty switches that can handle the high currents.

- Split lighting loads among circuits to reduce the current per switch.

Until all ballast manufacturers design their products to limit inrush currents, specify with caution for large group-switched fluorescent lighting applications.

7.2.5 Reliability

High failure rates among early generations of electronic ballasts made reliability a major concern among users. To counter reliability concerns, manufacturers have been identifying failure modes and addressing them with improved production techniques.

Electronic ballasts can fail well before their rated lifetime for several reasons. Many large-scale failures have been attributed to bad batches of components, so manufacturers are now choosing their suppliers more carefully. Defects can also occur in ballast assembly; this failure mode is being combatted by a shift to automated production lines. Another step now commonly taken is to "burn in" the ballasts by running them at full power before shipment, a step that at least keeps some "infant mortality" failures in-house. According to both manufacturers and industry analysts, many ballast failures are caused after shipment by poor grounding, dropping, miswiring, incompatibilities with adjacent equipment, or other installation-specific problems.[10]

Many ballast failures are caused after shipment by poor grounding, dropping, miswiring, incompatibilities with adjacent equipment, or other installation-specific problems.

Where Electronic Ballasts May Not Work

While electronic ballasts are almost always the best choice, in some cases they may not be appropriate, due to existing power quality problems, sensitive equipment, or operating conditions (for example, very high or low temperatures). In such cases, hybrid ballasts are usually the next best choice. Here are several situations where electronic ballasts may not be the best choice.

Security systems. Some security systems use a high-frequency magnetic field to detect products at entry points to a space, such as libraries and retail stores. The systems send out a low-intensity field that strikes a magnetized tape. The tape returns an echo indicating its presence. Most electronic ballasts emit a magnetic field of a similar frequency, causing such systems to shut down to protect themselves.

Manufacturers of detection equipment suggest that electronic ballasts not be installed within 20 feet of their systems to avoid this problem, but field experience shows that this recommendation may not be sufficient. Since magnetic fields penetrate concrete, the problem may be caused by ballasts in the ceiling of the floor above or below the security system, *or by power wiring feeding an electronic ballast* more than 20 feet away, because that wiring may also be emitting such a field. Another lighting designer recommends a 30-foot radius.[11]

Some security systems can be retrofitted with a notch filter to reduce their vulnerability to stray fields. Systems made after 1992 are typically immune to the problem. The system's service mechanic should be consulted prior to installing electronic ballasts in such areas. If any doubt exists, magnetic ballasts should be used instead of electronic ballasts.

Control systems. Some control systems that rely on powerline carrier (PLC) technology—especially residential systems—are vulnerable to the low level of interference caused by most electronic ballasts (as well as many other sources of radio frequency interference) and may fail to respond to commands from their central panel. This problem can sometimes be remedied by adding filters on the powerline to the central panel, but filtering may also be needed at the system's controllers, making this correction an expensive option. A better approach is to examine any existing control systems and check with the manufacturer about potential problems before installing electronic ballasts. A related but different problem can also occur. Filtration circuitry in electronic ballasts can "eat" the signals that powerline carrier systems use to communicate.

Recording studios. Wiring passing among amplifiers, recording equipment, and microphones can act like antennas, picking up radio frequency interference from many sources. While most electronic ballasts sold in the U.S. meet the FCC requirements for suppressing such emissions, compliance with such standards may be insufficient to avoid interference when the wiring surrounds or passes near light fixtures. The effect of this problem: a low-frequency hum is heard by the recording engineer when no sound is entering a microphone. Caution suggests the use of high-quality magnetic ballasts near sensitive audio equipment.

Data from the University of California lends credence to the "batch failure" theory, and generally supports the manufacturers' claims of reduced failure rates. The University installed over 32,000 high-frequency electronic ballasts involving 17 models and 3 manufacturers between 1987 and 1990. Field records revealed that while the overall failure rate was 1.46 percent,[12]

- seven models had failure rates below one percent,
- five models had failure rates of 1 to 1.4 percent,
- two models had failure rates of 2.7 to 4.3 percent, and
- three models, all from one manufacturer, had failure rates of 6.2 to 10.5 percent. Some of these ballasts were returned mistakenly by maintenance staff, who thought their slow starting meant they were defective.)[13] This manufacturer is no longer producing ballasts.

Lighting users and researchers indicate that trust in constant-output electronic ballasts has been solidified by good performance over the last couple of years. Doubts remain, however, about the reliability of dimming ballasts. Reliability data is difficult to come by, and while most manufacturers are now reporting failure rates of less than 1 percent, these figures are frequently averages for entire product lines, so they give little indication of the reliability of a particular model, such as a dimming ballast.

Reports of significant ballast failures continue, even with recent-vintage electronic ballasts. The experience of a maintenance engineer for a small university tells a story of success and failure (**Table 7-8**). Over three thousand electronic ballasts were installed at Augustana College in Sioux Falls, South Dakota between 1988 and 1992. All of the first batch of 500 ballasts failed and were replaced. Only 0.5 percent of the second batch failed. Even with this experience, the engineer notes that overall failure rates were higher with magnetic ballasts than with electronics; magnetic ballast failure rates can range up to 10 percent or more in a given batch. Nearly all failures occurred immediately or within a month of installation. The engineer reports that failure rates for electronic ballasts have consistently declined with successive generations of products.[14]

Table 7-8

Electronic ballast failure rates at Augustana College

Experience with electronic ballasts at this facility has been mixed. Several batches of ballasts had failure rates of 1.5 percent or less, but three batches had unacceptably high failure rates.

Date	Quantity of ballasts	Ballast manufacturer and model number	Failure rate (%)	Comments
1988	500	Advance R2P32TP	100.0	Complete changeout
			0.5	Replacement ballasts for above
1989	780	Magnetek Triad B432I120	1.5	
	360	Magnetek Triad B432I277	1.5	
1990	750	Advance REL4P32TP	7.5	
	1,250	Advance VEL4P32TP	28.0	Complete changeout
1992	1,250	Advance VEL4P32 (RH) TP	1.0	Replacement ballasts for above

Source: Augustana College [8]

At Columbia University, retrofitters have installed 41,000 electronic ballasts since early 1990, of which fewer than 100 failed.

These failures must also be tempered with the experience of users who have installed large quantities of ballasts with very few failures. At Columbia University, retrofitters have installed 41,000 electronic ballasts since early 1990, of which fewer than 100 failed, and most of those within a couple of weeks of being installed. This represents a failure rate of less than one quarter of one percent. The sample includes a wide variety of high-frequency electronic ballasts, including units for 1, 3, 4, and (mostly) 2 lamps, running at 120 and 277 volts, with normal and low ballast factors, and from manufacturers EBT, Magnetek, Motorola, Osram, and Valmont.[15]

Nearly all electronic ballasts now come with a five-year limited warranty. Caution is suggested, however, when employing the first model of any new technology since the labor and inconvenience costs inherent in replacing a ballast may greatly exceed the usual $10 labor allowance given by ballast manufacturers when they replace a defective product.

7.2.6 Ballast factor and light output

The lighting design of a space can be fine-tuned by selecting ballasts with low ballast factor (BF) that underdrive lamps or high-BF ballasts that overdrive lamps. Manufacturers vary ballast factor by adjusting the amount of electrical current used to drive the fluorescent arc.

Low BF (also called "low-power") ballasts, defined here as those units with BF below 0.88, are available in both rapid- and instant-start varieties. Instant-start versions can wear out lamps faster, since the combination of reduced arc current and no cathode heating is hard on the cathode's emissive coating. In most cases, however, the reduction in lamp life is more than paid for by the energy savings of the instant-start operation.

Although ballasts with BF up to 1.28 are available, high ballast factors also reduce lamp life. As discussed above (Figure 7-7), high current loading degrades the phosphors, resulting in accelerated lumen depreciation. For this reason, lamp manufacturers suggest a maximum ballast factor of 1.20.[16]

7.2.7 Ballast recommendations

With the wide variety of available ballast products, some general guidelines for ballast selection can be helpful. Our recommendations are as follows:

■ The combination of T8 lamps and nondimming high-frequency electronic ballasts has excellent energy efficiency, well-established performance, and a wide variety of suppliers. The majority of general illumination applications can make use of this simple prescription. Since all available 265 mA T8 lamps have good phosphors, reasonable light quality is virtually assured. With this lamp-ballast combination, technological concerns give way to questions of lighting design—mostly involving the distribution of light.

■ Instant-start ballasts with parallel operation provide the best efficacy and lighting utility. In applications for which no instant-start products exist, or where duty cycles are shorter than three hours, rapid-start ballasts are the logical choice.

■ For applications that suggest emerging technologies, the best advice is to try out products on a small scale. While inspecting other installations can be helpful, there is no match for trying the goods under your specific conditions. If there are problems, they will usually make themselves known quickly. The case study in **7.3.1** (page 150) gives one example of the lessons this approach can provide.

■ Developing a rigorous specification is an excellent way to protect yourself and your organization. **APPENDIX D** gives one example of a lighting "spec" that weeds out the pretenders. Table 7-9 summarizes technical recommendations.

■ Third-party technical evaluations are standard procedure for lighting equipment. Any reputable supplier should be able to provide independent reports, such as the common ETL lab tests,[17] for their products. **APPENDIX E** provides contact information for independent lighting laboratories.

Table 7-9

Recommended characteristics for electronic ballasts

Appendix D includes a complete sample specification for electronic ballasts and other lighting equipment.

	Recommendation
Power factor	Minimum 0.95
Current crest factor	Maximum 1.7
Total harmonic distortion	Maximum 20%
Frequency	Minimum 20 kHz
Transient protection	Category A, IEEE 587
Sound rating	"A"
Flicker factor	Maximum 5%
Ballast factor	Minimum 0.88, unless exceptionally low light is required. Maximum 1.20
History	Two years of production and documented performance, or satisfactory performance in an on-site pilot project.
Warranty	Minimum 3 years, including a $10 per unit labor allowance for failed ballasts.

7.2.8 Power reducers

Typically consisting of a transformer, resistor, and capacitor (and perhaps wave-smoothing electronics), power reducers—also commonly called current limiters—cut lamp current, generally dropping both wattage and light output by 30 or 50 percent, depending on the model. Some vendors of these devices claim, however, that they reduce wattage more than light output. Actual improvements in efficacy are usually very small, and are attributable to a reduction in fixture temperature, which can raise the light output of magnetically ballasted lamps (see Figure 7-15).[18]

Power reducers are installed either between a lamp and one of its sockets, or are hardwired in series with an existing magnetic ballast. Power reducers are not compatible with electronic ballasts, nor are they compatible with several types of magnetic ballasts. This severely limits the potential for incorporating power reducers in an energy efficient lighting system.

Sufficient problems exist with power reducers to limit their use to unusual circumstances. In some cases, for example, lamp and ballast warranties may be voided with their use. "Energy-saving" (or cathode-cutout) lamps and ballasts may fail to work properly, or suffer noticeably shortened life, depending on their type, when used with power reducers. While cheaper than new ballasts, power reducers are not acceptable in locations that are not overlit to begin with, and make no sense for new construction, where high-efficiency components can be installed to provide proper illumination levels. They may be appropriate, however, in overlit spaces where light levels can be significantly reduced. Other options, such as delamping and low-BF electronic ballasts, should also be considered for these situations.

As with other retrofit products (particularly reflectors, discussed below), it is important to separate improved efficiency *attributable to the device* from improvements due to additional procedures carried out during its installation, such as relamping and cleaning.

7.3 FLUORESCENT FIXTURES

Fluorescent fixtures bridge the gap between the technical discipline of lighting efficiency and the more artistic trade of lighting design. Efficiently distributing light is as important as efficiently generating it, and can dramatically affect visual comfort. Until recently, most fluorescent light fixtures for general area lighting were ceiling-mounted *direct* troffers that used painted white metal reflectors to distribute light and plastic lenses to even out the fixtures' surface brightness.

While cheaper than new ballasts, power reducers are not acceptable in locations that are not overlit to begin with, and make no sense for new construction, where high-efficiency components can be installed to provide proper illumination levels.

An increasing number of lighting applications now use *indirect* (or indirect/direct) fluorescent fixtures. These units, which come in a wide variety of styles and functions, usually bounce light off of walls or ceilings. Indirect (or indirect/direct) lighting creates a softer overall illumination level and reduces glare, but typically requires higher ceilings, more design expertise, and more costly (though sometimes fewer) fixtures than conventional direct troffers. Indirect lighting often can deliver better visual performance than direct lighting, with lower ambient illumination (and lower energy use) because it distributes light more uniformly. Choosing between the two types of lighting is largely a design decision; in either case, the same high-efficiency lamps and ballasts are available. Section **2.1.1** discusses some of the benefits of indirect illumination.

Despite the frequent superiority of indirect lighting, a large majority of commercial space is lit by direct fluorescent downlights. We now look in some detail at three aspects of those fixtures that affect lighting efficiency: reflectors, lenses, and louvers.

7.3.1 Reflectors

Reflectors are specially shaped retrofittable metal sheets (**Figure** 7-21) that improve (or attempt to improve) the efficiency and distribution of conventional ceiling-mounted fluorescent downlight fixtures. With higher reflectivity and more directional control than

Figure 7-21

MetalOptics reflector design

Well-designed reflectors improve fixture efficiency, and can widen or narrow light distribution. This fixture also shows a small-cell paracube louver, which reduces glare but absorbs a lot of light.

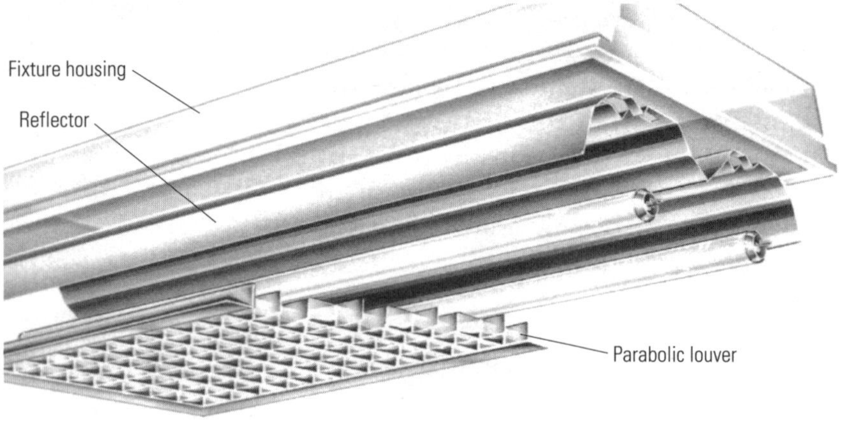

Source: MetalOptics

the white paint that lines most existing fixtures, reflectors can significantly decrease the internal losses of fixtures and widen or narrow their light distribution.

Reflectors are now available from over two dozen manufacturers, many using the same materials, the same computer-aided design programs, and even the same brake presses for bending the metal. Innovative designs, mounting methods, and usable lamp combinations have improved performance and simplified installation, and many millions of square feet of space have been successfully upgraded using them. The proper specification and use of reflectors can solve many lighting problems resulting from poor existing equipment.

Reflectors are recovering from a bad image. In the 1980s, manufacturers developed several types of films, metal reflectors coated with films, and polished aluminum reflectors to improve the efficiency of ceiling-mounted fixtures, often with widely ranging results. In some cases, the films were not designed to handle the heat and ultraviolet emissions from a nearby lamp, or the reflectors altered light distribution, creating dark and light areas where previously a uniform level existed. Other problems included optimistic performance claims and the confusion between benefits derived from the reflector product and those from fixture cleaning and relamping.

The careful use of reflectors, however, is an important tool for the lighting retrofitter. Understanding reflector technology is simplified by remembering a few rules:

- Reflector design—not material—is the key to the best performance.

- Highly efficient reflectors often require relocation of existing lamps to new positions to avoid glare and provide optimal light distribution.

- No single retrofit reflector kit will work well in all situations. Even in a building with one type of fixture, several reflectors may be used, each designed for the best pattern of light distribution for its position in the room.

- The potential to improve fixture efficiency depends mostly on the geometry and efficiency of the *existing* fixture, not that of the reflector kit.

- Savings analysis of proposed retrofits is best done on a room-by-room basis, not by fixture type throughout a large or diverse facility.

Performance and light levels. Delamping is the heart of reflector energy savings. A frequent claim is that reflectors will allow 50 percent delamping with little or no reduction in the fixture's light output. Without counting the temporary boost of higher lumen output of newly replaced lamps, this is a difficult goal to achieve, and is less a function of how good the reflector is than how bad the existing fixture is. In any application, delamping should be considered carefully, bearing in mind that it may also be possible to delamp *without* a reflector.

Because of the large population of existing 2 x 4 fixtures with four 4-foot lamps, removing two of these lamps is a common proposal.

This can be a realistic goal if one or more of the following conditions are present:

■ The space is measurably overlit, or a reduction in light level will be difficult to notice.

■ The existing fixture efficiency is extremely poor, and cannot be corrected by cleaning or lens replacement.

■ A proposed reflector will provide better light distribution, allowing more uniform illumination at a lower light level.

■ Replacement of the lamp/ballast combination increases the average light output per lamp.

■ A better lens, diffuser, or louver is also installed.

■ Existing lamps were very close together, raising their average temperature while shading each other.

Many reflector vendors design retrofits to maintain a fixture's light output without checking if doing so continues an already inappropriate room light level. While some may not feel that correcting an existing *low* light level is their job, doing so can increase savings when a higher ambient light level eliminates the need for incandescent task lighting that may be installed by occupants to correct such deficiencies. In other cases, *high* existing light levels may no longer be appropriate, either because the task has changed or because the space was overlit. Maintaining high light levels forfeits additional savings that would result by establishing an acceptable (lower) range and designing to meet it.

If the desired goal is a relatively even maintained 50 footcandles, for example, a typical room equipped with high-efficiency fixtures or good-quality specular reflector retrofit kits will use about one watt per square foot with T8 lamps and electronic ballasts. If the existing installation already approaches that light level using, say, 1.5 watts per square foot, it is likely that wattage cannot be reduced more than 33 percent ($[1.5 - 1] \div 1.5 = 0.33$) without also reducing lighting levels. In other words, taking out half the lamps and installing a reflector won't work in this case—unless the occupant will accept less ambient light or supplement it with efficient task lighting.

Some specular reflectors narrow the beam spread of a fixture, resulting in noticeable variations in light levels when measured between and under the retrofitted fixtures. This problem is especially obvious in partitioned open-plan offices with low ceilings: reflectors may cut off light that previously shone into one worker's cubicle, while greatly increasing the level in cubicles directly below fixtures. Since the human eye is most sensitive to *differences* in light levels, creating dramatic variations in illuminance is a quick way to demonstrate a lack of lighting competence.

Since the human eye is most sensitive to differences in light levels, creating dramatic variations in illuminance is a quick way to demonstrate a lack of lighting competence.

Retrofit kits commonly include new socket bars that relocate sockets (and thus lamps) to positions that optimize light output and distribution with the new reflectors. Once the lamps are repositioned, much of the apparent darkness down the center of the fixture is eliminated. Where a lens or diffuser is employed, slightly darkened areas may be visible as gray stripes when the fixture is viewed from an angle, even if the entire lens appears bright when viewed from directly below the fixture. Experience with many reflector installations, however, shows that these variations are noticeable primarily to lighting designers, not occupants, and are usually accepted as normal within a week or two after installation. Some striping is also common in fixtures equipped with lenses even *before* a retrofit takes place, and often becomes acute with delamping in the absence of a reflector.

With reflector retrofits, the photometrics of the finished space are completely under the control of the designer/specifier. Even in a building with only one type of fixture, a good retrofit may use half-a-dozen reflector designs to provide specific light distribution patterns for different functions or room conditions.

Reflector materials. While the quality of reflector designs is still highly variable, most of the questions concerning reflector materials have been resolved. There are four basic types of reflector materials:

- mylar-based silver film, attached to steel or aluminum,
- acrylic-based silver film, attached to steel or aluminum,
- anodized aluminum metal, and
- highly polished sputter-coated aluminum metal.

Nearly every material manufacturer claims that its product is best, but many of the "comparisons" are irrelevant to actual conditions met in the field. There are few significant differences among reflector materials; all work reasonably well, though anodized aluminum has the lowest initial reflectance. Film-surfaced products can be vulnerable to harsh conditions (such as metal shops and dirty mechanical rooms). Metal-surfaced products last longer under rough conditions, but may provide less light, depending on their surface treatment. Anodized aluminum is usually the least expensive, and polished aluminum the most expensive, with silver films generally falling between them.

Reflector material claims (and critiques) should be examined with skepticism. Some tests purporting to show the low resistance of silver film to cleaning solutions border on the ludicrous: few people, for example, would clean specular reflectors by submerging them overnight in concentrated cleaning fluid normally diluted for use as a surface washing agent.[19] Claims of resistance to ultraviolet radiation may be based on unrealistically harsh test conditions that could only occur

CASE STUDY **REFLECTOR RETROFIT**

In 1986, the Facilities Management office of Columbia University renovated its ceiling lighting to improve energy efficiency. The retrofit included delamping and adding specular reflectors to enhance the light reaching the workplane. One lamp was removed from each of the three-lamp 2 x 4 recessed fixtures, and the single-lamp magnetic ballast serving it was disconnected. One of the remaining 40-watt T12 lamps was exchanged for a Thrift/Mate® lamp equipped with a current reducer that cut light output and remaining wattage by 30 percent. The built-in task lighting fixtures (each using a single F40 T12 lamp) were not altered.

Ambient light levels in the open-plan office area dropped from about 35 footcandles (fc) to 20 to 25 fc, making the subdivided spaces appear somewhat dim. Levels under the task lights just above each desk remained at 120 to 150 fc. Some occupants found this contrast so great that they kept the task lights off, and used incandescent desk lamps instead.

A more careful upgrade was pursued in late 1992. All lamps were replaced with 3,500 K T8s, two-lamp instant-start electronic ballasts replaced the magnetic units, and better specular reflectors were installed. Task light fixtures were equipped with plug-in power reducers that reduced light output and wattage by 50 percent.

As a result, ambient light levels were raised to 30 to 35 fc while task light levels dropped to 60 to 70 fc. The better color rendering of the T8 lamps improved the overall appearance of the space and its furnishings, and the reduced contrast between task and ambient alleviated complaints of glare and excessive brightness. Total wattage was cut another 20 percent. Overall payback was slightly over three years, and occupants were quite pleased with the change.

when using HO or VHO lamps, which are best avoided anyway. Reflectors on the market today have sufficient UV resistance for normal applications.

The real world of reflector retrofits. Some vendors take light meter readings within a few hours of completion of an installation, avoiding the issues of lamp and fixture depreciation. Since catalog ratings of lamp output are based on measurements taken after 100 hours of burning—during which time lamps may lose 10 percent of their initial output—taking immediate measurements with brand-new lamps is one way to inflate the apparent performance of the retrofit. Normal lumen depreciation will remove another 10 to 20 percent of light output over the lamp's life. And as fixtures age and dust builds up on new reflectors, light output may drop by another 10 percent or more.[20]

Note that this kind of lumen depreciation and dust buildup will occur without reflectors as well. In fact, some reflector kits are well sealed against dust and, because they draw less wattage and build up less heat, may reduce the convection currents that draw dust into the fixture.

Taking immediate measurements with brand-new lamps is one way to inflate the apparent performance of a retrofit.

A good general rule for lighting retrofits—with or without reflectors—is to overdesign by 20 to 25 percent to compensate for these losses (or more, in dirty locations, such as industrial or mechanical facilities). Combined with the initial "excess" output of a new lamp, such an installation may be 30 percent brighter on its first day of operation than it will be at the end of the first lamp's rated life. For design levels of 50 footcandles, 60 to 65 footcandles may thus be required immediately after the retrofit. Another approach is to require light measurements after a week or two of operation, or at least 100 hours of burn time.

CHECKLIST FOR REFLECTOR RETROFITS

- If a vendor is involved, examine the proposed lighting audit sheet. It should include spaces for existing light levels, fixture heights and spacing, surface reflectances, and typical tasks being performed, as well as lamp, ballast and fixture data. When an audit lacks such information (due to incomplete data collection or a poor form), problems—and missed opportunities—are practically guaranteed.

- Choose retrofit kits that are listed by Underwriter's Laboratory, which means that using them will not violate the existing fixture's UL listing. Such a violation may invalidate fire insurance coverage. UL visits the facilities of kit manufacturers to ensure that their materials and mounting methods do not compromise safety. In Canada, changes to luminaires may invalidate the fixture's CSA (Canadian Standards Association) approval. If so, each building being retrofitted must have the luminaire equipment inspected and approved by local authorities.[21]

- Carefully consider the mounting methods of proposed reflectors. Many cheaper kits do not relocate sockets, and are attached by self-tapping screws through the reflector into the fixture body. Since reflectors often act as ballast covers, screwing them into a fixture body may complicate future access to the ballast. Other kits include new socket bars that also hold the reflector with tabs, slots, or wing nuts, making it simple to remove the reflectors to access the ballast, clean the reflectors, or change them should the lighting design be altered in the future. Such mechanisms also reduce the likelihood of reflector damage during installation, since no drilling occurs close to the reflector's polished surface. Designs that require one hand to hold the reflector while drilling and mounting with the other are particularly susceptible to misalignment, which can alter the fixture's light output dramatically.

CASE STUDY **LENS REPLACEMENT**

In the McVickar School for Social Work at Columbia University, energy staff measured light levels of 60 to 70 footcandles, but the fixtures appeared to cast a dingy yellow light. Occupants complained of poor visibility and asked that conditions be improved as part of the 1993 lighting upgrade. Upon close examination of the existing lenses, technicians found the culprit—light measurements showed a 33 percent loss due to dirt, yellowing, and poor quality plastic. As **Figure 7-22** indicates, even vigorous cleaning left room for improvement. The upgrade included replacing all lenses with good-quality acrylic lenses with a transmittance between 92 and 94 percent, and switching to high-CRI T8 lamps. The resulting improvement in light quality due to the combination of better lenses and lamps was so great that a general *reduction* in light levels to 45 to 55 footcandles appeared to the occupants to be an *increase* in light level.

Figure 7-22

Light loss through fixture lenses

Old and dirty lenses can absorb as much as one third of the light that tries to pass through them.

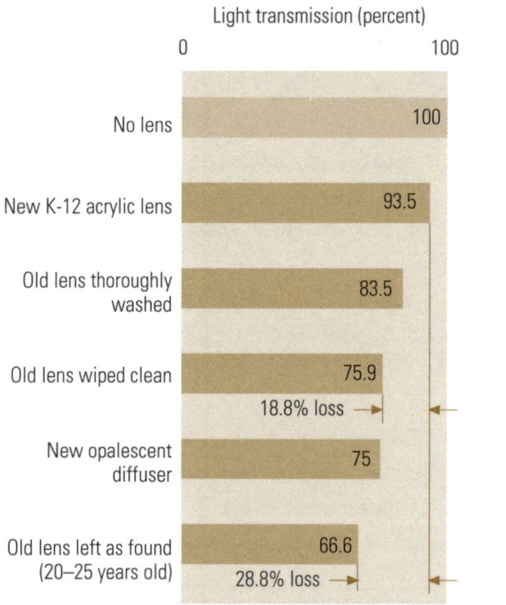

Source: Data from measurements taken at Columbia University on various lenses taken from McVickar School of Social Work, NYC, 1993.

7.3.2 Diffusers, lenses, and louvers

Diffusers and lenses are similar devices: thin plastic sheets that cover recessed fluorescent fixtures at the ceiling level. They differ dramatically in efficiency, however. *Diffusers* are milky-white translucent sheets that hide lamp images and disperse light nearly equally in all directions. They are notoriously inefficient (approximately 70 percent) because of their high light absorption. *Lenses*, on the other hand, redirect light rather than diffuse it, and their efficiency can be as high as 95 percent. Lenses can be differentiated from diffusers by their clear material and micro-patterns of surface prisms.

Most lenses and diffusers are made from one of three types of plastic: acrylic, polycarbonate, or polystyrene. While rarely used today, many older units were made from *polystyrene* because it is the cheapest of the three. Unfortunately, it also has the highest flame spread and smoke production, making it unacceptable to most building codes. Polystyrene lenses also yellow more quickly than other plastics, making them good candidates for replacement during a lighting upgrade. *Polycarbonate* is stronger than acrylic, making it ideal for outdoor and institutional fixtures (such as prisons) subject to abuse. While exhibiting high transmittance, it is also the most expensive of the three choices and yellows with age and exposure to UV radiation. *Acrylic* exhibits low flame spread and smoke production, can be stabilized against UV degradation and costs less than polycarbonate, making it the choice for most light fixtures.

Three-dimensional troffers/lenses. In extreme cases, three-dimensional lenses that project below the ceiling plane can be used to cast light further out of the ends and sides of fixtures. This technique can be particularly useful if the existing fixtures are spread too far apart. Three-dimensional lenses cast light onto the ceiling, reducing contrast and potentially allowing a lower light level in the space. In a hallway of the East Campus building at Columbia University, for example, it was possible to reduce the number of fixtures and widen the beamspread of each remaining fixture while ensuring that an even light level was maintained (**Figure 7-23**). Doing so cut the cost of a retrofit since fewer ballasts and retrofit kits (and thus less labor) were needed.

Polarizing lenses. Another type of lens promoted as an energy-saving option polarizes light, ostensibly improving contrast and clarity while minimizing glare. Proponents claim that doing so allows a lowered light level, and thus a reduction in lamp wattage, without sacrificing an occupant's ability to perform a task.

When one looks through a polarizing lens, reflected glare from a specular surface such as water is noticeably reduced, making it easier to see through that surface. The effect is much less noticeable when the light is polarized *before* it is reflected, however, and much of the light we see is reflected, not direct. In an attempt to resolve some of the controversy surrounding this subject, the Lighting Research Center (LRC) at Rensselaer Polytechnic Institute conducted tests of polarizing lenses.[22] Recognizing that most lenses and diffusers polarize light to varying degrees, LRC measured polarization through a polarizing panel and two materials not explicitly designed for polarizing: a translucent diffuser, and a prismatic lens. **Figure 7-24** shows that the degree of polarization increases for all three from essentially zero directly below a flat ceiling fixture to higher levels at greater angles from the luminaire.

Since polarization involves the absorption of light with specific orientations of vibration, it stands to reason that polarizing lenses have lower total transmissivity than non-polarizing lenses. The LRC tests found comparable fixture efficiency in a polarizer-equipped luminaire and one with a diffuser; both efficiencies were lower than is typically found in luminaires that use high-transmissivity prismatic lenses.

The real question is whether polarizing lenses can be used to save energy by reducing illuminance levels without adversely affecting visual performance. LRC concluded that "there is little basis for the Illuminating Engineering Society of North America to change its recommended illuminances for spaces with multilayer polarizer panels" and thus "does not consider multilayer polarizer panels to be energy-saving devices." To quote from the LRC research:[23]

> Increased contrast can allow reduced illuminance without affecting visual performance, and multilayer polarizers can increase contrast in some applications. However, the degree to which contrast improvements can be realized with multilayer polarizer panels depends upon several factors.
>
> **Task orientation.** For ceiling mounted multilayer polarizer panels, contrast is not substantially improved for vertically positioned tasks such as VDT screens. The greatest improvement in contrast occurs for horizontally positioned tasks.

Figure 7-23

Retrofit of three-dimensional lenses

Using lenses that project below the ceiling plane allowed these fixtures to be spaced farther apart than normal.

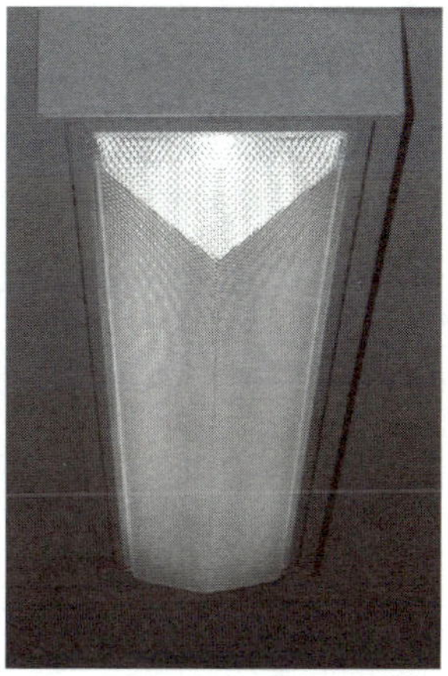

Source: Columbia University

Figure 7-24

Degree of polarization of light transmitted through three materials

All lenses and diffusers polarize light to varying degrees.

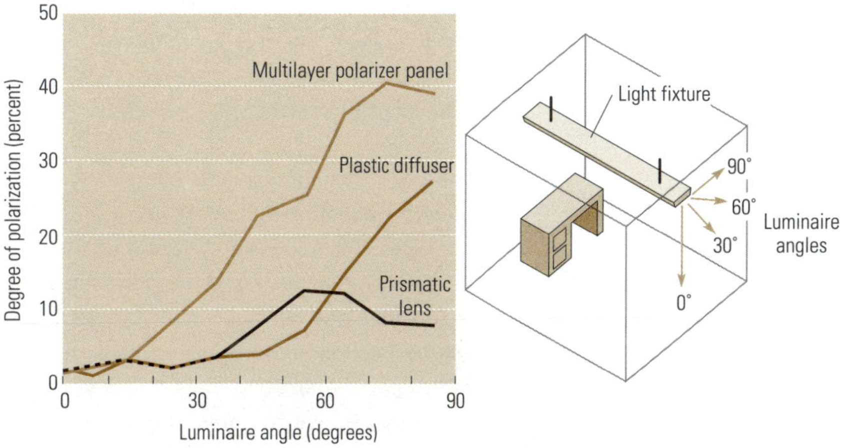

Source: LRC [18]

For more information on independent research on polarizers, see the following three reports from the Lighting Research Center at Rensselaer Polytechnic Institute:

"Multilayer Polarizer Panels" (August 1993) by Mark Rea

"Full Spectrum Polarized Lighting: An Evaluation" by P.R. Boyce, J.M. Gutkowski, and R.W. Strobel (September 30, 1992)

"A Supplementary Evaluation of Full Spectrum Polarized Lighting" (in press)

Available from:
National Lighting Product Information Program
Rensselaer Polytechnic Institute
Troy, NY 12180-3590
tel 518-276-8716
fax 518-276-2999
web www.lrc.rpi.edu

Task specularity. Matte materials, such as typing paper, are not greatly affected by multilayer polarizer panels mounted in any position. Multilayer polarizer panels reduce specular reflections from glossy materials such as glass or drafting film.

Lighting geometry. The incremental benefits of illumination from multilayer polarizer panels compared with that from more conventional luminaire lenses are maximized when the task is positioned so as to form a specular reflection between the observer and the luminaire. Little or no contrast benefits result from multilayer polarizer panels when specular reflections are not present.

Degree of polarization. The benefits of polarized light emitted from a multilayer polarizer panel are angle dependent, even at the specular angle. . . . Typical viewing angles for horizontal reading tasks are between 10 and 30 degrees. Because the degree of polarization [at such task angles] is so small, multilayer polarizer panels of the type tested by NLPIP would have little effect on reducing specular reflections for typical task geometries.

Absolute contrast. Visual performance is nearly constant for medium to high contrasts at a given background luminance. Contrast improvements within this range have very little effect on visual performance. . . . Where low-contrast tasks, such as pencil writing on paper, are performed, slight improvements in contrast can result in significant improvements in visual performance.

The contrast improvements of multilayer polarizer panels are greatest when all of the [above] conditions are met simultaneously. If one or more of these conditions is not met, the contrast improvements will be minimal or nonexistent. Thus, for multilayer polarizer panels to be very effective relative to conventional luminaires, an office would need to use horizontally positioned, low-contrast, specular task materials under fixed lighting geometries seen at extremely large specular angles. These conditions are not likely to be common in typical commercial environments.

Overall, LRC's research suggests that polarizers make sense in some applications but not all. **Figure 7-25**, from the LRC report quoted above, summarizes the kinds of conditions that affect the applicability of this technology.

Louvers. Specular parabolic cube louvers (**Figure 7-26**) can avoid much of the light losses of translucent lenses. By employing highly reflective surfaces shaped to send light down to a task (instead of bouncing it back and forth between parallel surfaces, as occurs with "eggcrate" style louvers), these devices block lamp images, reducing the potential for reflected glare in computer screens or other shiny materials. The literature includes recommendations for the angle at which louvers should cut off light output.[24]

Figure 7-25

The value of vertically polarized light under different conditions

The value of polarized light varies with task and viewing conditions.

	Will help	Won't help
Task orientation	Horizontal	Vertical
Task specularity	Glossy	Matte
Lighting geometry	Specular angle	Other angles
Degree of polarization	High	Low
Absolute contrast	Low	High

Source: LRC [18]

Figure 7-26

Parabolic cube louvers in fluorescent lighting fixture

Louvers reduce glare and improve light distribution compared to lensed fixtures, but may create a sense of darkness in the space.

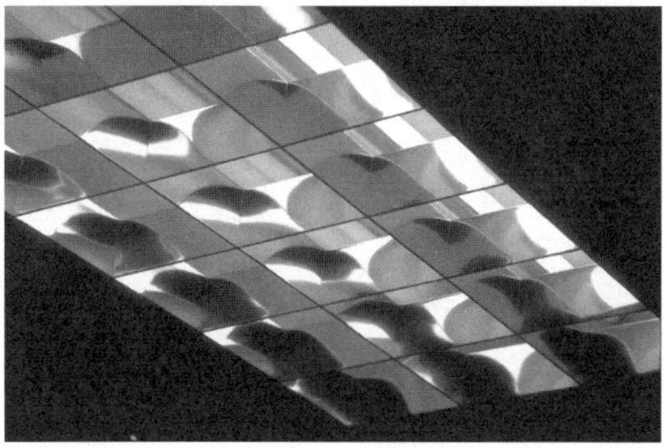

Although small paracube louvers (with openings smaller than 1 inch) are effective at reducing glare, they absorb a great deal of light and dramatically reduce fixture efficiency. Appropriate louver cube sizing (over 1.5 inches) typically requires a deeper fixture and a different method of attachment than conventional shallow fixtures can provide. Larger (or "deeper") cell louvers are typically made of metal, which is also expensive to shape into a parabolic section. These options are usually available only in new fixtures, and not as a retrofit kit. When applied properly, however, such large-cell parabolic louvers create an acceptable appearance and distribution, adding value to a lighting upgrade. They are especially useful in minimizing VDT screen glare.

The downside of "paracube" louvers is that they may make ceilings and nearby walls appear dark, especially in spaces where furnishings are also dark and do not reflect much light between them. The same fixture equipped with a standard acrylic lens appears to brighten up a space because the surfaces in an occupant's field of view are bright. Paracube louvers may thus make a space appear "dark" or cave-like even though measured task light levels can actually be higher than with lensed fixtures. Correction of this increased contrast level could involve wall-mounted fixtures, such as low-brightness sconces or wall washers that would add just enough indirect light to brighten the ceilings or walls.

The introduction of specular paracubes has also led to their inappropriate use by interior designers who like their "high-tech" look and employ them in areas where low brightness and sharp cutoff are unnecessary. When, for example, they are used in corridors, a cavern-like ambience may be created, which to many is depressing.

While specular paracubes thus have their place, they are prone to misuse. Whatever they can do can in many cases be better accomplished with a good direct/indirect lighting design (see **CHAPTER 3**).

1 Rick Wilson, personal communication (July 2, 1997), Linear Fluorescent Product Manager, Osram Sylvania, 100 Endicott Street, Danvers, MA 01923, tel 508-750-2158, fax 508-750-2432, e-mail wilsonr@osi.sylvania.com.

2 Mark S. Rea, ed., *Lighting Handbook: Reference & Application,* 8th Edition (New York: Illuminating Engineering Society of North America, 1993), p. 13.

3 List prices from GE Lighting, mid-1997. Discounts are usually available.

4 Rea [2], p. 212.

5 Barbara Ponce, personal communication (February 1994), Advance Transformer, Rosemont, IL, tel 708-390-5000; G.D. Garbowicz, "An Improved Hybrid Fluorescent Lamp Ballast," *Journal of the IES*, v. 21, no. 2 (Summer 1992), pp. 42–45.

6 Rea [2], p. 209.

7 N. Bleeker and W. Veenstra, " The Performance of Four-Foot Fluorescent Lamps as a Function of Ambient Temperature on 60 Hz and High-Frequency Ballasts," Paper #45A, Illuminating Engineering Society Conference (August 1990).

8 Elmer Smolnisky, personal communication (March 21, 1994), Maintenance Supervisor, Physical Plant, Augustana College, Sioux Falls, SD 57197, tel 605-336-4326, fax 605-336-5299.

9 Robert Sackow, Johnson Controls, internal memo dated November 22, 1994.

10 Rudy Verderber, "Electronic High-Frequency Ballasts (Past, Present, and Future)," LBL-28792 (March 1990), Lawrence Berkeley National Laboratory, Berkeley, CA tel 510-486-4000, fax 510-486-7000.

11 Lori Garcia, personal communication (May 1994), Illumination Engineer Advisor, British Columbia Buildings Corporation, 3350 Douglas Street, Victoria, BC, V8W 2T4, Canada, tel 604-775-0035, fax 604-660-1980.

12 Raul Abesamis, Paul Black, and Jeffrey Kessel, "Field Experience with High-Frequency Ballasts," *IEEE Transactions on Industry Applications*, v. 26, no. 5 (September/October 1990), pp. 810–811.

13 Jeff Kessel, personal communication (June 21, 1994), University of California, Office of Physical Resources, Berkeley, CA 94704, tel 510-642-6194.

14 Smolnisky [8].

15 Lindsay Audin, 1 Everett Avenue, Ossining, NY 10562, tel/fax 914-762-4939, e-mail energy-wiz@aol.com.

16 Jim Hagar, personal communication (July 5, 1994), Product Development Engineer, Magnetek Corp., 900 East State Street, Huntington, IN 46750-2954, tel 219-356-7101 x551, fax 219-356-1250.

17 Intertek Testing Services, 3933 U.S. Route 11, Industrial Park, Cortland, NY 13045, tel 607-753-6711, fax 607-753-9891, e-mail ima@itsqs.com, web www.itsqs.com.

18 Lighting Research Center, "Specifier Reports: Power Reducers" (March 1992), National Lighting Product Information Program, Rensselaer Polytechnic Institute, Troy, NY 12180-3590, tel 518-276-4835, fax 518-276-2999, web www.lrc.rpi.edu.

19 This claim was made in a 1990 print ad by a vendor of polished aluminum reflectors (1990).

20 Rea [2], p. 398.

21 Lori Garcia [11].

22 Mark Rea, "Multilayer Polarizer Panels," *Lighting Answers* (August 1993), Lighting Research Center, Rensselaer Polytechnic Institute, Troy, NY 12180-3590, tel 518-276-8716, fax 518-276-2999, web www.lrc.rpi.edu.

23 Rea [22].

24 "RP-24: IES Recommended Practice for Lighting Offices Containing Computer Visual Display Terminals," Illuminating Engineering Society, New York, NY (1990), p. 13.

8 COMPACT FLUORESCENT LAMP TECHNOLOGIES

Continued on next page

Continued from previous page

8 Compact Fluorescent Lamp Technologies

Compact fluorescent lighting is a key tool for achieving energy efficiency in many common applications formerly dominated by incandescents. While the technology can provide energy savings of up to 75 percent and slash maintenance costs, the real challenge is to obtain excellent visual quality with these compact sources. The experience of lighting designers, lamp and fixture manufacturers, and system users provides valuable guidance in how to apply—and how not to apply—compact fluorescents. These lessons are especially salient as federal programs like Green Lights and Energy Star expand the use of compact fluorescents in the commercial and residential sectors.

8.1 Understanding the Basics

Compact fluorescent lamps differ from full-size fluorescent lamps in several key respects. They are narrower; most CFLs of 13 watts and less are T4s (four-eighths of an inch in diameter) and most over 13 watts are T5s (five-eighths of an inch in diameter). Most CFLs are shorter and connect to their power supply through a base or socket at one end of the lamp, as opposed to the single or bi-pin bases at both ends of full-size fluorescents. Compact fluorescent lighting systems consist of a lamp (often with a starter built into the base), a lamp holder, and a ballast.

Compact fluorescent lamps use the same preheat and instant-start initiation and operating techniques used in full-size fluorescent lamps, that is, electron-emitting filaments, mercury gas, and phosphors coating the inside of glass tubes (see **7.1** for a description of fluorescent operation). Their narrow diameter makes it

Highlights

- Product offerings have expanded rapidly, with CFLs now available from 5 to 28 watts in screw-in versions, and up to 55 watts in hardwired models.

- Early claims about light output, energy savings, and lifetime were somewhat exaggerated, because they ignored the effects of temperature, cycling, mounting position, fixture performance, lumen depreciation, power quality, and variations in efficacy with lamp size. If such issues are not properly dealt with, they can collectively reduce CFL light output by more than half relative to the manufacturer's rating and substantially reduce lamp life.

- One of the most important yet underrated benefits of CFL technology is the dramatic reduction in labor and relamping costs it offers compared to incandescents.

- Lighting manufacturers have responded to utility concerns about poor power quality by developing a growing array of CFLs with high power factor and low harmonic distortion.

- Dimming and three-way switchable CFLs are now available from numerous manufacturers, and dedicated CFL fixtures can be purchased in a wide variety of sizes, styles, and price points.

Figure 8-1

Relative efficacy of major light sources

CFL efficacies vary considerably with wattage and ballast type. Small lamps driven by standard magnetic ballasts are the least efficient.

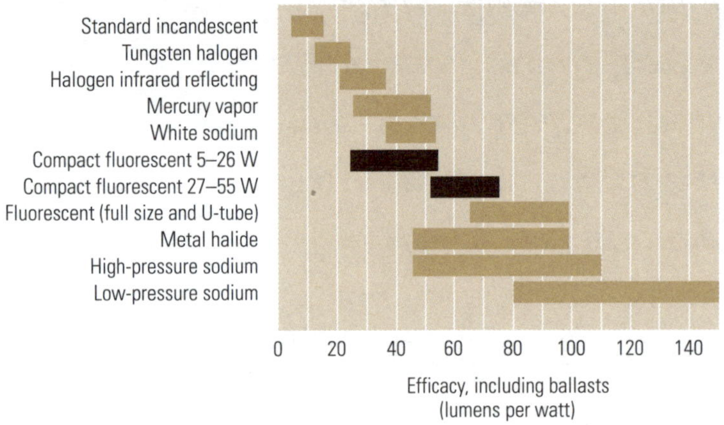

Figure 8-2

CFL lamp-ballast systems

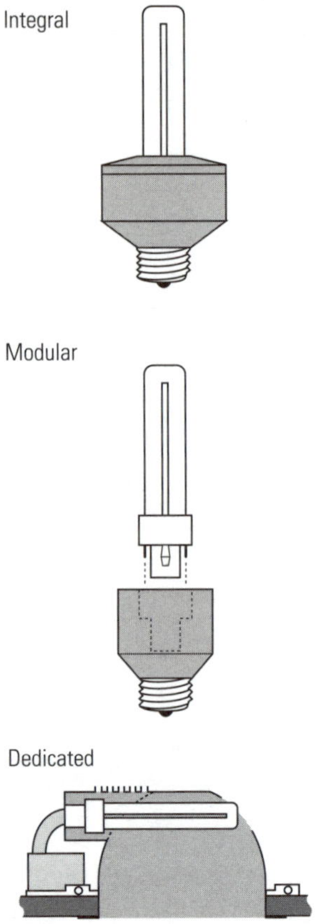

economical to use the same high-quality rare earth phosphors used in modern T8 lamps (though the mixtures are not always identical), which gives these lamps high efficacy and good color rendering.

CFLs are generally available in four color temperatures: 2,700 K (similar to the warm tone of standard incandescent lamps), 3,000 K, 3,500 K, and 4,100 K. Some 5,000 K and 6,500 K models are also available in certain sizes. Because CFLs often replace incandescents with color temperatures around 2,700 K, CFL color choice is important when trying to maintain an existing ambiance. Studies indicate that at lower light levels, warmer colors are preferred (see **2.2.3**).

CFL lifetime ranges from 8,000 to 20,000 hours at three hours per start, but can be considerably less with more frequent cycling. Lengths range from 4.5 to 22.5 inches, lamp wattage from 5 to 55 watts, and light output from 250 to 4,800 lumens.

Most of the early CFLs used magnetic ballasts, but most screw-in products now employ high-frequency electronic ballasts, which are smaller, lighter and more efficient. Dimming ballasts for some CFLs are available, but they are still fairly expensive.

One fact that is not widely recognized about CFLs is that efficacy (lumens/watt) varies considerably with lamp wattage and ballast type and quality (**Figure 8-1**). A 5-watt lamp on a low-quality magnetic ballast, for example, can be as low as 27 lumens per watt. At the other extreme, two 39-watt CFLs powered by a single high-quality electronic ballast deliver nearly 73 lumens per watt, compared to a typical incandescent lamp's 15 lumens per watt (see **8.3** for more on efficacy).

8.1.1 The CFL family

There are three types of CFL lamp-ballast systems, illustrated in **Figure 8-2**: integral, modular, and dedicated.

- *Integral units* combine a lamp, ballast, and adapter (the Edison screw base—or other standard base in some countries—which fits the incandescent lamp socket) in a single sealed assembly, which must be discarded when the lamp burns out.

- *Modular units* plug a separate lamp into an Edison or other style of adapter/ballast. This design offers the economic advantage that when the lamp burns out the entire assembly need not be discarded. Instead, a relatively low-cost replacement lamp can be installed in the same ballast base, which typically lasts for 40,000 to 60,000 hours. Lamps for modular units have either two or four pins in their bases. Those of 9 watts or less use the same ballast and pin configuration. CFLs above 9 watts generally use different bases and ballasts for each wattage, so they can't be interchanged. Since lamps of the same wattage can have different, incompatible bases, users and specifiers must pay careful attention to base configurations.

- *Dedicated systems*—also called *hardwired systems*—consist of a ballast and fluorescent lamp socket that are permanently wired into a fixture by the fixture manufacturer or as part of a retrofit kit. As with modular systems, the lamp can be replaced with another pin-based compact fluorescent when it burns out.

8.1.2 Lamp configurations

Compact fluorescent lamps come in a wide array of shapes and sizes which are difficult enough to keep track of without the added confusion of brand-specific terminology for the various configurations. The various configurations are described in **Figure 8-3**, along with the names that different vendors have applied to such products.

Figure 8-3

The CFL family

Twin-tube

First introduced to the U.S. market in 1981 under the Philips tradename of "PL,"such designs consist of two straight, parallel, miniature fluorescent tubes (or one U-shaped tube with parallel legs) both ending side-by-side in a plug-in base that contains an integral starter. General Electric calls its family of such lamps Biax (short for "biaxial") and many people use this term generically. Sylvania (before it was bought by Osram) called them "twin tubes." Osram Sylvania uses the original Osram trade name "Dulux." All of these terms describe the same two legged style of lamp. To minimize confusion, we use the term "twin-tube" or simply "twin" in this report because it is the most descriptive of the various names in common use.

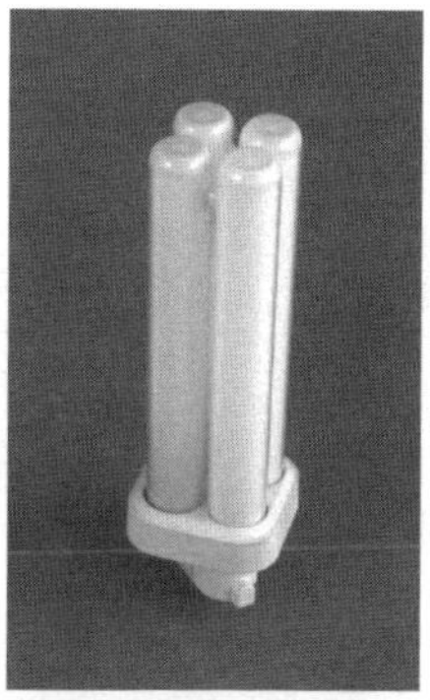

Quad-lamp

Quad-lamp, quad-tube, double biax, or in Philips Lighting's parlance, PL "clusters" or "PLC" lamps, pack four parallel tubes (or two U-shaped tubes with parallel legs on the same base as twin tubes, and thus provide nearly the same light output as twin tubes of equivalent wattage in only half the length.

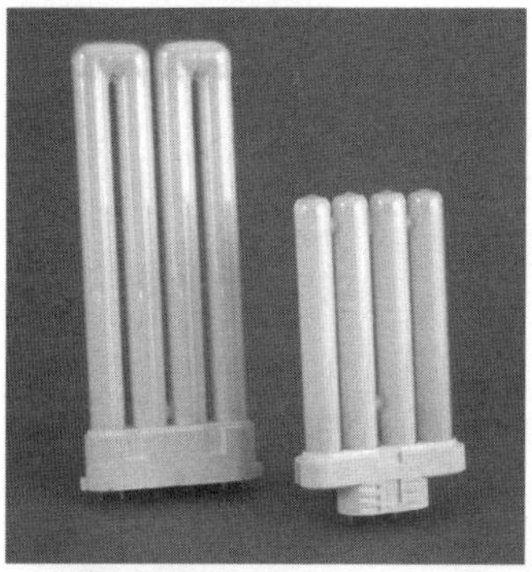

F-lamp

The "F-Lamp" features two twin tubes aligned in a single plane, shoulder to shoulder. F-Lamps are well suited to task lights and low-profile surface and recessed fixtures and are currently manufactured by Panasonic and Osram Sylvania.

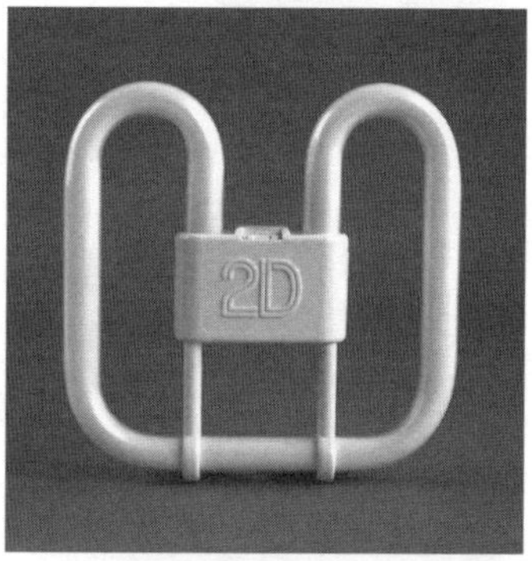

2-D

A relatively new addition in North America to the CFL family is the "2-D" or "double-D" lamp, first introduced in Europe in the mid-1980s and originally available only as an import. General Electric is now the sole U.S. manufacturer of this flat 4-pin configuration, which comes in five models between 10 and 38 watts. A 55 watt model is still pending. This lamp configuration is ideal for low-profile surface and recessed fixtures.

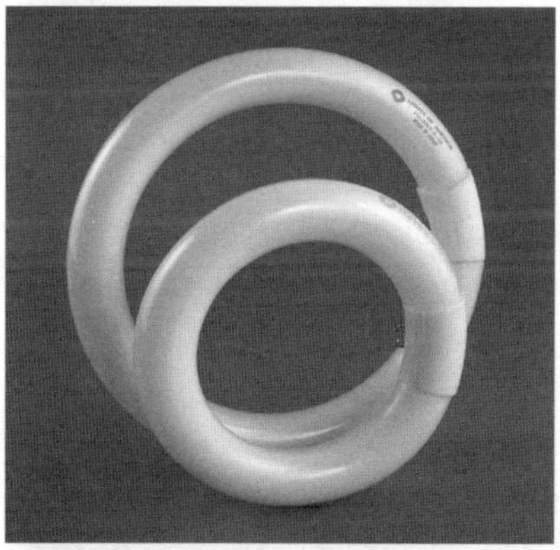

Circline

Circular (also known as "Circline") fluorescent lamps are not universally viewed as members of the CFL family. They came into use in the 1940s, and many of them utilize the same cool white phosphors as standard T12 lamps. However, at least one manufacturer—Lights of America—produces circular compact fluorescents with rare earth tristimulus phosphors and instant-on electronic ballasts in warm and cool color temperatures. The 20-, 22-, and 30-watt versions from Lights of America are among the few compact fluorescents that can replace 100- to 150-watt incandescents with comparable light output and still fit in residential reading lamps. Philips offers Circlines with rare earth phosphors but does not sell them with ballasts.

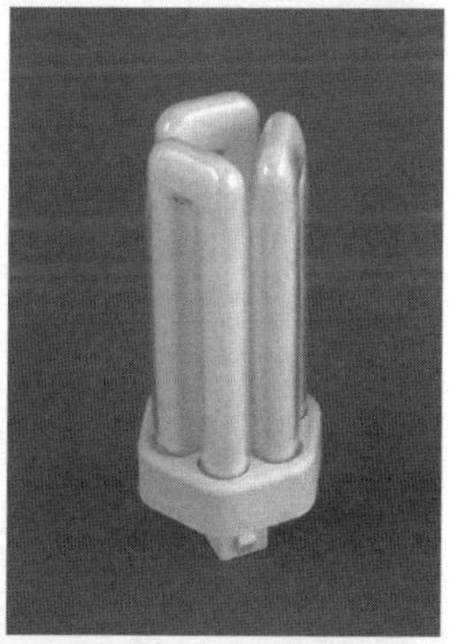

Triple-twin

To generate even more light from a shorter lamp, various manufacturers offer CFLs with three twin tubes. GE Lighting and Osram Sylvania cluster their triple tube lamps in a triangular or "delta" configuration. Other manufacturers, including Philips Lighting, offer triple tube lamps with three parallel arch-shaped curved tubes. These products are often referred to as "triple biax" lamps.

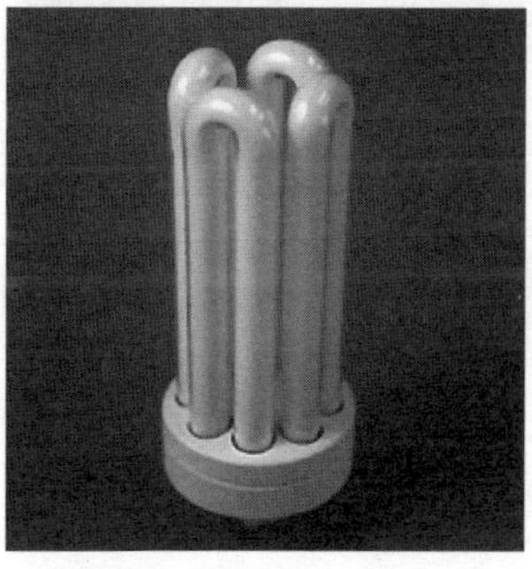

Oct lamp

The quadruple twin configuration, also known as "Performance Biax" or "Oct lamp" by General Electric, is available in 28-watt screw-in and 42-watt plug-in models. Though it increases light output per inch of lamp length, its design also traps a great deal of light inside, yielding diminishing efficiency returns.

Helical

Perhaps the most unusual new twist in CFL lamp types is the helical shape being introduced by Duro-Lite under the name Spiral-Lite. This shape closely approaches the light distribution and physical dimensions of incandescent lamps by packing more luminous surface area into a tighter space. The Duro-Lite lamps are available in 11, 15, 20 and 23 watt models. GE Lighting introduced a helical CFL in Europe in 1995 under the name Heliax, but the product has been discontinued and will not be introduced in the U.S.

Lighting designers have four choices when using CFL technology:

- Bare lamps, either integral or modular.
- Screw-in retrofit kits that include modular lamp and ballast plus a reflector and/or lens.
- Hardwired retrofit kits that allow permanent modification of an existing fixture, sometimes altering its light distribution as well as its light source.
- New fixtures designed to hold only CFLs.

8.1.3 CFL sales

Sales of CFLs have risen dramatically since their introduction in the early 1980s (**Figure 8-4**). Total U.S. sales rose from 200,000 units in 1982 to over 35 million in 1994.[1] World CFL sales rose from 45 million units in 1988 to 240 million units in 1995.

8.2 APPLICATIONS AND ECONOMICS

Compact fluorescents can be used in an impressive array of commercial and residential applications traditionally served by incandescents. They have been incorporated into a wide variety of fixture types, including recessed downlights, wall washers, desk lights, table lights, wall sconces, pendants, undercabinet fixtures, landscape lights, and floodlights. Their long life makes them particularly appealing for hard-to-reach places, where the costs and risks of frequent lamp changes are high. **Table 8-1** lists the suitability of CFLs to various standard applications.

8.2.1 Application tips and notes of caution

While CFLs have many uses, there are some applications where they just don't make sense or where extra care must be taken to achieve satisfactory performance:

- Most CFLs do not produce enough light for high-ceiling (above 12 feet) ambient lighting applications. This has been overcome in some fixtures by using multiple lamps. Other alternatives include higher-color-rendering HID sources and halogen infrared lamps, which are often more effective in such applications because of their higher light output.

Figure 8-4

CFL sales on the rise

U.S. CFL sales rose from 200,000 units in 1982 to over 35 million units in 1994.

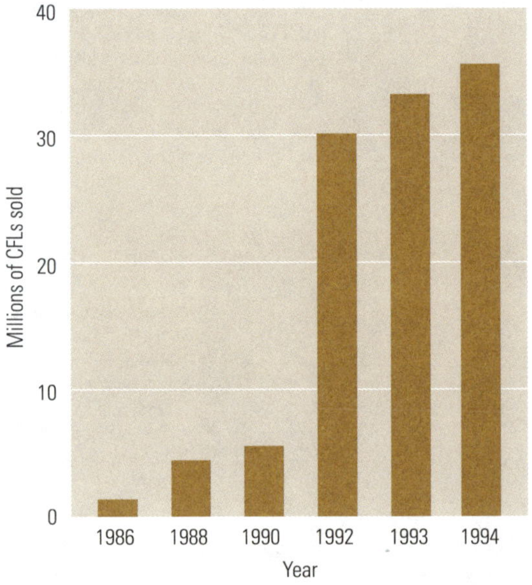

Source: EPRI (1986, 1988 and 1990 data), and U.S. Census (1992, 1993 and 1994 data) [1]

World CFL sales rose from 45 million units in 1988 to 240 million units in 1995.

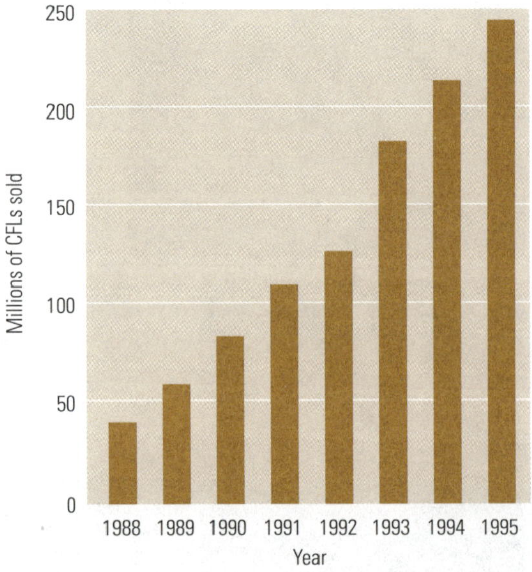

Source: Worldwatch Institute [1]

Table 8-1

Compact fluorescent lamp-luminaire application matrix

At least one type of CFL is a superior substitute for incandescents in many applications.

● Superior choice ◉ Suitable choice ○ Poor choice

	Downlights	Surface lights	Pendant fixtures	2' x 2' fixtures	Sconces	Illuminated signage	Floodlights	Table and floor lamps
T-4 twin tube	● [a]	◉	◉	○	●	●	●	○
T-4 quad tube	● [a]	◉	◉	○	●	●	●	◉
T-5 twin tube	○	◉	○	●	◉	○	◉	○
Circline	○	●	◉	○	◉	○	○	●
2-D	●	●	◉	○	◉	◉	○	●

Note: a: Superior if used with high-quality reflector

Source: EPRI, *Advanced Lighting Guidelines* [2]

■ Because they are linear diffuse light sources, CFLs are not appropriate for accent lighting or glittery retail display, where a tight beam or sparkle is needed. Halogen or some of the new compact HID sources are better in such instances.

■ Bare CFL replacements should be limited to fixtures that do not use a reflector, such as sconces, globes, and chandeliers; downlight retrofits should always use kits or lamps with built-in reflectors.

■ The reflectors in many fixtures designed for incandescents may not provide the desired quantity or distribution of light when retrofitted with a CFL. Many retrofit reflector kits are available to adapt existing fixtures to deliver quality light with a CFL source.

■ Theft-resistant devices are available to protect CFL installations from vandals.

■ Many CFLs are too long to fit in some fixtures originally designed for incandescents—some models may protrude below the ceiling line when retrofitted into downlights, or they may stick up above the shade of a table lamp, causing unacceptable glare (**Figure 8-5**). Given the wide variety of configurations now available, such problems are avoidable by selecting a shorter lamp (such as some of the new triple-tube lamps) or by using the various products available for adapting the fixture.

■ The screw-in adapter/ballast bases of some CFLs are too wide to fit in some luminaires. This problem can be avoided by selecting a different model of CFL, by using fixture adapters, or by using a luminaire designed for CFLs.

■ Light output and lifetime of CFLs are reduced by operation in ambient temperatures that are either too hot or too cold. Care is thus required for outdoor applications where temperatures drop below freezing and in unvented indoor fixtures where heat may build up. For more on the effects of temperature on CFLs and how to avoid problems, see **8.4.5**.

■ Some magnetically ballasted CFLs flicker for a few seconds when starting and produce an audible hum. This may make some lamps unacceptable in quiet settings such as homes and libraries. Some magnetically ballasted CFLs are also fairly bulky and heavy and can tip over small table lamps designed for much lighter incandescents. All these problems can be avoided by using lighter, silent, and instant-starting electronically ballasted units.

■ CFLs cannot generally be installed on existing dimming circuits, though some new CFL systems (such as the Advance Mark X) can be dimmed with standard dimmer switches or switched to low, medium and high light levels on three-way switches. Only four-pin CFLs with special electronic ballasts can be dimmed. Continuous dimming ballasts and lamps can cost $40 or more for the package, though prices are expected to continue dropping (see **8.8** for more information).

Figure 8-5

When a CFL is too much

Most misapplications of CFLs aren't as blatant as the one we staged for this photo, but we've seen a few that come close. Like any tool, a CFL has to be used properly to do its job.

■ Frequent on-off cycling can dramatically reduce the operating life of CFLs, so they shouldn't be used with occupancy sensors or where frequent cycling is expected. Some electronic timers can also shorten CFL life by maintaining a trickle current through the lamp when it is off, which ruins starters and cathodes.

■ Some electronically ballasted CFLs emit infrared radiation that can interfere with the remote controls for televisions, stereos, and video players. Care should be taken when specifying CFLs for such settings. For more on infrared interference, see **8.7**.

As the preceding application highlights illustrate, compact fluorescent lighting technology presents something of a two-edged sword. CFLs offer dramatic energy and labor savings, and are now available in enough product configurations to give designers and users many choices, but they require greater sophistication than incandescents to use properly. For example, if a higher light level is needed, a higher wattage incandescent can be used in the same fixture (up to the fixture's thermal limit). Most CFL fixtures, however, have a socket and ballast that accommodate only one lamp wattage. Moreover, in directional fixtures like downlights, distribution of light from the CFL is limited by the fixture, whereas with incandescent fixtures, lamps with different beamspreads can be interchanged.

Lighting maintenance personnel, previously content with knowing only the shape and wattage of a lamp, are now faced with variables such as multiple pin configurations with the same wattage, several sizes of replacement adapters, competing claims regarding lamp and retrofit kit performance, different lamp color temperatures, and potential lamp and ballast theft.

Table 8-2

Economics of CFLs versus incandescents

	Baseline incandescent 75W	CFL Option 1 21W 2D	CFL Option 2 26W quad
Performance			
Initial light output (lumens)	1190	1350	1800
Design light output (lumens)	1071	972	1296
Lamp lifetime (hours)	1,000	10,000	10,000
Power input (W)	75	21	26
Efficacy (lumens/W)	15.9	64.3	69.2
Energy			
Annual energy use (kWh/year)	187.5	52.5	65
Annual energy use ($/year)	$15.00	$4.20	$5.20
Annual energy savings ($/year)	n/a	$10.80	$9.80
Maintenance			
Lamp cost ($/lamp)	$1.00	$20.00	$20.00
Relamping labor cost ($/lamp)	$5.00	$5.00	$5.00
Annual lamp cost	$2.50	$5.00	$5.00
Annual labor cost	$12.50	$1.25	$1.25
Annual maintenance savings (lamp+labor) ($/year)	n/a	$8.75	$8.75
Summary			
Annual operating cost (energy+maintenance) ($/year)	$30.00	$10.45	$11.45
Annual operating cost savings ($/year)	n/a	$19.55	$18.55
Payback on first CFL installed (years)	n/a	1.3	1.3

Table notes:
- Design light output for incandescent lamp based on 10% lumen depreciation at 40% of rated life.
- Design light output for CFLs based on .90 ballast factor and 20% lumen depreciation at 40% of rated life.
- Annual operating time=2500 hours
- Lamp costs: incandescent=$1; CFL=$20
- Lamp replacement labor cost=$2.50
- Electricity cost=$0.08/kWh

8.2.2 Economics

The first cost of CFL equipment may appear high compared to incandescents, but the energy savings and reduced relamping costs cause them to pay back very rapidly in medium to high duty factor applications (over 2,000 operating hours per year). The longer the annual operating hours, the more attractive the economics of CFLs becomes, because more incandescent relamping costs are being avoided per year.

Prices of CFL products vary considerably among regions. Most fall into the following ranges, depending on wattage and purchased quantity:

- modular CFL lamp and ballast: $5 to $20
- integral CFLs: $10 to $30
- retrofit kits: $25 to $50 including reflector
- new CFL fixtures: $30 to $250+.

Table 8-2 illustrates energy, lamp, and labor savings from replacing a 75-watt incandescent lamp with either a 21-watt 2D lamp or an 26-watt quad lamp. (These lamps come close to a 3:1 wattage ratio—the "rule of thumb" discussed in **8.4**). The annual savings for a $20 CFL are about $10 in energy costs and $9 in maintenance. Including labor for the installation, the first CFL in this application pays back in just over a year. Similar calculations can be performed for most other CFL retrofits.

8.3 Efficacy

The efficacy of CFLs (light output in lumens per watt of input power) varies with lamp size, configuration, and ballasting (**Table 8-3**). Two basic rules of thumb hold true:

■ Higher wattage lamps are generally more efficient than similar lower wattage lamps, in part because their ballast losses are a lower percentage of the total lamp-ballast power requirement.

■ Electronically ballasted units are generally about 20 percent more efficient than comparable magnetically ballasted models.[2]

8.4 Light Output

Many a CFL has been substituted for an incandescent lamp using the rule of thumb that a CFL uses only 25 percent of the wattage to deliver the same light output. Sometimes this is true, but many manufacturers' product literature exaggerate CFL performance by "rounding up." For example, a CFL may be advertised as a replacement for a 75-watt, 1,200 lumen incandescent, but may only produce 1,000 lumens. A more accurate description would put the CFL midway between the light output of 60- and 75-watt incandescents, but advertising tends to err on the side of exaggeration. The result in some cases has been customers dissatisfied with a retrofit that reduced lighting levels.

Table 8-3

Typical efficacy of CFL lamp/ballast combinations

The typical values shown here illustrate that higher wattage lamps are generally more efficient than lower wattage ones and that electronically ballasted systems are more efficient than magnetically ballasted systems.

	System efficacy (lumens per watt)	
Twin-tube CFL	Magnetic ballast	Electronic ballast
5W	27	n/a
7W	32	n/a
9W	37	44
13W	50	53
18W	49	61
27W	63	67
40W	68	74

The most serious misrepresentations of light output occur with reflector lamps. The reflector turns a lamp into a directional source whose output is more appropriately expressed in terms such as candlepower, beam angle, and total reflector efficiency. Nevertheless, most CFL reflector lamp manufacturers rate their products in terms of bare lamp lumens, a relatively meaningless figure for this application.

In addition, many such products are advertised as the "equivalent" of incandescent "R" lamps. No fluorescent reflector unit can deliver the beam candlepower or light distribution of a reflector flood lamp, however (see Figure 6-2). Nor do any CFLs approach the photometric performance of incandescent reflector spot or ellipsoidal reflector (ER) lamps. Yet most CFL reflector-lamp products are promoted as direct replacements for incandescent lamps, leading to many highly unsatisfactory (and often inefficient) applications.

A three-to-one wattage equivalence between CFLs and incandescents is a fairly safe rule of thumb for most applications.

Although CFL reflector lamps don't provide the same photometric performance as their incandescent counterparts, there are many cases in which reflector CFLs are ideally suited to replace incandescent spot and flood lighting. The key is to match the technology to the requirements of the application, and to avoid misleading comparisons between different types of light sources.

Disappointments with light output have been a big problem in residential applications. New England Power Service Company found from evaluations of its DSM programs that the primary reason residential customers gave for removing CFLs and reinstalling incandescents was "insufficient brightness."[3] Because customers are less likely to continue using a

Figure 8-6

Realistic comparison of light output between incandescent and compact fluorescent lamps

This chart deducts 15 percent from published CFL light output data for a more realistic comparison with incandescents. The de-rating accounts for ballast factor, temperature, and position effects, all of which usually reduce the light output of CFLs. The wattages listed for CFLs are for the lamps only; actual wattages will be higher because of ballast losses.

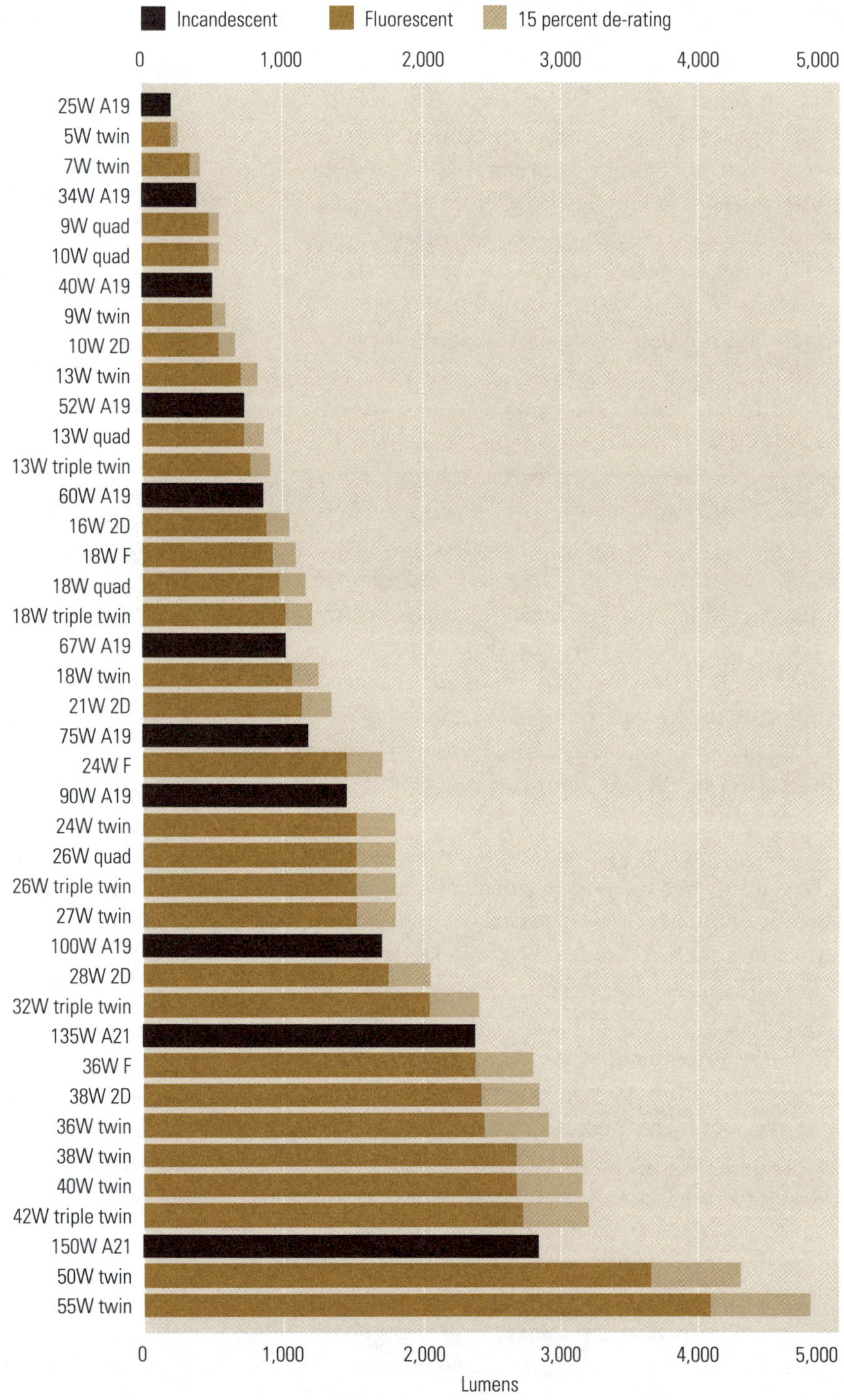

new technology if their early experience with it is negative, utilities have been anxious to avoid such experiences. Consequently, some utilities have taken a more conservative approach to light output equivalence between CFLs and incandescents in their DSM programs' literature. A *three*-to-one wattage equivalence between CFLs and incandescents is a fairly safe rule of thumb for *most* applications. Rather than claiming that a 20-watt CFL replaces a 75-watt incandescent, the 20-watt CFL should be said to replace a 60-watt incandescent.

This more conservative rule of thumb is more realistic than the often-used 4:1 ratio, for five basic reasons:

- Manufacturers' sales literature often overstate light output.
- Efficacy of CFLs varies with wattage, configuration, and ballasting, as described in **8.3**.
- As CFLs age, their light output diminishes, sometimes more so than incandescents.
- CFLs' light output is sensitive to temperature, mounting position, and other factors.
- System performance depends on fixture performance as well as lamp performance.

Note also that the greater the number of "legs" in a lamp, the lower its light output compared to fewer-legged lamps of the same wattage, because the extra legs shade more of the lamp's own light output. Hence, a 9-watt twin-tube produces 600 lumens while a 9-watt quad delivers only 575 lumens.

Figure 8-6 shows both rated and realistic light output for most of the CFL lamps now on the market, and how they compare to incandescent light output. For this comparison, we assume that the CFLs provide 85 percent of their rated output because of ballast factor (see **8.4.2**) and temperature and position effects (see **8.4.5**).

8.4.1 Accuracy of light output ratings

When the National Lighting Product Information Program tested two dozen 13-watt CFL lamps with rated light output of 860 lumens, they produced eight to 16 percent less light than the manufacturers said they would (**Table 8-4**).[4] These tests were conducted at standard rating conditions with a "reference ballast" (ballast factor = 1.0), so this result is separate from the effects of ballast factor, temperature, and position discussed below. Although this test was limited to 13-watt lamps, it raises the concern that manufacturers' ratings may overstate CFL light output in other size classes as well. Specifiers and utilities can minimize the risk of such "lumen inflation" by requiring independent test reports before purchasing lamps.

8.4.2 Effects of ballast factor

Ballast factor describes how hard a ballast drives its lamp: it is the ratio between a lamp's output with the ballast being measured and the lamp's output with a standard reference ballast. Bare-bulb lumen ratings for CFLs assume the lamp is driven to its full rated output. In fact, many ballasts underdrive lamps, although the ballast factors of modular CFL ballasts are rarely published. The National Lighting Product Information Program tested 11 magnetic ballasts for 13-watt CFLs, finding that ballast factors ranged from 0.79 to 0.94, with an average of 0.87.[5]

8.4.3 Light output stabilization

Unlike incandescent lamps, which reach full light output immediately, compact fluorescent lamps' light output reaches a peak after a few minutes, then descends again before reaching a stable peak after several hours (**Figure 8-7**). For the first five hours of operation, a typical CFL will put out between 85 and 95 percent of its rated light output. Only after five hours of continuous operation does it reach full output. The lamp's duty cycle can thus have a significant effect on the light levels that are actually delivered. If installed in an application where it will be turned on only for brief periods, the lamp will not have time to reach its

For the first five hours of operation, a typical CFL will put out between 85 and 95 percent of its rated light output.

Table 8-4

Measured light output of CFLs

For all the lamps listed below, manufacturer-rated light output is 860 lumens. Every lamp tested produced less light than was indicated by the manufacturer's rating. Three separate lamps were tested from each manufacturer, all with a "reference ballast," (ballast factor = 1.0).

Manufacturer	Trade name	Catalog number	Light output from 3 separate tests (lumens)[a]
GE Lighting	F13DBX	18844	787
			774
			752
OSRAM	F13DTT/27K	DD090	773
			773
			773
Philips	PL-C 13W/USA	245464	750
			745
			732
Sylvania	F13DTT	20451	745
			738
			726

Note: a: All lamps are NEMA-designation CFQ13W/GX23 lamps with a CCT of 2,700 K. These results were obtained under standard ANSI conditions, including stable light output operation at 25°C (77°F) still air in a vertical, base-up position on a reference ballast.

Source: NLPIP [4]

Figure 8-7

Light output stabilization of a CFL

A CFL takes some time to reach full light output.

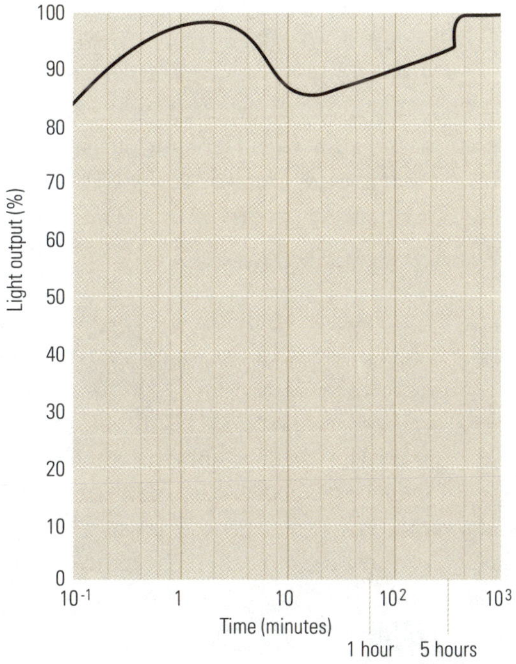

Source: Osram Sylvania

full output. This phenomenon should be considered in determining what wattage CFL to use to replace a given incandescent.

8.4.4 Lumen depreciation

Lamps are generally marketed on the basis of their initial light output. Light output degrades over time, however, from the natural aging of the lamp phosphors. Most CFLs lose between 10 and 30 percent of their light at 40 percent of rated life.[6] In general, smaller lamps have higher percentage losses than larger lamps. Because CFLs are often specified on the basis of their initial rated lumens, installations that start out producing adequate light levels may not do so over their full operating life. Incandescent lamps also lose about 20 percent of their light output over their rated life. For these reasons, New England Electric Service requires that the light output of CFLs eligible for its DSM programs and their incandescent equivalents be measured at 40 percent of lamp life (see box on NEES CFL criteria, page 177). **Figure 8-8** illustrates the comparative effects of lumen depreciation on CFLs and incandescents.

8.4.5 Effects of temperature and position on light output

The lumen output listed in lamp catalogs is determined under laboratory conditions where the ambient temperature is held at 77°F and the lamp is oriented in the base-up position. Because CFL light output is sensitive to both temperature and orientation, variations in these conditions may reduce output up to 30 percent.

CFL light output depends on the mercury vapor pressure within the lamp. This vapor pressure, in turn, is related to the minimum lamp wall temperature (MLWT)—the temperature at the coldest spot on the lamp's glass envelope. During operation, free mercury vapor condenses at the "cold spot"—typically at the point furthest from the cathode—controlling vapor pressure throughout the envelope. Lamp construction, lamp wattage, ballasting, ambient temperature, luminaire design, lamp orientation, and air circulation all affect the MLWT.

Below the optimal mercury vapor pressure, there are not enough mercury atoms in the discharge to produce maximum light output. At higher-than-optimum mercury vapor pressure, there are too many mercury atoms in the discharge, also limiting light output. The maximum light output and luminous efficacy for most CFLs is obtained at ambient fixture temperatures between 68° and 77°F—the temperatures at which photometric tests are usually conducted. This leads to a MLWT of about 113°F. By comparison, temperatures inside downlight fixtures are often as high as 122°F, which can lead to a 149°F or higher minimum lamp wall temperature.

Lamp orientation also affects mercury vapor pressure and light output. All lamp manufacturers publish rated lamp lumens based on vertical base-up position, which is the optimal burning

Figure 8-8

Lamp lumen depreciation for CFLs and incandescents

Incandescent lamps typically lose 20% of their light output by the end of their rated life. CFLs lose from 10% to 30%, with the greatest losses in the smaller lamps.

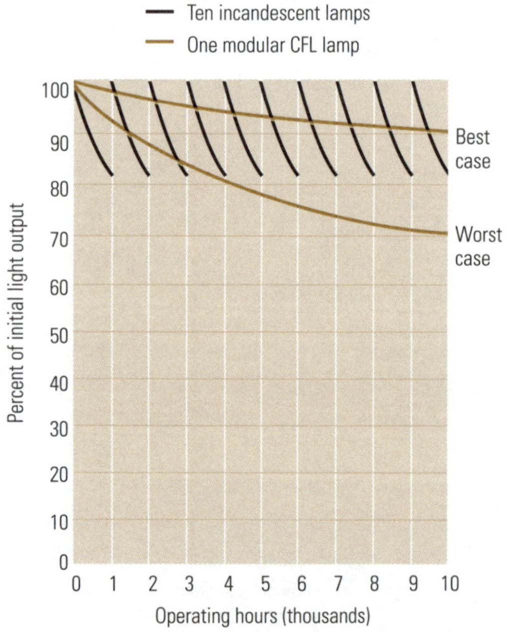

Source: Adapted from NLPIP [4]

orientation. In horizontal and base-down orientations, light output typically decreases, because the MLWT rises. For a 13-watt twin tube at 77°F, the light output is about 13 percent less in the base-down orientation relative to base-up operation, and more than 5 percent less in a horizontal position (**Figure 8-9**). But by tilting the lamp head downward just a few degrees from the horizontal, most of the loss in light output can be mitigated.

There are various options for minimizing these effects in field applications. These include:

- convective venting,
- thermal bridges, and
- mercury amalgams.

Convective venting. One of the easiest ways to reduce the temperature inside CFL fixtures is convective venting. By promoting passive ventilation through the fixture (from the room cavity to the ceiling plenum in a commercial downlight application, for instance), minimum lamp wall temperatures can be greatly reduced, and lumen output and efficacy can be increased 10 to 20 percent, depending on the number of lamps and type of ballasting (**Figure 8-10**).

Carefully designed fixtures can keep lamps cooler and minimize optical losses. Lawrence Berkeley National Laboratory (LBNL) has developed and tested numerous venting strategies for downlight fixtures, including various front and back vent geometries in fixtures with horizontally mounted lamps (**Figure 8-11**).[7] The appropriate venting configuration will depend on the thermal conditions of the particular application. Front vents remove heat directly from the lamp head area,

Because CFL light output is sensitive to both temperature and orientation, variations in these conditions may reduce output up to 30 percent.

Figure 8-10

Venting improves light output

Double 13-watt lamp fixtures tested with and without convective venting show that unvented lamps lose light output due to higher operating temperature.

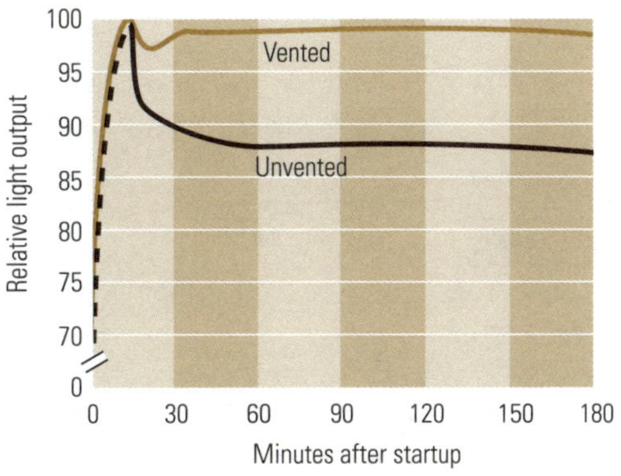

Source: Lawrence Berkeley Laboratory

Figure 8-9

Lamp orientation and temperature affect light output

Base-up oriented CFLs emit the most light under normal fixure operating temperatures.

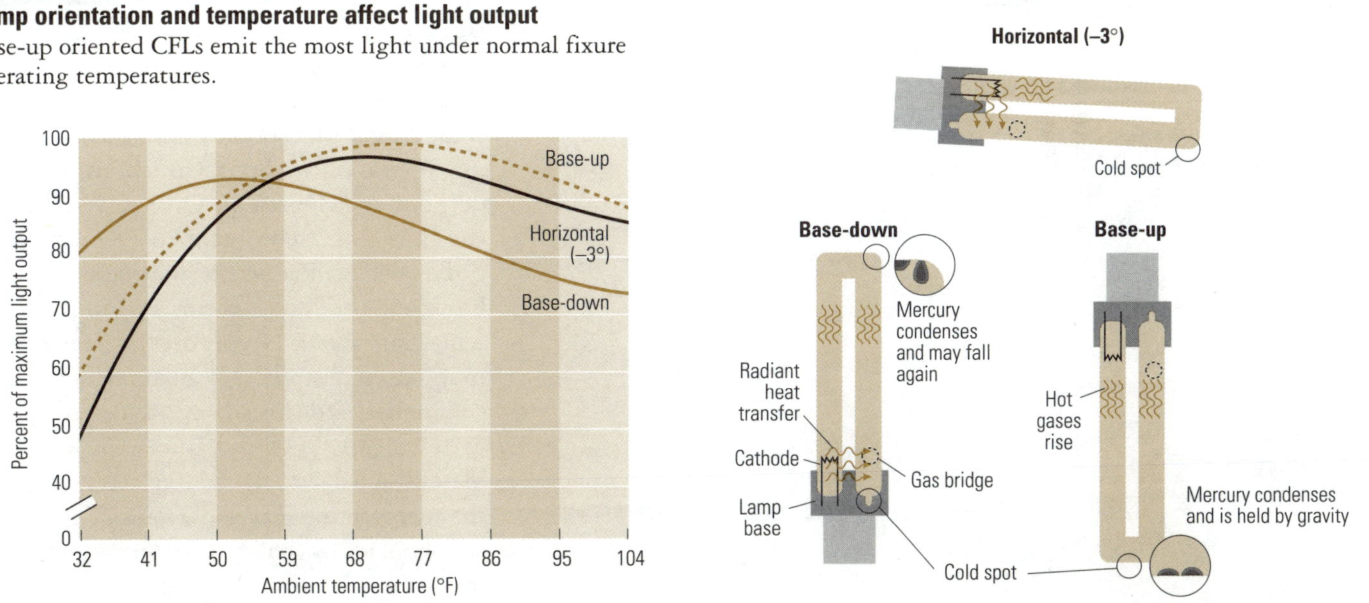

Source: Osram Dulux Compact Fluorescent Handbook; Lawrence Berkeley National Laboratory

while back vents induce a secondary convective airflow over this area in addition to directly venting the socket region of the lamp, which may extend lamp life. Very high temperatures may cause heat-sensitive components in the lamp base to fail prematurely, or may damage the physical connections between plastic base and metal contact pins.

Regardless of venting configuration, vented fixtures typically accumulate less dirt than nonvented ones and hence experience less loss in light output due to dirt build-up. This self-cleaning cycle is inhibited, however, in enclosed vented fixtures—including some lensed downlights and reflectorized screw-in CFLs. The light source attracts and traps insects, and dust settles inside the lens because there is inadequate air circulation to flush it out.

Figure 8-11

Thermal venting designs for CFL downlight fixtures

The appropriate venting configuration depends on the thermal conditions of the application.

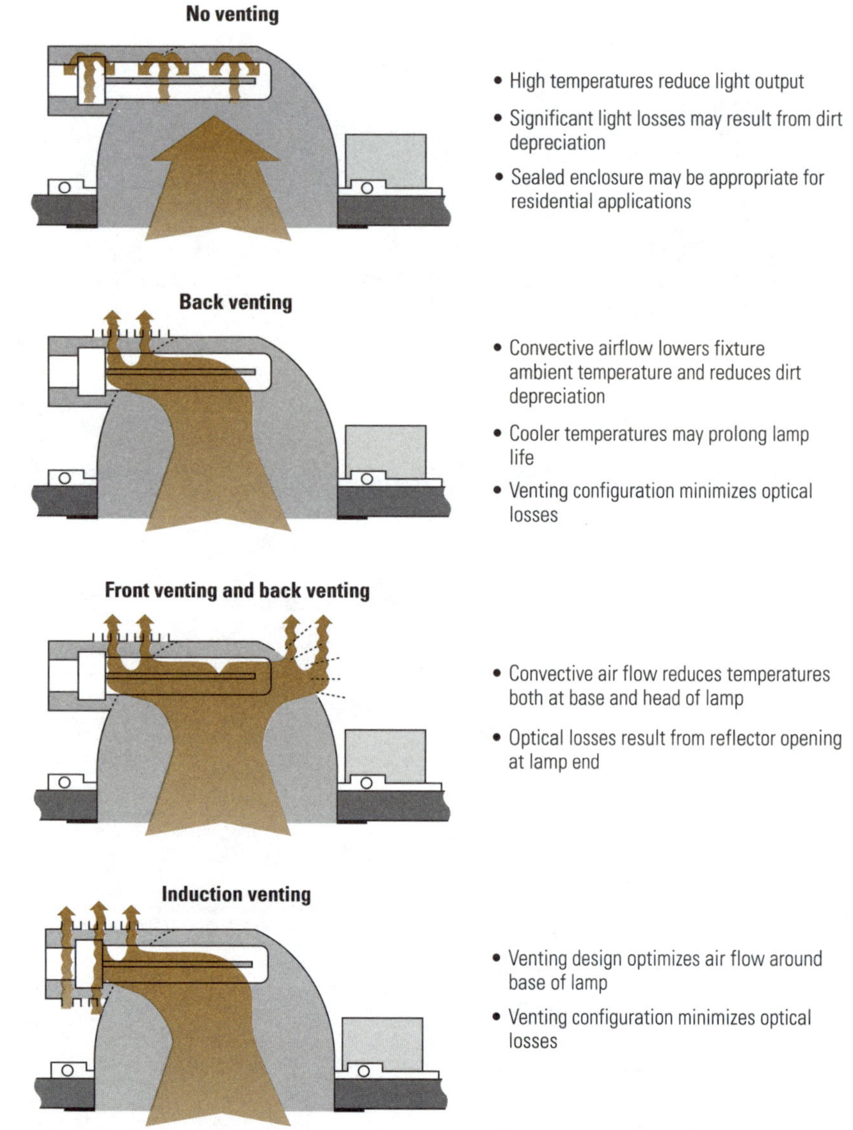

Source: E SOURCE, Lawrence Berkeley National Laboratory

Thermal bridge technology. Thermal bridges can also be used to conduct heat away from the lamp, thereby maintaining lower temperatures. This approach may offer the best solution for enclosed fixtures, where venting is less effective.

A variety of experimental heat-sink devices have been developed by LBNL for use with screw-in compact fluorescent reflector lamps equipped with integral reflectors and lenses. **Figure 8-12** illustrates one such design, which uses a simple metal heat-sink device that attaches to the lens and contacts the end of the CFL by means of a heat-conducting spring. The metal contact cools a small area at the tip of the lamp, transferring heat through the lens to three radial fins.

Installing this device on a 13-watt screw-in CFL reflector lamp operated in a recessed downlight, LBNL researchers found that its light output, after reaching equilibrium, dropped by only one to two percent from its maximum value as the lamp stabilized.[8] By comparison, light output from the same product operated without the thermal bridge declined by 21 percent. Optical losses due to the heat sink were found to be only two to three percent. Because of the thermal bridge, however, the lamp warms up more slowly, requiring about 10 minutes to reach maximum light output. No manufacturers have commercialized this concept to date.

Mercury amalgams. By adding certain alloys to the mercury in CFLs, light output reductions at high and low ambient temperatures can be minimized. The amalgam stabilizes the mercury vapor near its optimum vapor pressure over a wider temperature range, helping to

Figure 8-12

Relative light output for screw-in CFL with and without heat sink

The heat sink increased light output by about 20%, but caused the lamp to reach maximum output more slowly.

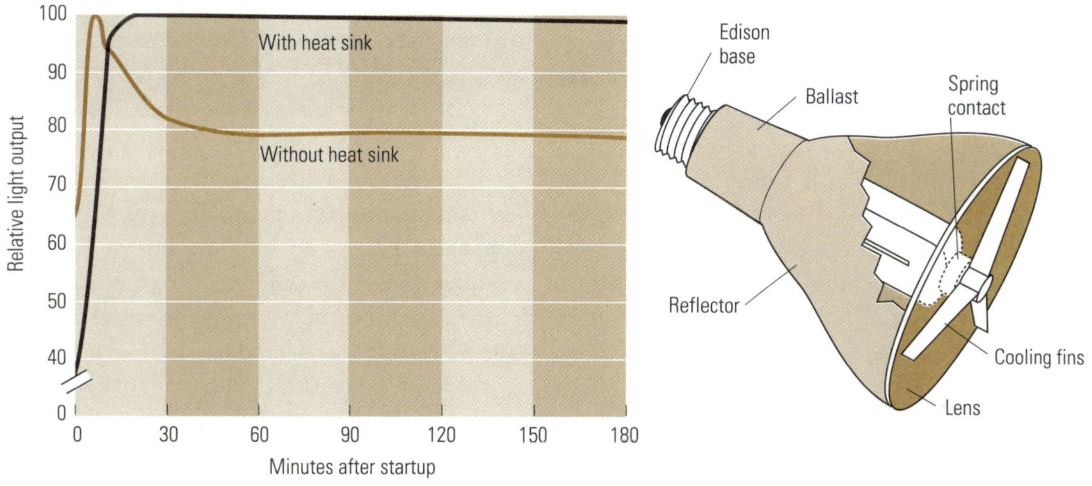

Source: Lawrence Berkeley National Laboratory

maintain maximum light output. In one experiment, light output was maintained between 88 and 100 percent of maximum at ambient temperatures ranging from -5° to 149°F, compared to nonamalgam lamps which dropped to 10 percent light output at -5°F and to 65 percent light output at 149°F (see **Figure 8-13**).[9] Perhaps equally important, mercury amalgams also reduce or eliminate the lamp orientation sensitivity discussed above. Mercury amalgams are used in several Philips, Osram-Sylvania, and Panasonic products (**Table 8-5**).

Most manufacturers are approaching amalgams cautiously. Disadvantages associated with amalgam lamps include low initial light output, slower run-up, and added cost. They are also harder to start, and require a special preheat starting cycle and higher starting voltage to avoid sputtering of the emissive coating. This stress on the ballast can cause some flicker. The slow buildup of initial light output may be countered by the addition of an auxiliary amalgam. In this case, the primary amalgam shifts the high-efficiency thermal operating window for the lamp while the auxiliary amalgam speeds the time it takes for the lamp to attain full brightness.[10] Though magnetic ballasts can be used to operate amalgam lamps (as in the Panasonic line), they will not

Figure 8-13

Relative light output vs temperature in amalgam and non-amalgam CFLs

Mercury amalgams maintain consistent output over a wide range of temperatures.

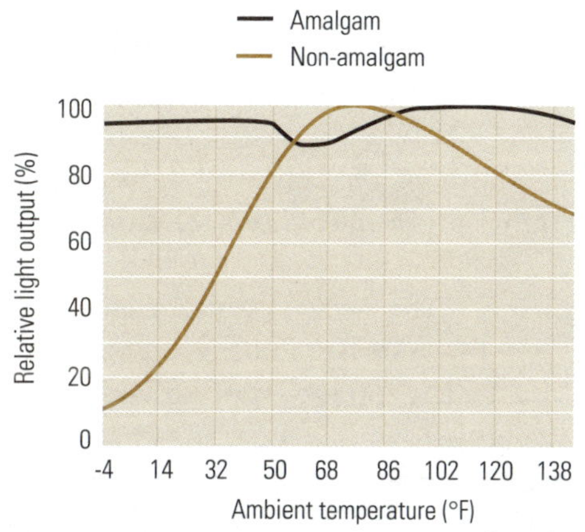

Source: Serres [10]

Table 8-5

Commercially available mercury amalgam CFLs (spring 1994, U.S. market)

Relatively few CFLs contain mercury amalgams.

Manufacturer	Lamp	Wattage	Description
Philips	SLS Earthlight	15, 20, 23	Electronic, integral, triple tube
	SL Earthlight	17, 18	Electronic, integral
	PL-T	26, 32	Electronic, modular
Panasonic	EFG	16	Electronic, integral
	EFT	16	Electronic, integral

reliably start the lamps much below 32°F. For colder starts, the higher voltages available only from electronic ballasts are required.

Proper venting would eliminate the need for amalgams in most interior applications, at lower cost and without requiring the more sophisticated ballasts amalgam lamps require. In addition, if a fixture is designed for an amalgamated lamp and doesn't have good venting, a maintenance person in the future may install a nonamalgam lamp with the same base design, which might then overheat and suffer significantly reduced light output.

8.4.6 Cold-temperature operation of CFLs

Contrary to popular belief, many compact fluorescent lamps work well in cold conditions. However, specifiers must carefully look at the application and product offerings to ensure satisfactory results. There are no universally applicable guidelines available to assist specifiers in making these choices, but there are two main issues to be aware of:

- the ability of the lamp to start in cold temperatures, and
- the ability of the lamp to produce significant light output in cold temperatures.

Ballasting is the most important element in cold weather starting. During normal operation, the ballast initiates starting with a high-voltage pulse to ionize the mercury and strike the arc. The colder it gets, the harder this becomes. If an arc is not initiated, the lamp will keep trying to restrike until an arc is established or the emissive coating on the lamp's electrode wears out, killing the lamp.

The ballast must provide adequate open-circuit voltage to ionize the mercury. Though many standard magnetic ballasts will start CFLs in subzero (°F) temperatures, particularly lower wattage T4 lamps (five, seven, nine watt) and higher wattage T5 quad lamps (22 and 28 watt), most electronic ballasts are superior in such situations because they provide higher open-circuit starting voltage. This holds true only for electronic ballasts with starting modes incorporating preheat or modified instant-start circuitry. Those with rapid-start circuitry will not work satisfactorily in the cold. Electronically ballasted, instant-start, screw-in CFLs (seven to 28 watt) and hardwire CFLs (seven to 55 watt) are now available which will start at or below 0°F.

Because lamp manufacturers cannot control ballast selection for modular lamps, they are conservative in rating their lamps for cold temperatures. Most CFL manufacturers specify a minimum operating temperature (MOT) of 0°F, although many of these lamps will work in colder temperatures. Philips Lighting rates their Earthlight series for an MOT of -10°F, and GE rates a few of its screw-based products down to -10°F. Panasonic specifies an MOT of -22°F for their entire product line. Manufacturers of hardwired luminaires, however, select a specific combination of lamp and ballast and thus can safely rate their luminaire for colder temperatures.

Once started, CFL light output depends on lamp wall temperature. Full light output will only be attained if the heat given off by the CFL can adequately warm the luminaire enclosure. This is facilitated in luminaires which are tightly constructed, recessed, and small. Lamp orientation also affects light output, with base-down installations performing better than base-up ones in cold operating environments. Note that this is just the opposite of warm-environment applications, where base-up installation is preferable. This approach could also degrade summertime performance. Another way to ensure adequate cold environment performance is to use mercury amalgam lamps, which are less sensitive to ambient temperature or orientation.

To apply CFLs in cold conditions, check both with manufacturers and end users who have "weathered" a cold winter using the products in question; confirm outdoor ambient temperature extremes, luminaire construction, lamp orientation, and ballasting, and perhaps do small-scale testing at your local meat locker.

8.5 Ballasts and Power Quality

Ballasts for CFLs fall into two main categories: electronic and magnetic. Key features of these two ballast types are summarized in Table 8-6.

Table 8-6

Comparative features of magnetic and electronic ballasts for CFLs

Characteristic	Magnetic	Electronic
CFL base compatibility	Mostly 2-pin	2- or 4- pin
Noise level	Slight hissing hum	Very quiet
Cost: Edison-base hardwired	$10–$25 $7–$20	$15–$35 $25–$45
Number of lamps powered	1 or 2	1, 2, or 3
Dimmability	No	Available
Expected lifetime: Edison-base hardwired	 5–7 years 10–15 years	 5–7 years 15–20 years

The importance of CFL power quality has been debated within lighting industry and utility circles for several years. The two primary issues are the power factor (PF) and total harmonic distortion (THD) of the ballasts. Typical ranges for various ballast types are shown in Table 8-7.

Table 8-7

Power factor and total harmonic distortion of CFL ballasts

Utility DSM program requirements are driving the market toward high-power-quality electronic ballasts.

	Ballast type: Magnetic		Ballast type: Electronic	
	Normal PF	High PF	Normal PF	High PF
Power factor	0.4–0.7	0.8 to > 0.9	0.4–0.7	>0.9
THD (%)	6–18	15–27	75–200	16–42

Source: NLPIP CFL *Specifier Report, Advanced Lighting Guidelines*

Both magnetically and electronically ballasted CFLs are available with normal-power-factor (NPF) or high-power-factor (HPF) ballasts. Because utility DSM programs have targeted incentives at higher power quality CFLs, more of these products are entering the market. While they cost more, the increased demand and competition stimulated by utility programs are bringing their price premium down.

Specifiers often have a choice between normal-power-factor and high-power-factor ballasts in hardwired CFL fixtures as well. HPF ballasts are generally more efficient and provide higher lumens in such applications.

Some argue that CFLs are such small loads that their power factor and THD should not be of major concern, especially since many other end-use devices on the power grid—personal computers, copiers, laser printers, microwave ovens, televisions, stereos, variable-speed motor controls, and others—also degrade power quality in varying degrees and typically use far more power per unit. The experts disagree among themselves, however, as evidenced by the differing results from three recent studies designed to predict or measure the power quality impacts of large numbers of CFLs.

A residential neighborhood served by a single power feeder saturated with electronically ballasted CFLs found little measurable power quality effects.

New England Electric System designed a computer model to judge the effects on a typical distribution feeder line of two CFLs with 130 percent THD operating simultaneously in each of about 2,500 households served by the line. The model concluded that this would cause enough voltage distortion to exceed the five percent limit recommend by IEEE. However, measured data gathered by Lawrence Berkeley Laboratory showed "only minor effects from a building totally outfitted with electronically ballasted fluorescent lamps (not only compact fluorescents)." And an experiment conducted by Sacramento Municipal Utility District in which a residential neighborhood served by a single power feeder was saturated with electronically ballasted CFLs found little measurable power quality effects.[11]

Although the technical debate is still unresolved, the issue is receding because a number of utilities are setting power quality standards for the CFLs to be eligible for their incentive programs. PG&E, for instance, has two tiers of CFL rebates, and the higher incentives require power factor over 0.9 and THD below 33 percent. New England Electric (see box on page 177) set an even more stringent THD standard of 25 percent and power factor greater than 0.9. Manufacturers are responding with an ever-increasing array of high-power-quality products.

A sampling of screw-based CFLs with low THD and high power factor is listed in **Table 8-8.** They are drawn from a list of products eligible for NEES DSM programs and from the Green Seal certification list. Since new products are entering the market on a regular basis, this list should be viewed as illustrative, not comprehensive.

In addition to the rapidly proliferating selection of high-power-quality screw-in CFLs, the number and variety of high-power-quality electronic ballasts for hardwire applications are increasing as well. **Table 8-9** lists manufacturers of U.S. hardwire electronic CFL ballasts on the U.S. market as of early March 1994. Virtually all of these products have high power quality.

Table 8-8

Selected screw-base CFLs with high power quality on U.S market (spring 1994)

All of the listed products meet New England Electric System requirements of power factor ≥ 0.9 and THD < 25 percent.

Lamp manufacturer	Wattage	Lamp model #	Ballast model #	Type
Panasonic	18	FDL18LE	Elba ECL-18-SC-B	2-pc electronic
			LOA 2004-MPF	2-pc electronic
GE	20	FLE20TBX	built-in	1-pc triple biax
LOA	20	2020TP-MPF	sold with lamp	2-pc electronic circline
Panasonic	20	EFD20LE.UHD	built-in	1-pc electronic
Philips	20	SLS 20/RH	built-in	1-pc electronic
Panasonic	22	FDL22LE/D	Enertron 3900 HPF	2-pc magnetic
			Prolight PL222HPF	2-pc magnetic
			Tek-Tron or Feit E22RH	2-pc magnetic
LOA	22	2022TP-MPF	sold with lamp	2-pc electronic circline
Panasonic	25	EFD25LE.UHD	built-in	1-pc electronic
Panasonic	27	FDL27LE	LOA 2027-MPF	2-pc electronic
Panasonic	28	FDL28LE/D	Tek-Tron or Feit	2-pc magnetic
			10028HPF	2-pc magnetic
LOA	30	2030TP-MPF	sold with lamp	2-pc electronic circline

Source: NEES, Green Seal

Table 8-9

Key to manufacturers of hardwire electronic ballasts for compact fluorescents

AD Advance Transformer
10275 West Higgins Road
Rosemont, IL 60018
tel 800-372-3331
or 847-390-5000
fax 847-390-5109
web www.advancetransformer.com

E Energy Savings
2266 North Palmer Drive
Schaumburg, IL 60173
tel 847-925-8400
fax 847-925-8490

ET ETTA Industries
5680 Central Avenue
Boulder, CO 80304
tel 303-444-2244
fax 303-444-1222

H Hatch Transformer
5403 West Gray Street
Tampa, FL 33609
tel 813-288-8006
fax 813-288-8105
e-mail hatch@interaccess.net
web www.hatchtrans.com/hti

L Lightolier
100 Lighting Way
Secaucus, NJ 07096
tel 508-679-8131
fax 508-674-4710
web www.lightolier.com

LA Lights of America
611 Reyes Drive
Walnut, CA 91789
tel 909-594-7883 or 800-321-8100
fax 909-598-5931
web www.lightsofamerica.com

LU Lutron Electronics Company
7200 Suter Road
Coopersburg, PA 18036-1299
tel 610-282-3800
fax 610-282-0298
web www.lutron.com

MA Magnetek
26 Century Boulevard, Suite 600
Nashville, TN 37214
tel 615-316-5340
fax 615-316-5186
web www.magnetek.com

MI Mitor Industries
418 South 2nd Street, 3rd Floor
Mankato, MN 56001
tel 507-387-1599
fax 507-387-7491
web www.light-link.com/mitor

MO Motorola Lighting
887 Deerfield Parkway
Buffalo Grove, IL 60089
tel 708-215-6300
fax 708-215-6311
web www.motorola.com/AECG/MLI

O Osram Sylvania
100 Endicott Street
Danvers, MA 01923
tel 508-777-1900
fax 508-750-2992
web www.sylvania.com

P Power Lighting Products (formerly Valmont Electric)
1770 Commerce Park Drive
El Paso, TX 79912
tel 915-877-4380
fax 915-877-1499

PR Prescolite Controls
1251 Doolittle Drive
San Leandro, CA 94577
tel 510-562-3500
fax 510-577-5022

R Robertson Transformer
13611 Thornton Road
Blue Island, IL 60406
tel 708-388-2315
fax 708-388-2420

Table 8-9

Manufacturers of hardwire electronic ballasts for compact fluorescents

In response to utility demands for high-performance CFLs, most of the major ballast manufacturers have developed electronic ballasts for hardwire applications. In almost all cases, these products are rated at power factor of 0.9 and THD of 33 percent or less. This table lists manufacturers by abbreviation; see the adjoining list for full manufacturer names and contact information.

Lamp configuration	Lamp wattage	1 lamp	2 lamp	3 lamp	120 V	277 V	Other voltage
Twin tube	5	E[a]	E		E		
	7	E[a]	E		E		
	9	E[a], LA, MA	E, MA		E, LA		MA (230 V)[b]
	13	E[a], ET, LA, MA, PR	E, ET, L, MA, MO, PR	ET	E, ET, L, LA, MA, MO, PR	E, ET, L, MA, MO, PR	ET (220/240 V), MA (230 V)
	18	E, ET, H, MA, R	E, ET, H, L, MA[e], MO, R	ET	E, ET, L, MA, MO, R	E, ET, L, MA[e], MO, R	ET (220/240 V), MA (230 V)[b]
	24/27	E, ET, H, L, MA, MO, R	E, ET, L, MA[e], MO, R	ET	E, ET, L, MA, MO, R	E, ET, L, MA[e], MO, R	ET (220/240V), MA (230 V)[b], R (347 V)
	36/39	AD, E, ET, H, L, LU, MA, R	AD, E, ET, L, LU, MA, R	ET, LU, MA	AD, E, ET, L, LU, MA, R	AD, E, ET, L, LU, MA, R	ET (220/240 V), MA (230 V)[b]
	40	AD, E, ET, L, LU, MA, O[e], P, R	AD, E, ET, L, LU, MA, O[e], P, R	ET, LU, MA, O, P	AD, E, ET, L, LU, MA, O, P, R	AD, E, ET, L, LU, MA, O, P, R	ET (220/240 V)
	50	AD, E, ET, L, LU, P, R	AD, E[a], ET, LU, MA, P, R	ET	AD, E, ET, L, LU, P, R	AD, E, ET, L, LU, P, R	ET (220/240 V)
	55	E, ET, L, MA, O[e], R	E[a], ET, MA, O	ET	E, ET, L, O, R	E, ET, L, O, R	ET (220/240 V), MA (230 V)[b]
Quad	9	ET, MA	ET	ET	ET	ET	ET (220/240 V), MA (230 V)[c]
	10	H, MA, MI	H		MI		MA (230 V)[c]
	13	E, ET, H, LA, MA, MI, PR, R	E, ET, H, L, MA, MO, PR, R	ET	E, ET, L, LA, MA, MI, MO, PR, R	E, ET, L, MA, MO, PR, R	ET (220/240 V), MA (230 V), R (347 V)
	18	E, ET, H, LA, LU, MA, MI, O, PR, R	AD, E, ET, H, L, LU, MA, MO, O, PR, R	ET	AD, E, ET, L, LA, LU, MA, MI, MO, O, PR, R	AD, E, ET, L, LU, MA, MI, MO, O, PR, R	ET (220/240 V), MA (230 V), R (347 V)
	26	E, ET, H, LU, MA, MI, MO, O, PR, R	AD, E, ET, H, L, LU, MA[e], MO, O, PR, R	ET	AD, E, ET, L, LU, MA, MI, MO, O, PR, R	AD, E, ET, L, LU, MA[e], MI, MO, O, PR, R	ET (220/240 V), MA (230 V)[b], R (347 V)
	27	LA			LA		
Delta	18	E, H, MA, MI, O, R	E, H, MA, MO, O, R		MA, MI, O, R	MA, MI, O, R	E (347 V), R (347 V)
	26	AD, E, H, L, MA, MI, MO, O, PR, R	E, H, L, MA, MO, O, PR[e], R		AD, L, MA, MI, MO, O, PR, R	AD, L, MA, MI, MO, O, PR, R	E (347 V), R (347 V)
	32	AD, E, H, L, LU, MA, MI, MO, O, PR, R	E, L, O, PR[e], R		AD, L, LU, MA, MI, MO, O, PR, R	AD, L, LU, MA, MI, MO, O, PR, R	E (347 V), R (347 V)
	42	AD, E, L, MI, PR	E[a], L, PR[e]		AD, L, MI, PR	AD, L, MI, PR	AD
2D	10	E, ET, H, MI[d]	ET, H	ET	E, ET, MI[d]	E, ET	ET (220/240 V)
	16	E, ET, H, MA, MI[d]	ET, H	ET	E, ET, MA, MI[d]	E, ET, MA	ET (220/240 V)
	21	E, ET, H, MA, MI[d], R	ET, H, R	ET	E, ET, MA, MI[d], R	E, ET, MA, R	ET (220/240 V)
	28	E, ET, H, MA, MI[d], R	ET, H, R	ET	E, ET, MA, MI[d], R	E, ET, MA, R	ET (220/240 V)
	38	E, ET, H, MI[d]	ET	ET	E, ET, MI[d]	E, ET	ET (220/240 V)
Circline	20	ET, LA	ET	ET	ET, LA	ET	ET (220/240 V)
	22	ET, H, LA, R	ET, H	ET	ET, LA, R	ET	ET (220/240 V)
	30	ET, LA	ET	ET	ET, LA	ET	ET (220/240 V)
	32	ET, H, R	ET	ET	ET, R	ET	ET (220/240 V)
	40	ET, R	ET	ET	ET, R	ET	ET (220/240 V)
Dulux F	18	H, MA, O[e], R	H, MA, O[e], R		MA, O, R	MA, O, R	MA (230 V)[b]
	24	H, MA, O[e], R	H, MA, O[e], R		MA, O, R	MA, O, R	MA (230 V)[b]
	27	LA			LA		
	36	AD, H, LA, MA, O[e], R	AD, MA, O[e], R		AD, LA, MA, O, R	AD, MA, O, R	MA (230 V)[b]

Notes: a: Available by December 1997.
b: These 230 V ballasts are offered in Europe only.
c: Offered in NPF (normal power factor) only.
d: To be introduced summer 1997.
e: Available fall 1997.

8.6 LIFETIME

CFL lifetime is highly sensitive to lamp cycling. Lamps are rated on the basis of three-hour duty cycle, which means three hours on, 20 minutes off. Most CFLs have rated lives of about 10,000 hours, though this can be dramatically reduced by frequent cycling or extended to 20,000 hours or more in continuous operation.[12] **Figure 8-15** illustrates the reduction in lifetime typically encountered with short cycling times. The impact of cycling on lamp life should be factored into the economic analysis of occupancy sensors which can lead to frequent cycling in areas that are alternately occupied and unoccupied many times a day.

GREEN SEAL STANDARD

The Environmental Standard for Compact Fluorescent Lamps, developed by Green Seal, is a rigorous performance screen that was adopted for a time in the early 1990s by a number of utilities. Green Seal is a nonprofit organization that develops environmental standards, certification and "green buying guides" for a wide range of products. The Green Seal CFL standard includes a full range of environmental criteria, including energy performance, the materials used in packaging, and product labeling information. The standard was released in July 1992 and is being currently being considered for revision by Green Seal. The full standard is available from Green Seal or its Web site (www.greenseal.org). Highlights are summarized in **Figure 8-14**.

A number of manufacturers have sought and received Green Seal certification for their products, including GE and Lights of America (LOA). While it is strictly voluntary, Green Seal's certification program requires re-certification every year; both GE and LOA have elected not to re-qualify for the seal in 1997.

For information on Green Seal or its programs, contact:
Green Seal
1770 Rhode Island Avenue NW
Washington, DC 20036
tel 202-331-7337
fax 202-331-7533
e-mail greenseal@grnseal.cais.com
web www.greenseal.org

Figure 8-14

Summary of the Green Seal standard

Minimum average efficacy

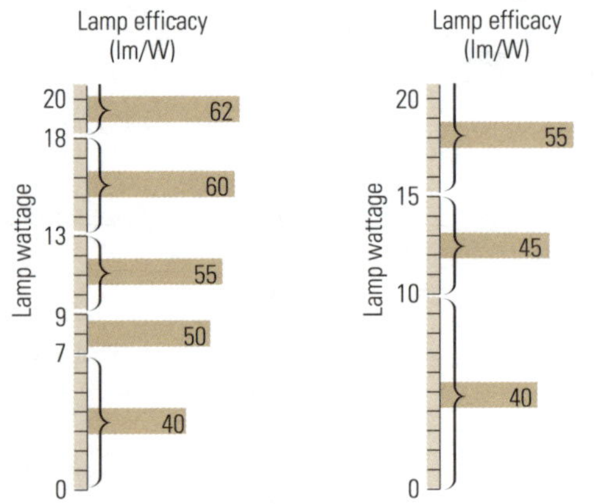

Product safety, performance, and lifespan

Mercury
Minimum average of 15 mg.

Radioisotopes
Zero

Starting
Start smoothly within 4 seconds at minimum rated operating temperature

Power quality
Products can be designated "Class A" if they meet these criteria:
Power factor >0.9 leading or lagging
Harmonic distortion <33% in triplets
Color rendering index of no less than 80

Life
Self-ballasted and modular CFLs: 8,000 hours at 3 hours per start
Ballast adaptors: four 8,000-hour lamp life cycles

Source: Green Seal

Figure 8-15

CFL lifetime as a function of switching cycle

CFL lifetime ratings are based on 3-hour switching cycles. More frequent cycling reduces lamp life. Less frequent cycling can extend service life to 2 or more times the rated life.

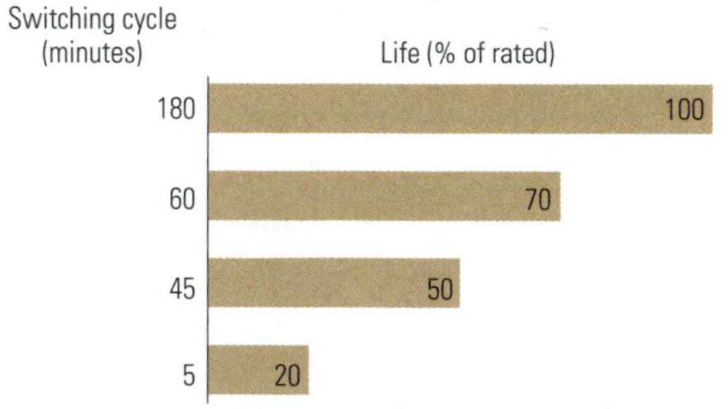

Source: Dulux CFL Handbook

8.7 CFL INTERFERENCE WITH REMOTE CONTROLS

Compact fluorescent lamps—particularly those that use high-frequency electronic ballasts—have the potential to interfere with other electronic devices in two ways.[13] The first type of interference is radio frequency and line interference, which in the U.S. falls under the jurisdiction of the Federal Communications Commission (FCC). FCC's Class A standard applies to commercial applications; most CFL products meet this standard. The FCC judges residential or consumer applications to have greater potential for

NEES CFL SCREEN CRITERIA

New England Electric System has been a pioneer in the utility DSM field for the past decade. NEES aggressively pushed CFLs in its early programs but became concerned about several problems with the technology, particularly light output and harmonic distortion.

Evaluations of residential lighting programs revealed that the primary reason program participants gave for removing CFLs was insufficient brightness. NEES was aware that CFL manufacturers tend to overstate the incandescent "equivalence" of their products on a lumen basis. The utility was also concerned about the total harmonic distortion produced by electronically ballasted CFLs, which often exceeds 100 percent. High THD can cause a range of problems, from transformer overheating to consumer equipment malfunctions. Although individual CFLs are small loads, NEES believes that large numbers of them with high THD could deleteriously effect the quality of power it delivers to customers.

Wanting to promote good CFL technology, in 1992 the utility designed a testing protocol that all screw-in CFLs must pass to be eligible for inclusion in NEES DSM programs. The utility also initiated a similar program for CFL fixtures in 1994. As of 1997, NEES has joined forces with four other northeast utilities and endorsed the performance standards specified by the EPA Energy Star Residential Fixtures program (see **APPENDIX B**).

One of the most impressive accomplishments of this program was that it helped to move the market to develop higher-power-quality products. Until recently, only magnetically ballasted CFLs could meet the 25 percent THD standard, but, not wanting to be left out of a major DSM program, a number of lighting manufacturers developed electronically ballasted CFLs that meet the NEES screen.

interference with televisions, radios and cordless phones, so its Class B standard is more stringent. Since few electronically ballasted fixtures were sold for the residential market in the past, ballast manufacturers had paid little attention to earning Class B certification. However, EPA's new Energy Star specification (see **APPENDIX B**), specifically calls for Class B certification, since qualifying products will be used in a residential setting. Ballasts that are already FCC Class B certified (such as Valmont's REF-213-120), will thus be in greater demand in the future.

The major lamp manufacturers are redesigning their integrally ballasted CFLs to operate above 56 kHz and avoid interference with remote controls.

The second type of interference occurs with infrared remote controls for televisions and VCRs, which send packets of information either in the 33- to 40-kHz range or at approximately 56 kHz. These frequencies overlap with the lamp excitation frequencies of electronically ballasted CFLs, which typically range from 20 to 65 kHz. Harmonics (multiples) of these CFL frequencies can also cause interference, so the most promising solutions involve shifting electronic ballast frequencies above the operating range of remote controls (harmonic frequencies are higher than the fundamental frequency). The major lamp manufacturers are redesigning their integrally ballasted CFLs to operate above 56 kHz and avoid interference with remote controls.

The National Electrical Manufacturers Association (NEMA) recommends that consumer electronics manufacturers use the 33- to 40-kHz range for remote controls, since the 56 kHz units tend to be more susceptible to disruption by compact fluorescents.[14] For problems with existing CFLs, possible remedies include shading the lamp, moving it farther away from the TV, or switching to a different model CFL. A more drastic solution is to switch to low frequency, magnetically ballasted CFLs, but these can cause other problems with hum, flicker and slow start times, while also yielding lower efficacy.

8.8 DIMMING

One of the conspicuous drawbacks of CFL technology has been the lack of cost-effective and reliable dimming options. The main reason for the lack of dimming CFLs is not technical, but economic. Any four-pin lamp can be dimmed as long as cathode voltage is maintained. However, it costs about the same to incorporate dimming circuitry in the ballast for a single 13-watt lamp as in a double 32-watt, four-lamped fixture, so manufacturers have not aggressively pushed dimming technology across the full range of CFLs.[15] A number of attractive approaches are beginning to emerge, mostly in units of 18 watts or more, that promise to extend the range of CFLs into settings like restaurants, theaters, hotels, and other places where dimming is a common feature.

Use caution in interpreting claims about dimming products, however. A number of claims have been made over the years by manufacturers saying that new dimming products were imminent, but some of these never materialized, and others encountered availability or reliability problems and were pulled from the market. Real progress is being made, but slowly. Options available on today's market and some of the current claims of imminent product introductions are described in this section.

There are various ways to achieve a dimming effect, including the following:

- Intelligent design with a mix of light sources, including some that don't dim and others that do.
- Multiple lamps on separate controls.
- Step dimming.
- Continuous dimming.

A dimming effect doesn't necessarily require that all light sources in a space be dimmable. In settings such as conference rooms, on/off CFLs can be used for general ambient lighting and dimmable HIDs, linear fluorescents, or incandescents can be used to ramp light levels up and down during presentations. This can often be much less expensive than using dimming CFL technology throughout.

Another option is to wire CFL downlights in every-other-fixture patterns and place them on separate switches so that each switch controls alternating fixtures. Since many CFL fixtures yield wider and more even light distribution than most incandescent installations, this approach can work well as a "dimming" strategy without creating hot spots and dark zones.

Three-way switching for table or floor lamps can be achieved with screw-based CFL sources designed for this task. The Lights of America "MultiLite" uses a circline lamp powered by a three-way electronic ballast to provide one-third, two-thirds or full brightness. The company claims a 75,000-hour operating life for the ballast. The end-user cost of the MultiLite is about $20. Payback times can be very attractive for such products, not only because the power savings at the highest level are approximately 116 watts, but also because three-way incandescent bulbs are short-lived and expensive. Lights of America hopes to extend this technology into a series of integrated circuit-based dimming CFL products in late 1997 or 1998.

ProLight offers a dedicated table lamp priced at about $40 to $45 to end users (**Figure 8-16**). Its CL2022E lamp includes two concentric 20-watt and 22-watt Circline lamps mounted on a screw-in base that will fit in three-way table lamps. The user can select either lamp alone or both lamps simultaneously, to achieve 900, 1,150, or 2,050 initial lumens. This three-way CFL provides light output comparable to 60- and 75-watt incandescents, and, in its highest setting, about 15 percent more lumens than 100-watt standard incandescent lamps. These are magnetically ballasted lamps, with color temperature of 3,000 K, CRI of 82, power factor of 90 percent, and THD of 33 percent. Demand for this product has not been sufficient for distributors to keep it in stock, but orders can be placed directly with ProLight.

Multilamp step-dimming CFL downlight fixtures are also available. Prescolite and Louis Poulsen Lighting offer three-level switching fixtures which have three 18-watt or 26-watt CFLs. A lower wattage lamp can be specified for the center lamp depending on the light level desired.

Step dimming can also be built into single lamps that operate at several light levels. Lumatech is developing and hopes to release by early 1998 a screw-in dimming unit that can work with Osram-Sylvania's 18-watt triple-twin lamp. These will be reflector-type units with high-power-factor, low-total-harmonic-distortion ballasts. A dedicated dimmer will be included with the system.[16]

Figure 8-16

ProLight's three-way circline lamp

This three-way CFL provides light output comparable to 60- and 75-watt incandescents, and, in its highest setting, about 15 percent more lumens than 100-watt standard incandescent lamps.

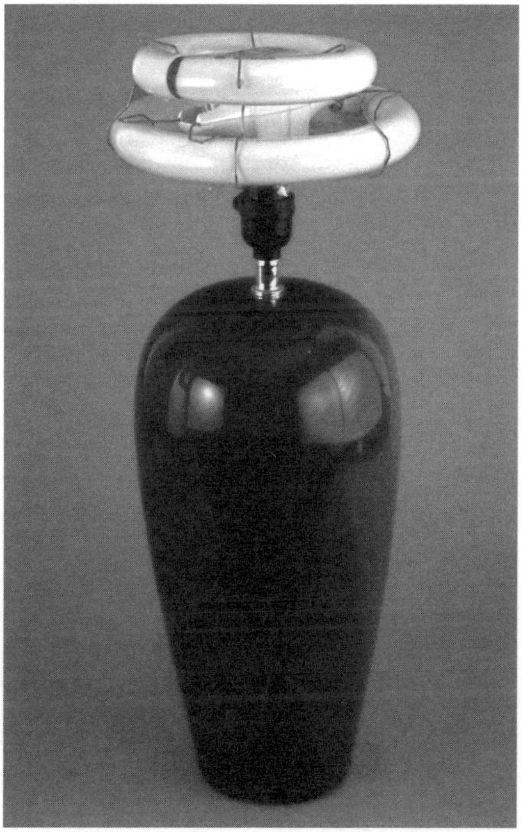

Though manufacturers once felt their greatest challenge in developing dimming CFLs was technical, it now appears that marketing is the primary issue with these products.

Continuous dimming of CFLs (smooth control between full output and minimum output) is limited to electronic ballasts driving four-pin rapid-start lamps of 13 watts or more. Below 10 or 15 percent light output, stability can be a problem with electronically-ballasted CFLs, while magnetically-ballasted CFLs cannot be dimmed at all. Only a handful of full-range dimming ballasts and controls for CFLs are currently available, and they tend to be expensive: adding dimming capabilities can easily double the price of a luminaire. Competition has yet to bring down dimming ballast prices as rapidly as initially expected. Dimming CFL products available include:

- Lutron Electronics offers the Hi-lume line of dimming ballasts that can dim 18 and 26-watt four-pin CFLs from 5 to 98 percent light intensity, with power factor greater than 0.9 and THD under 20 percent. The Hi-lume requires line voltage power plus separate control wiring, a dedicated controller, and a compatible slide switch or other means of controlling the light level.

- Prescolite produces the Intelect electronic full-range dimming ballasts for 13-, 18-, 26-, 32-, and 42-watt CFLs. Available as an option to their line of downlight fixtures, these ballasts can dim from 100 to 5 percent light output with greater than 0.9 power factor. The Prescolite system requires low voltage control wires in addition to line voltage power. The dimming option adds about 30 percent to the fixture cost.[17]

- The Advance Mark X dimming ballast can control a pair of 26-watt quad lamps between 5 and 100 percent from a thyristor (incandescent-type) wall dimmer. While initially designed to operate only with Advance's Mark X 500-watt two-wire dimmer, compatible dimmers should be available from a number of manufacturers by mid-1997. The ballast operates above 42 kHz, reducing the potential for interference with infrared remote controls. It maintains a power factor of more than 0.9 and THD of less than 20 percent throughout its dimming range.

- Solium offers two-wire dimming ballasts for 18-, 26-, and 32-watt compact fluorescents. The achievable dimming range is 20 to 100 percent and no proprietary dimmer is required. End-user cost is approximately $25 to $30. At the time of this writing, the future of Solium's business and products is uncertain.

Though manufacturers once felt their greatest challenge in developing dimming CFLs was technical, it now appears that marketing is the primary issue with these products. Manufacturers that have offered dimmable or three-way switchable fixtures have faced difficulty primarily because the limited number of fixture styles they have been able to offer have only appealed to a small number of buyers. Instead, companies are beginning to offer hard-wired modules that are installed in lieu of the Edison base—for example, a pin-based socket that can only accept a quad-lamp CFL—on whatever fixture style and type the buyer chooses. These modules are particularly popular in hotel/motel applications, since they are theft-resistant and usable in a wide variety of decorative styles.

8.9 CFL FIXTURES

The best CFL is worthless if installed in a poor fixture. As discussed above, CFL performance is highly sensitive to temperature and orientation, factors which are often determined by the fixture in which the CFL is installed. Other aspects of fixture performance are critical also.

Most kinds of pendants, desk lights, exit signs, incandescents (including sconces, globes, dishes, drums, downlights, and tracklights, among others), have been duplicated in CFL versions, and retrofit kits are available to convert most kinds of existing incandescent fixtures to CFL technology. For a unit to be considered a "CFL fixture," it should have a built-in ballast and a dedicated pin base that can only accept CFL lamps; fixtures with screw bases and screw-in CFLs can too easily be refitted with high-wattage incandescent lamps.

8.9.1 Outdoor CFL fixtures

Outdoor fixtures typically have long burn hours that make them attractive candidates for CFL use. The array of outdoor fixtures with dedicated pin bases for compact fluorescent lamps continues to grow (**Table 8-10**). Home improvement centers like Home Depot and Home Base now carry dedicated outdoor CFL fixtures and replacement pin-based lamps, making the technology available to most residential and small commercial customers at affordable prices.

The Environmental Protection Agency's Energy Star program for outdoor fixtures (see **APPENDIX B**) will encourage the use of two additional control strategies: photocells and motion sensors. In order to receive the Energy Star rating, outdoor fixtures will need to turn off automatically during daylight hours.

8.9.2 Indoor CFL fixtures

Until recently, indoor light fixtures designed to use pin-based compact fluorescent sources were found almost exclusively in commercial and high-end residential projects. But with the development of more diverse lamp shapes and sizes and improved ballasts, CFL technology has spread to a wide range of fixture styles and prices for the residential and commercial marketplace. A few trends are evident in today's CFL fixtures:

■ *Design evolution.* CFLs are no longer thought to be a design liability. The newest fixtures are designed to exploit CFLs' advantages of flat geometry and diffuse light output. This is particularly evident in the slender profiles of sconces: to meet the requirements of the Americans with Disabilities Act (ADA) these fixtures must project no more than four inches the wall, making them ideal candidates for compact fluorescent sources.

Table 8-10

Representative outdoor CFL fixtures

Lights of America 611 Reyes Drive Walnut, CA 91789 tel 800-321-8100 fax 909-594-6758	Surface-mount fixtures and floodlights using twin-tube and F-lamps	
Technical Consumer Products 23632 Mercantile Road, Suite A Cleveland, OH 44122-5916 tel 800-324-1496 fax 216-765-0336 web www.yan.com	"Coach light" with 13W quad lamp and incandescent-bulb diffuser	
Sun Dagger Design P.O. Box 4355 Santa Fe, NM 87502 tel 800-889-7109 fax 800-320-1136	Southwest-style ceramic sconces with twin-tube lamps	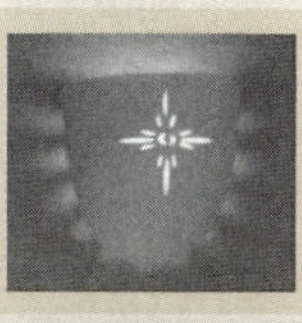
Brownlee Lighting 3071-K N. Orange Blossom Trail Orlando, FL 32804 tel 407-297-3677 fax 407-297-3705	Post lights, floodlights, and pathway lights using twin-tube lamps	

■ *Light source choices.* Many manufacturers offer fixtures with a modular design—they can be ordered with incandescent, halogen or compact fluorescent sources in the same basic housing. By separating decor choices from light source choices, these fixtures remove one of the principal obstacles to more efficient residential lighting.

■ *Wider array of lamps and ballasts.* At one time, virtually all compact fluorescent fixtures used magnetic ballasts and 13-watt twin tube lamps. This inexpensive though rudimentary solution was often plagued by noticeable ballast hum and insufficient or uneven light output. With the arrival of 2D, linear quad, triple tube, and other high output light sources, however, fixture designers find themselves with a much wider range of options. This increases efficiency and can reduce ballast cost. In addition, electronic ballasts are becoming more compact and versatile with each passing year, making it easier to accommodate them in a wide range of fixtures.

■ *Competitive performance.* Fluorescent fixtures have traditionally competed on price, and thus had to survive on thin margins and tight parts budgets. The new generation of fixtures, though more expensive, is increasingly competing on value, style and performance. Such fixtures are more profitable to sell, more efficient, and more likely to satisfy customers. They also help to dispel the myth that fluorescent lighting is ugly, by demonstrating that fixture aesthetics are more a function of fixture design and lamp/ballast matching than of any inherent "ugliness" of fluorescent sources.

One example is shown in **Figure 8-17**, where 160-watt fixtures provide as much light output as 600-watt incandescents with excellent light distribution and appearance. These fixtures use high-output twin-tube lamps to yield high ambient light levels. Because the sources are hidden behind a diffuser, the fluorescent versions offer superior aesthetics, since their incandescent alternatives show "hot spots" on the diffuser.

Nearly two-thirds of the incandescent downlights in the U.S. are candidates for retrofit or replacement with compact fluorescent lamps.

Other multilamp fixtures are now available for replacing even high-bay HID fixtures, providing savings through higher photometric and ballast efficiencies. Sportlight's CFL industrial luminaires, for example, use six or eight 26-watt CFLs in a fluted reflector to replace HIDs and incandescents in high-bay industrial and warehouse lighting.

The Energy Star certification and labeling program for residential light fixtures will help buyers select good-quality CFL fixtures for their homes. Fixtures that meet the Energy Star specifications for efficiency, optical performance, and power quality can carry the Energy Star logo on their packaging. See **APPENDIX B** for a complete listing of these specifications.

Recognizing that CFL options are available for most types of fixtures, we focus our fixture discussion on the downlight, which is one of the most common kinds of fixtures for CFLs, in both retrofit and new installations. Many of the same issues apply to other fixture types as well.

8.9.3 Compact fluorescent downlights

Downlight fixtures—often referred to as "recessed cans"—are among the most common lighting fixtures in commercial and residential settings. As such, they represent one of the most promising targets for improving lighting energy efficiency. We estimate that downlight fixtures account for more than one-fifth of the 2.8 billion incandescent lighting sockets in the United States, where over 30 million new downlight fixtures are sold each year.

With existing and foreseeable technology, nearly two-thirds of the incandescent downlights in the U.S. are candidates for retrofit or replacement with compact fluorescent lamps (CFLs) or fixtures. The remaining one-third, however, are unlikely to ever be replaceable with CFL technology because of constraints on light output, lighting quality, size, and cost-effectiveness of CFL alternatives. In many cases, other efficient alternatives, such as HIDs or halogen IR lamps, can replace conventional incandescents when CFLs are not appropriate.

Fully utilizing the compact fluorescent lamp's efficacy advantage in downlighting systems requires that consideration of thermal environ-

Figure 8-17

CFL pendant fixture

Each of the pendants in this picture use four 39-watt twin-tube CFLs powered by two electronic ballasts using only 160 watts to deliver the same light output as 600 watts of incandescent lighting. They provide strong uplight and soft downlight through a diffuse lens.

The four CFLs in each fixture are arranged as shown

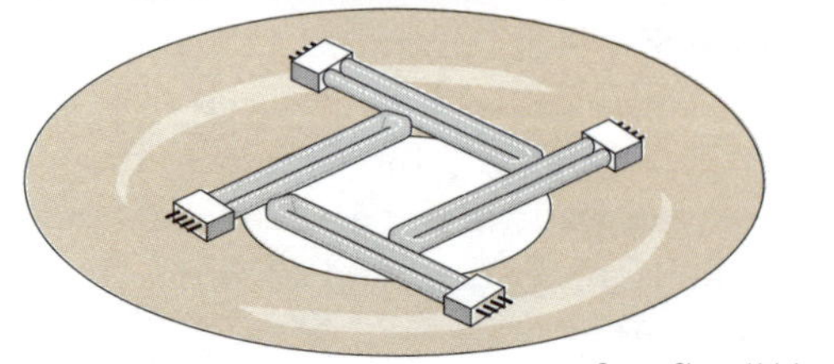

Source: Shaper Lighting

ment, optical geometries, and ballasting be integrated into the fixture design. Many compact fluorescent downlights sold today are simply incandescent fixtures relamped with CFL sources. These fixtures' efficiencies are poor, often falling below 50 percent. In addition, the CFL lamps in such fixtures typically operate at higher than optimal temperatures, reducing light output, lamp life, and ballast life. Poor lamp/ballast matches (low power factor, high total harmonic distortion, and low ballast factor) can also contribute to low light levels and/or poor power quality. Moreover, while "fixture efficiency" ratings may be used to compare the relative performance of different fixtures, other factors must also be considered to ensure good lighting quality, ranging from distribution of light output to glare control. The bottom line: simply specifying "CFL downlight fixtures"—whether in utility rebate programs, corporate purchasing policies, or building specifications—is not sufficient to ensure high-performance, energy efficient downlighting.

Most CFL downlights are sold on the basis of their high luminous efficacy relative to incandescent alternatives. Yet making direct efficiency comparisons among various CFL downlight fixtures is still more art than science. While standards for photometric testing of bare CFLs have been established by the Illuminating Engineering Society (IES), no specific standard exists for testing CFL luminaires. CFL fixtures are now tested according to procedures used for other indoor fluorescent fixtures, but which may not take into consideration all of the factors particular to CFL performance in downlight luminaires. As a result, uncertainties persist about the reliability and comparability of test data provided by manufacturers and even by independent laboratories. Since ambient temperature and lamp orientation have substantial effects on light output, test procedures that vary only slightly may yield substantially different results.

The best advice for those specifying, purchasing, or evaluating CFL downlight fixtures is *caveat emptor*—let the buyer beware. To evaluate performance of CFL downlights, we recommend that the specifiers secure samples from standard production stock, and use performance data from independent laboratories. While photometric data can be off by five percent due to extra polishing of reflectors and other efforts to "test the best" product, thermal and optical differences between test and production fixtures can alter performance values by as much as 10 to 20 percent.

Photometric data can be off by five percent due to extra polishing of reflectors and other efforts to "test the best" product.

Lamp orientation. As discussed above, light output from bare CFLs may vary by up to 15 percent at different orientations because of the effect that lamp orientation has on mercury condensation inside the tube. Not all labs use the same orientation for calibration of the flux output from the bare lamp. Testing labs use relative, as opposed to absolute, photometry in which a factor is determined to "calibrate" the flux output (lumens) of the bare lamp to the rated value. Any inconsistent test protocol during calibration would make it impossible to compare fixture efficiency values from different manufacturers, since the calibrated flux value establishes the denominator for fixture efficiency. In general, fixture efficiency values based on vertically calibrated lamps will be lower than those based on horizontally calibrated lamps.

Confounding the problem, lamp manufacturers generally publish flux output data based only on tests conducted with the lamp in the vertical, base-up position. This approach conforms with recommended practices established by IES for bare lamps. For fixtures with lamps in a horizontal orientation, however, designers sometimes mistakenly combine fixture photometry data based on calibration in the horizontal position with lamp manufacturers' lumen ratings based on calibration in the vertical position. To detect such inconsistent use of data, buyers should check photometric reports for the lamp orientation used during calibration. Lumen ratings based on horizontal calibration can be obtained from some, but not all, lamp manufacturers.

Temperature effects. As also discussed above, light output from CFLs is sensitive to ambient temperature and air circulation. At the high operating temperatures common for most fixtures, light output may be compro-

Application correction factors for CFL fixtures

To help designers to relate photometric data based on standard test conditions to the real world, the Advanced Lighting Guidelines, sponsored by the California Energy Commission, U.S. DOE, and EPRI provide "application correction factors" (AFs) for fluorescent sources. These factors are intended to give designers a more accurate estimate of the installed power consumption and lumen output of a luminaire based on its design, mounting, typical thermal conditions, and lamp orientation. For specific combinations of lamp, ballast, and luminaire, the factors can be multiplied by the fixture coefficient of utilization (CU) and fixture efficiency to obtain values that more accurately reflect installed performance. The Advanced Lighting Guidelines' application correction factors for CFL fixtures (including, but not limited to downlights) are presented in **Table 8-11**. The application factors range from 0.8 to 0.9 depending on the lamp and ballast type, fixture design, and mounting.

Table 8-11

Application correction factors for CFL fixtures

By multiplying the application correction factors shown in the table by the fixture's photometric report ratings for coefficient of utilization and fixture efficiency, lighting designers can obtain more realistic estimates of installed power consumption and lumen output of CFL fixtures. The application correction factors are only accurate when applied to the specific installations described briefly in the column headings and more completely in the notes below.

	Recessed unvented base-up[a]	Recessed vented base-up[b]	Recessed unvented horizontal[c]	Recessed vented horizontal[d]	Surface closed uninsulated[e]	Surface closed insulated[f]	Wall open uninsulated[g]	Wall closed jar[h]	Free air ANSI table[i]
9-watt twin-tube lamps, 120-volt standard reactor ballast, ballast factor : 0.85									
Application correction factor	0.82	0.84	0.81	0.84	0.83	0.80	0.83	0.83	0.85
Input watts: 1 lamp (1) ballast	13	13	13	13	13	13	13	13	13
2 lamps (2) ballasts	26	26	25	26	25	25	26	26	26
13-watt twin-tube lamps, 120-volt standard reactor ballast, NPF, ballast factor : 0.90									
Application correction factor	0.87	0.89	0.85	0.89	0.88	0.84	0.88	0.88	0.90
Input watts: 1 lamp (1) ballast	16	16	16	16	16	16	16	16	16.5
2 lamps (2) ballasts	32	33	32	33	32	32	33	32	33
13-watt quad-tube lamps, 120-volt standard reactor ballast, NPF, ballast factor : 0.90									
Application correction factor	0.87	0.89	0.85	0.89	0.88	0.84	0.88	0.88	0.90
Input watts: 1 lamp (1) ballast	16	16	16	16	16	16	16	16	16.5
2 lamps (2) ballasts	32	3	32	33	32	32	33	32	33
26-watt quad-tube lamps, 277-volt standard reactor ballast, HPF, ballast factor : 0.90									
Application correction factor	0.87	0.89	0.85	0.89	0.88	0.84	0.88	0.88	0.90
Input watts: 1 lamp (1) ballast	32	33	32	33	32	32	33	32	33
2 lamps (2) ballasts	65	65	64	65	64	64	65	65	66
26-watt quad-tube lamps, 277-volt electronic ballast, ballast factor: 0.90									
Application correction factor	0.88	0.89	0.86	0.89	0.88	0.85	0.88	0.88	0.90
Input watts: 1 lamp (1) ballast	24	25	24	25	24	24	25	24	27
2 lamps (2) ballasts	49	51	48	50	48	47	51	49	54

Note: Ballast factor correction. The ballast factors shown are industry averages from 1992. If the ballast factor for a given lamp-ballast combination is substantially different from the listed value, the input values can be determined by applying the following equation: Corrected input power = listed input power x $\frac{\text{intended ballast factor}}{\text{listed ballast factor}}$

Descriptions of luminaire installations

a: Recessed unvented downlight with vertical base-up lamps and nominal 7-inch aperture recessed in acoustic tile ceiling.
b: Same as "a" except luminaire is vented to improve thermal performance.
c: Recessed unvented downlight with horizontal lamps and nominal 7-inch aperture. Recessed in acoustic tile ceiling.
d: Same as "c" except the luminaire is vented.
e: Surface-mounted "vandal resistance" luminaire (enclosed) on uninsulated gypsum wallboard ceiling.
f: Same as "e" but insulated ceiling.
g: Open top wall sconce, surface mounted to uninsulated wall.
h: Enclosed and gasketed wall bracket, industrial "jelly jar" type luminaire.
i: Free air table lamp, essentially ANSI/catalog values.

Source: *Advanced Lighting Guidelines*

mised by 15 percent or more. During testing, it is important to allow the lamp to stabilize in order to represent accurately the thermal effects on light output at the ambient temperature specified in the testing protocol. Still, most laboratory tests are conducted at ambient temperatures significantly lower than those observed for recessed fixtures in most commercial buildings. The performance of CFL downlights in diverse real-world applications varies widely.

Visual comfort. No single performance metric, such as fixture efficiency, can adequately capture all of the performance characteristics that are important. For example, to get more light out of a fixture, the lamps can be moved lower within the fixture. While this will increase fixture efficiency, it may also increase glare. Hence, manufacturers must make trade-offs between various characteristics that are not reflected in any one quantitative measure of fixture performance. To avoid making decisions based on misleading data, verify that independent testing information and manufacturers' specifications for lamp location are the same. This will ensure that a manufacturer, for instance, has not tested a fixture, determined its efficiency, then decided to raise the lamp in the fixture to increase visual comfort.

Compact fluorescent downlight technology. Well-designed compact fluorescent downlight fixtures should ideally maximize light output, minimize glare, and provide a range of beam spreads, making it possible to replace a wide variety of incandescent sources and fixture types with much more efficient CFL alternatives. Most of the fixture options offered for incandescent downlights are available for CFL downlights, including round and square apertures, various lenses, specular and matte reflector finishes, 4-inch to 12-inch cross sections, and beamspreads from 20° to 86°. Wall-wash and adjustable "eyeball" CFL downlights are also available, although the latter generally have low optical efficiency.

The combined savings realized through reduced energy use, fewer lamp replacements, and lowered maintenance costs can quickly recover the marginal cost of high-performance CFL systems in many commercial applications. **Table 8-12** provides a simplified comparison of incandescent and fluorescent downlighting systems in terms of performance and cost. In this example, replacing incandescent "A" lamp fixtures with more efficient, electronically ballasted fluorescent downlights reduces unit power density (UPD) by 57 percent—from 3.0 to 1.3 watts per square foot. If the number of fixtures can also be reduced

Table 8-12

Performance comparison for selected downlighting systems

These lighting approaches will provide roughly 30 footcandles of illumination at the task plane. Other information must also be considered, however, such as glare, beamspread, spacing-to-mounting-height ratio, and the presence of a lens or louver.

		Lamp type	Fixture aperture (inches)	Fixture efficiency (%)	Fixture efficiency with AF[a] (%)	Rated lamp lumens	Input (W)	Coefficient of utilization[b]	Lamp/ballast efficacy (lm/W)	Unit power density (W/ft²)	Ft² per fixture	Energy cost[c] (\$/ft²-y)	Fixture cost[d] (\$/fixture)
Incandescent		1–150W R40	6 1/4	98	98	1,900	150	1.01	12.7	3.13	48	0.82	40
Incandescent		1–150W A	6 9/16	67	67	2,850	150	0.70	19.0	3.01	50	0.79	65
Fluorescent	Magnetic ballast	2–26W quad	7 3/8	53	48	3,600	65	0.49	49.8	1.65	39	0.43	150
Fluorescent	Magnetic ballast	2–26W quad	8	76	68	3,600	65	0.70	49.8	1.14	57	0.30	150
Fluorescent	Electronic ballast	2–26W quad	7 3/8	53	48	3,600	51	0.49	63.5	1.30	39	0.34	175
Fluorescent	Electronic ballast	2–26W quad	8	76	68	3,600	51	0.70	63.5	0.90	57	0.24	175

a: Rated fixture efficiency for CFL fixtures has been multiplied by an application factor (AF) of 0.9 to reflect typical performance of recessed vented fixtures with horizontally mounted lamps.
b: Coefficients of utilization were calculated assuming a Room Cavity Ratio of 1.0 and ceiling/wall/floor reflectances of 0.8/0.5/0.2 respectively. An application factor (AF) of 0.9 has been multiplied by the coefficient of utilization values.
c: Based on 6¢/kWh and 4,380 hours operation per year.
d: Based on contractor pricing for small quantities (with lamps).

through use of fixtures with a greater fixture efficiency, UPD can be reduced even further, to 0.9 watts per square foot. This 70 percent overall energy savings is indicative of the potential benefits of optimizing the overall lighting design.

Compact fluorescent light sources are increasingly being used in downlight fixtures for both new construction and retrofit applications. Hundreds of CFL downlight products are available; the majority are produced by a few dozen manufacturers. **Figure 8-18** illustrates several common configurations for such fixtures.

CFL downlights for new construction. CFL downlight fixtures represent a rapidly growing share of the market for new downlight fixtures. We estimate that in the U.S. alone, approximately 28 million downlights are sold annually into the commercial sector, while about five million are sold for residential use. Most downlight fixture manufacturers report that CFL products represent a significant and rapidly growing share of their sales. For one major manufacturer, sales of CFL downlights increased by 35 percent in 1992.

As with incandescent downlights, three grades of CFL downlights are available: specification, commercial, and economy. Each grade is distinguished by the fixture's construction, assembly materials, reflector finish, optical performance, and, often, visual comfort (**Table 8-13**).

While the specification grade is the most expensive option, it is not necessarily the most efficient. In fact, there is often not much correlation between fixture cost and efficiency. Some commercial grade and even economy grade downlight fixtures may have higher fixture efficiency ratings than comparable specification grade fixtures, though visual comfort may be compromised as a result of inferior glare control. Typical fixture efficiencies for new CFL downlights are 50 to 60 percent, but they range from 40 to over 80 percent.

Figure 8-18

Anatomy of CFL downlights

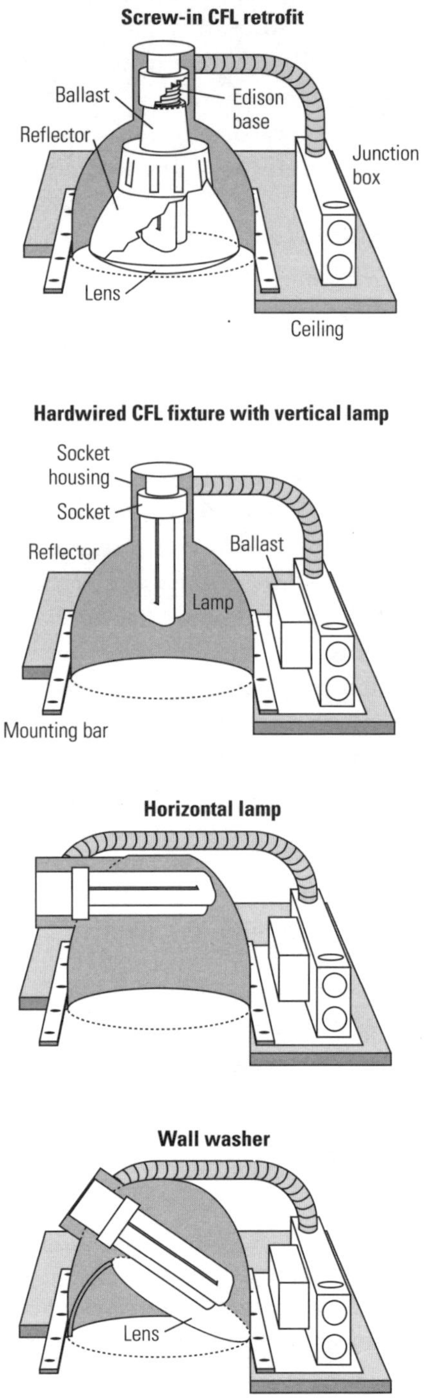

Downlight technology for remodel/retrofit applications. Given that more than 500 million incandescent downlights are already installed throughout the U.S., the market for CFL retrofit products is potentially much greater than that for new downlights. Retrofit technologies, however, must accommodate a myriad of existing fixture designs. For this reason, optimizing the performance of retrofittable fixtures is especially challenging.

There are two basic types of compact fluorescent retrofit systems for downlights: screw-in retrofits and hardwire retrofits.

■ *Screw-in retrofits.* Screw-in compact fluorescents are the simplest and least expensive retrofit option for incandescent downlight fixtures. Although a few are excellent, most of these are poor performers, so choose carefully when going this route.

The most basic—and all too common—option is to screw a bare CFL into a standard incandescent downlight. This is typically a light distribution disaster. Since the fixture's reflector was designed for a concentrated, symmetrical incandescent light source, not the long, diffuse CFL, much of the light remains trapped in the fixture. Moving the lamp farther down in the fixture will get more light out of the fixture but typically creates glare.

The other main screw-in option is a CFL retrofit kit that includes a screw-in ballast, a reflector that attaches to the ballast and serves in lieu of the original fixture reflector, and the lamp itself. In the world of these reflectorized CFL kits there are a few high-quality products and a lot of mediocre ones.

The reflector is the main feature that distinguishes product quality in this arena. Many products use for a reflector the same kind of glass bulb with a reflective surface that is used in incandescent R, PAR, and ER lamps. The bulb/reflector attaches to the screw-in ballast and fits inside the downlight. Because these bulbs are designed for incandescent sources, however, they are not very good at getting the light from a CFL out of the downlight can.

Several hundred models of such screw-in compact fluorescent reflector lamps are available on the U.S. market in 5- to 28-watt models. Available in magnetically or electronically ballasted integral or modular models, the "fixture efficiency" (the amount of light escaping the reflector/lens assembly divided by the light output from the lamp alone) for most of these ranges from 50 to 70 percent. When installed in relatively shallow fixtures (where the reflector extends close to the fixture lip), the efficiency of the entire system will be about the same as the screw-in fixture alone, but the lamp may create glare, since most products in this category have wide light distributions. Installing them in deeper fixtures reduces glare, but at the cost of trapping much of the light inside the fixture.

A higher-performance alternative is to use a screw-in retrofit kit that incorporates a metal reflector specifically designed for a compact

Table 8-13

Typical design features for the three grades of CFL downlight fixtures

	Frame and socket holder	Reflector	Cutoff design*	Ballast	Retail price
"Specification"	die-cast aluminum	high reflectance and optimized design	wide range of cutoff designs available, including deep cutoffs	electronic or high power factor (PF) magnetic ballast	$95–175
"Commercial"	stamped steel	high to medium reflectance and good design	moderate cutoff	electronic ballast or high or normal PF magnetic ballast	$45–95
"Economy"	stamped steel	medium to low reflectance and "generic" design	minimal cutoff	normal power factor magnetic ballast	$25–45

*Cutoff refers to the control of light dispersed at high angles relative to the vertical axes of the lamp. Deep cutoff designs minimize glare.

fluorescent lamp. The reflector attaches to the screw-in ballast and fits inside the existing downlight. Unlike the glass bulb reflectors described above, however, these designs are very effective at delivering light out of the downlight fixture. Several manufacturers make such products, including Lumatech, ProLight, QuadraTech, TekTron, and Eastrock Manufacturing.

The fact that screw-in CFLs can be easily removed is of particular concern in commercial settings where theft may be a problem. To ensure long-term savings, many utilities are encouraging installation of hardwire retrofits to make sure that CFLs are not subsequently replaced with incandescent lamps. Some manufacturers of screw-in reflector products have developed locking devices that make it difficult to convert retrofitted fixtures back to incandescents.

■ *Hardwire retrofits.* Hardwire downlight retrofit kits are designed to permanently upgrade existing incandescent fixtures so that all they can use are compact fluorescent lamps. Most of these use the existing downlight's mounting frame and junction box, to which a new lamp holder/socket housing, reflector, and ballast are installed. Both the lamp holder and reflector are specifically designed for the specified CFL. The retrofit installation generally takes less than 10 minutes and requires few tools because most parts clip or screw in. When the pin-based CFL burns out, it can only be replaced with a similar lamp—not a screw-in incandescent. One- and two-lamp retrofits are available with vertically or horizontally mounted lamps, offering the capability of replacing up to 150 watts of incandescent lighting with two 26-watt CFLs, for instance. A number of downlight manufacturers now offer such retrofit kits for their standard incandescent downlights. In addition, a handful of specialty manufacturers fabricate custom hardwire downlight retrofit kits. Once retrofitted, fixture efficiencies typically range from 50 to 80 percent.

One glare-free option for retrofitting incandescent downlights with CFLs is the surface-mounted drum fixture that attaches below existing downlights by clips or wires and connects electrically with an Edison base on a wire whip. Light distribution is changed from downward-focused to wide by the kit's diffuser. Alta offers such kits using the 21-watt General Electric 2D CFL (**Figure 8-19**).

Figure 8-19

CFL surface-mounted drum fixture

This CFL downlight retrofit avoids glare and provides wide light distribution by attaching a surface-mounted drum fixture to the bottom of the existing downlight. This design, from Alta, uses a 21-watt GE 2D compact fluorescent.

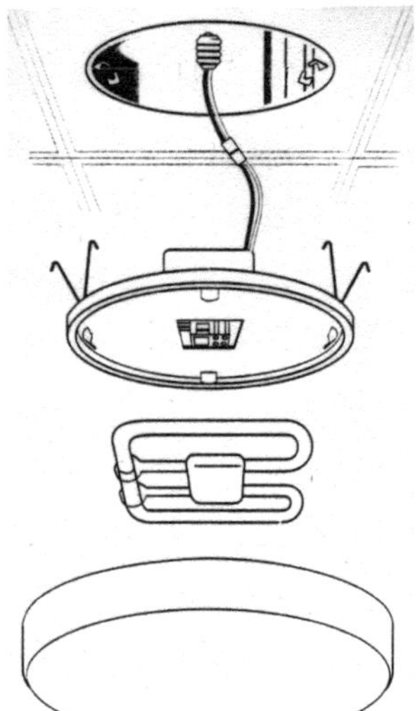

Source: Alta Illumination

Another type of "quasi-hardwire" retrofit kit screws a new lampholder and power supply assembly into the existing fixture cap (**Figure 8-20**). Also called a cap replacement kit, the new assembly contains a new cap with pin-based CFL lampholders, attached to a ballast, which in turn is attached to a medium screw base that screws into the old cap, which is pushed back out of sight into the recessed ceiling. Although such kits are not formally "hardwired," from below it appears that the fixture can receive only pin-based CFLs, minimizing the possibility of relamping with incandescents. Many cap replacement kits reuse the original fixture's reflector. For reasons discussed above, this can be a photometric disaster, putting out only half the light of the original incandescent fixture. To ensure a high-quality retrofit, use kits that include a new reflector designed for CFLs, and check the photometric features of the proposed replacement in advance.

Figure 8-20

Compact fluorescent cap replacement retrofit

While not technically a "hardwire" retrofit, from below it appears the fixture can receive only pin-based CFLs, minimizing the possibility of relamping with incandescents.

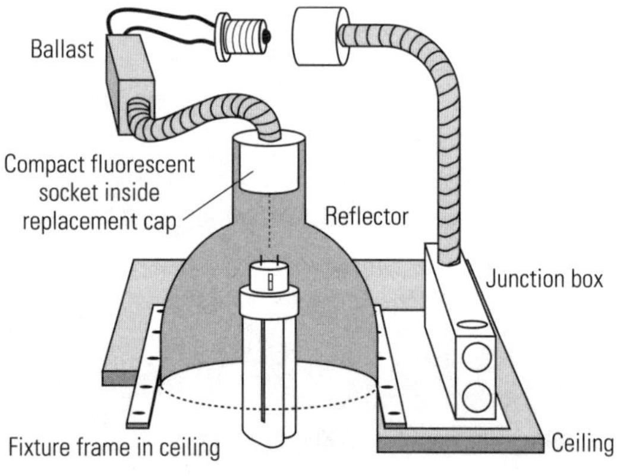

Technical limitations. While CFL downlights can replace incandescents in many cases, some downlight applications, such as architectural and display lighting, may require higher beam candlepower, a narrower beam, or a light that imparts more "sparkle" on specular or crystalline surfaces than CFL downlights can provide.

The highest center beam candlepower available with CFL downlights only approaches 1,500 to 1,600 (30° beamspread); the tightest beamspread available is 20°. By comparison, a 50-watt PAR 30 halogen-IR narrow flood has a 23° beamspread with a center beam candlepower of 4,000. A 50-watt halogen-IR narrow spotlight has a beamspread of only 7°, and a center beam candlepower of 19,500. Even the smallest compact fluorescent downlights may be too large for some space-constrained applications or where the appearance of a more compact fixture is desired. Finally, many downlight installations require continuous dimming capability, a feature which requires special wiring and significantly higher cost in CFL systems.

Factors affecting fixture performance. As discussed above, the overall performance of CFL downlights depends on fixture optics, thermal environment, and ballasting. Each must be designed and optimized specifically for the CFL source to maximize system efficiency. Since changing any one of these design variables will affect system performance, designers and specifiers must understand these interactions to insure that the overall system performance is optimized.

■ *Optical design.* As opposed to the "point source" filament of an incandescent lamp, the entire surface area of a compact fluorescent lamp emits light, making CFLs inherently more difficult to control optically than incandescent sources. Efficiently directing the CFL's output with a reflector requires a more sophisticated design, particularly in applications requiring high-angle cutoff for glare control.

Optical design is further complicated by the fact that the light emitted from the CFL's surface varies axially and radially (**Figure 8-21**). Several manufacturers are working on advanced reflector geometries which factor this phenomenon into the design. One manufacturer, ProLight, has recently introduced the Omni/Flood screw-in CFL reflector, which achieves a "fixture efficiency" of 72 percent for the lamp/reflector combination with no lens. A single lamp mounted in the vertical position represents the easiest reflector configuration to enhance optically due to its radial symmetry.

Figure 8-21

Luminous intensity distribution

The light emitted from CFLs varies along the axial and radial axes. This makes the design of effective reflectors for CFLs more challenging than reflectors for incandescent sources. This figure shows the luminous intensity of a 13-watt quad-lamp.

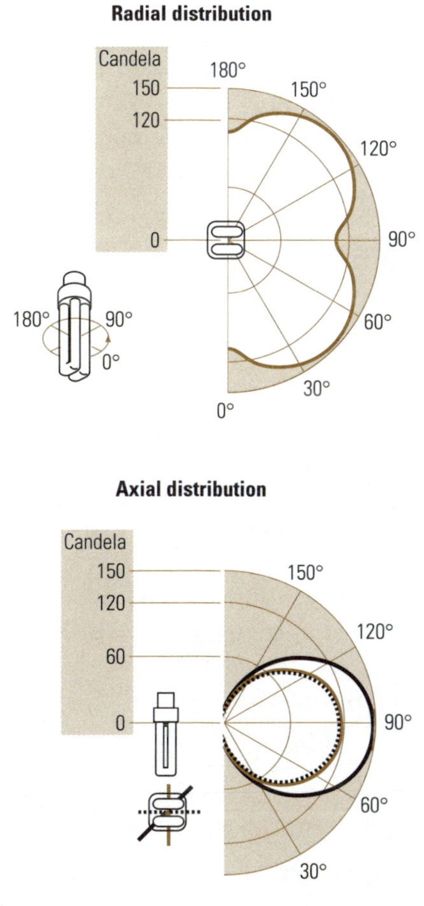

Source: Osram

Moreover, the source efficacy in a single-lamp downlight, as compared to a multilamp system of the same total wattage, will inherently be higher because the single lamp has a greater length of active discharge area per cathode. For example, it would be more efficient to specify a fixture with one vertically-mounted 26-watt quad tube than two 13-watt twin tubes if the desired candlepower distribution is available. (Horizontally mounted quads may be less efficient, however, than side-by-side twin tubes, because the four legs of the quad shade each other.) Achieving high light output and/or wide light distributions, however, often necessitates the use of multiple lamps and, in turn, more complex reflector designs. As many as three lamps are used in some compact fluorescent downlights; depending on fixture design, the lamps may be mounted in vertical, horizontal, or sloped orientations. When multiple lamps are used, however, some fixture efficiency is sacrificed due to occlusion from "lamp shadowing." In addition, the more complicated configuration of lamp socket assemblies typically results in higher optical losses.

■ *Reflectors.* To get as much light out of a fixture as possible, reflectance must be maximized and bounces per exit ray minimized. With specular reflectors, however, the greater the amount of light reflected downward, the darker the fixture itself appears when viewed from an oblique angle. In many applications, it may be esthetically desirable to be able to see where the light is coming from, rather than "dark holes" in the ceiling. At the same time, it is often desirable to balance brightness ratios on the ceiling. Excessive reflector luminance may result in glare, either directly or as reflected in computer screens. Therefore, reflectivity must be considered with regard to both performance and esthetics.

New reflector designs are continually being introduced to improve optical performance and glare control. Some manufacturers, including Lightolier and Kurt Versen, have developed double-reflector optical systems for their vertically mounted quad lamp fixtures to maximize performance (see **Figure 8-22**). While it is more expensive to use two separate reflectors than a single reflector, this design allows the upper reflector to be optimized to control the light distribution while the lower parabolic reflector is used to achieve glare control. Lenses or louvers are also commonly used where fixtures have large apertures and are mounted in high ceilings.

Figure 8-22

Dual reflector design

Vertical lamp, dual reflector designs are well-suited for delivering maximum candlepower in a narrow distribution for task lighting. The upper reflector gathers and concentrates lamp output to the desired spacing pattern. Lower cones, designed for glare comfort, can be clear or colored.

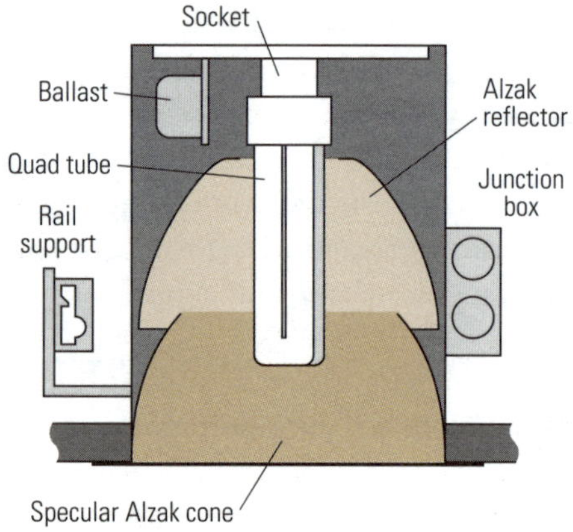

Source: Kurt Versen Company

To maximize reflector efficiencies in downlights using two horizontally mounted lamps, some manufacturers have incorporated a reflector contour at the top of the fixture that bends down between the lamps to maximize reflectance. Several manufacturers have developed fixture designs that rotate the long axis of the lamps approximately 45° to expose more of the lamp surface area to the reflector.

■ *Reflector finish.* Reflector finishes for CFL downlight products include coatings and chemical conversion processes, and are classified by the method of application and by the type of material applied. Anodized aluminum, created by an electrochemical conversion process, is the most commonly used reflector finish for CFL downlights. In the anodizing process, the aluminum is first cleaned and chemically treated. The aluminum substrate can be etched to produce a matte appearance or dipped to enhance reflectivity. Total reflectance can range from 60 to 90 percent.

When light strikes an anodized surface, some is reflected from the top surface of the anodic coating and some travels through the coating and is reflected by the aluminum substrate. Due to the difference in the optical paths these waves travel, they reach the eye at different times, producing an iridescent "rainbow" effect when triphosphor fluorescent sources are used. Aside from esthetic compromise, this iridescence can also cause glare. "Low iridescence" anodized reflectors are available but these are less reflective and more costly.

By comparison, most high-performance metallic coatings used for reflectors avoid iridescence, while achieving total reflectance of up to 96 percent. A number of companies offer coatings which involve the vacuum deposition of aluminum onto a metal or plastic base material. These reflectors absorb so little light that they appear dark from most viewing angles when operating, eliminating nearly all direct glare.

Reflector materials can be specified with a number of reflective characteristics and appearances to correspond to specific job requirements. Reflectors are typically offered in specular, semi-specular, and diffuse/matte finishes. A semi-specular reflector has a lower reflectance than specular material and exhibits more diffuse reflectance. These reflectors are offered in clear and colored finishes. Gold finishes are sometimes specified to "warm" the light output from the CFL downlights for esthetic reasons. Because a clear specular surface reflects more light in the blue part of the spectrum, it will not appear as warm. Note that colored finishes, however, typically result in reduced reflectance values and reduced light output. A similar effect can often be obtained without the light loss penalty by specifying a 2,700 K lamp color.

1 1982 and 1986 values from P.C.J. Gallagher, "The Revolution in Lighting Retrofits," *Proceedings Energy Technology Conference* (April 1987), Electronic Advanced Systems Corporation, West Conshohocken, PA. 1988 and 1990 data from "EPRI Survey and Forecast of Marketplace Supply and Demand for Energy Efficient Product: Phase 1 Report" (August 20, 1991). 1992 and 1993 data from U.S. Census. Worldwide data from Evan Mills & Nils Borg as cited in *Vital Signs 1994*, Worldwatch Institute, Washington DC, tel 202-452-1999.

2 *Advanced Lighting Guidelines*: 1993, p. 6-3 available from EPRI, report number TR-101022s, Revision 1 or from the California Energy Commission, 1516 Ninth Street, Sacramento, CA 95814-5512, tel 916-654-4287, e-mail energia@energy.ca.gov, web www.energy.ca.gov/energy.

3 "Compact Fluorescent Standards" (March 1993), p. 1, New England Electric System, Laura McNaughton, Manager of Residential Marketing, 25 Research Drive, Westborough, MA 01582-0099, tel 508-366-9011, fax 508-366-6551.

4 National Lighting Product Information Program, "Specifier Report on Screwbase Compact Fluorescent Lamp Products" (April 1993), Table 6, p. 25.

5 NLPIP [4], p. 6

6 NLPIP [4], p. 7.

7 Michael Siminovitch, personal communication (March–April 1993), Staff Scientist, Lawrence Berkeley Laboratory, 1 Cyclotron Road, Berkeley, CA 94720, tel 510-486-5863, fax 510-486-6940.

8 Siminovitch [7].

9 Anthony Serres, "A Method to Improve the Performance of Compact Fluorescent Lamps," 1992 IESNA Conference, Paper no. 42, Philips Lighting Co., Route 3, Box 505, Fairmont, WV 26554.

10 Serres [10].

11 Judy Jennings, "So Many Sockets, So Little Time," *Home Energy* (March/April 1993), p. 7.

12 Lindsay Audin, personal communication (February 1994), based on experience with hundreds of installations at Columbia University 1990–1993. One installation of 28-watt quad lamps has been operating continuously for more than 20,000 hours with less than a 1% burnout rate.

13 This discussion is condensed and updated from a longer treatment in the previous version of this book. See "Compact Fluorescent Lamp Technologies," E SOURCE *Lighting Technology Atlas*, Chapter 8 (1994), pp. 178–181.

14 National Electrical Manufacturers Association, "Interaction of Infrared Controls And Electronic Compact Fluorescent Lamps," National Electrical Manufacturers Association, 1300 North 17th Street, Suite 1847, Rosslyn, VA 22209, tel 703-841-3200, fax 703-841-3300.

15 Roy Sierleja, personal communication (February 1994), GE Lighting, tel 216-266-2074.

16 Hal Dohling, personal communication (September 1997), Lumatech, 5900 Christie Avenue, Emeryville, CA 94608, tel 510-654-4300.

17 Leslie Davis, personal communication (April 21, 1997), Prescolite, 1251 Doolittle Drive, San Leandro, CA 94577, tel 510-577-5374, fax 510-577-5022.

9 HIGH-INTENSITY DISCHARGE LIGHTING

9 HIGH-INTENSITY DISCHARGE LIGHTING

High-intensity discharge (HID) lighting sources are the primary alternative to high-wattage incandescent lamps wherever an intense point source of light is required. While HID lamps can provide high efficacy in a wide range of sizes, they have special requirements for startup time, restrike time, safety, and mounting position. Replacing incandescents with HIDs requires careful consideration of these issues, as well as the asymmetric output, lumen depreciation, and color shifting characteristics that many HID sources exhibit.

There are three basic types of high-intensity discharge lamps: mercury vapor (MV), metal halide (MH), and high pressure sodium (HPS). Although low-pressure sodium lamps are technically considered "miscellaneous discharge"—not HID—sources, we discuss their operation here as well. All these lamps usually require a separate ballast for starting and power conditioning. The 1992 EPACT legislation in the U.S. has minimal impact on HID lighting (see box on EPACT impacts, page 197).

HIGHLIGHTS

■ HID lamps can provide very high efficacy, offering energy savings of 50 to 90 percent when replacing existing incandescent sources. Efficiency upgrades within the HID family can provide smaller but significant savings of about 10 to 50 percent.

■ Mercury vapor lamps are approaching obsolescence as other HID sources (metal halide and high-pressure sodium) with better efficacy, color rendering, and lumen maintenance become available at lower cost, especially in larger lamp sizes. At the smaller end of the spectrum, mercury vapor lamps can be replaced with screw-in compact fluorescent lamps that do the same job for less than half the wattage and provide better color.

■ Small HID fixtures are appearing as the metal halide and high-pressure sodium families expand to include smaller reflector lamps, more lamps that can be safely used in open fixtures, and lower-wattage single-ended lamps.

■ As the price of color-corrected "white" high pressure sodium (HPS) sources drops, more fixture manufacturers will offer lower-cost equipment, creating greater demand for electronically ballasted versions.

■ If "instant-on, instant-restrike" technology filters down to lower-wattage lamps, HID lamps could replace incandescents in even more applications.

■ Electronic ballasts for HID lamps are an emerging technology. Unlike electronic ballasts for fluorescent lamps, energy performance is *not* greatly improved with these ballasts. The primary benefits of electronic ballasts are reduced size, weight, and noise, and better control of color and lumen maintenance.

HID sources—primarily mercury vapor lamps—became popular in the 1940s and 1950s for industrial applications because they lasted longer and had greater lumen output than incandescents. In the early 1960s, demands for better lumen maintenance, color rendering, and efficacy fostered development of metal halide and high-pressure sodium lamps. HID sources are now available in sizes of 35 to 2,000 watts from six primary manufacturers. More than 20 manufacturers offer HID fixtures for a variety of applications, from retail track lighting to industrial and sports lighting. **Figure 9-1** shows that HID light sources include the highest efficacies now available, while **Figure 9-2** shows that like other types of lamps, the best efficacies come from the largest sources.

Table 9-1 lists typical applications for HID lamps. HIDs operate well at temperatures down to -20°F, making them excellent for outdoor applications such as street lighting, billboards, and security illumination.

Figure 9-1

HID efficacy

HID light sources span a wide range of efficacies.

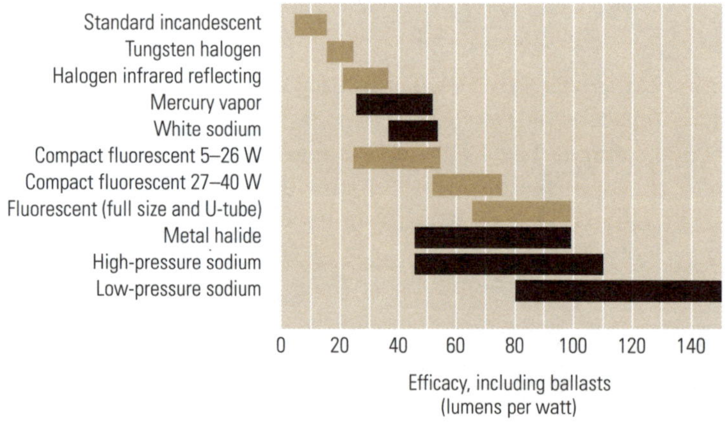

Figure 9-2

HID lamp efficacy vs size

Like other lamp types, HID sources are most efficient in larger sizes.

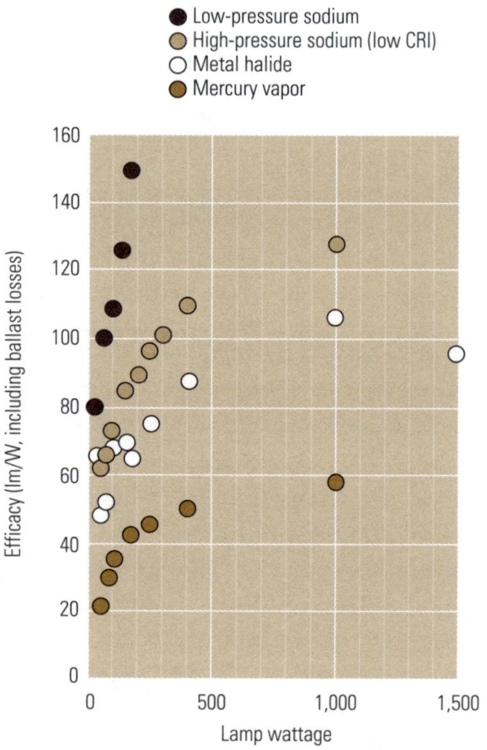

Source: Consolidated Edison Co. [14]

Table 9-1

HID light sources and applications

Excellent color properties and a wide range of sizes make metal halides the most versatile HID light sources.

Application	Metal halide	High-pressure sodium	White high-pressure sodium	Low-pressure sodium	Mercury vapor
Interior-decorative	●		●		
-downlights	●		●		●[a]
Parking areas	●	●		●	
General outdoor	●	●		●	
Roadway/tunnel	●	●		●	
Sports arenas	●				
High-bay spaces (hangars, warehouses, etc.)	●	●			●[a]
Low-bay spaces (supermarkets, light industrial shops, etc.)	●	●			
Outdoor signage	●				●[a]

Note: a: Where access is difficult or dangerous.

9.1 MERCURY VAPOR LAMPS

Mercury vapor lamps (**Figure 9-3**) operate by passing an electric current through vaporized mercury, similar to fluorescent lamp operation. As energized mercury atoms return to their normal condition, they radiate electromagnetic energy, primarily in the visible and ultraviolet (UV) ranges. The key difference: while fluorescent lamps contain mercury vapor at a pressure of only about 0.06 psi, mercury vapor lamps are pressurized at 30 to 60 psi. This higher pressure creates more visible light directly through mercury discharge. Mercury vapor lamps are available in sizes from 50 to 1,000 watts.

IMPACTS OF THE 1992 EPACT LEGISLATION ON HID LIGHTING

The National Energy Policy Act of 1992 did not create any specific performance targets for HID lamps. Instead, the U.S. Department of Energy was charged with making a determination of the potential costs and benefits of standards for HID lamp efficacy. That determination has not yet been made as of late 1997, but is expected sometime in the next year or two. Should that determination result in a rulemaking that requires performance standards for HID lamps, the DOE would take the following actions:

- Create lamp testing requirements.
- Create lamp efficiency standards based on the lamp tests.
- Create lamp labeling standards.
- Require compliance with the efficiency and labeling standards.

These actions would require about five years from the initiation of testing requirements to full compliance with standards, so EPACT will not affect the efficiency or availability of HID lamps until early in the next decade (if at all). The DOE is prioritizing the development of equipment efficiency standards by their potential nationwide energy impacts. Under this process, HID lamps are a relatively low priority compared to other devices such as clothes washers and water heaters.

Some mercury vapor lamps use phosphors to convert UV radiation to visible light, just as fluorescent lamps do. Others, however, use no phosphors, depending solely on the visible emissions from excited mercury atoms. The quality of such light is poor, and lamp efficacy is below that of most phosphor-coated lamps. Phosphor-free mercury lamps are usually identified in catalogs by the description "clear," and have lower lumen output than the "white" or "deluxe" phosphor-coated lamps of the same wattage. With or without phosphors, mercury vapor lamps emit more ultraviolet radiation than standard fluorescents.

Figure 9-3

Mercury vapor lamp

Mercury vapor lamps are similar to fluorescent lamps, but have lower efficacy.

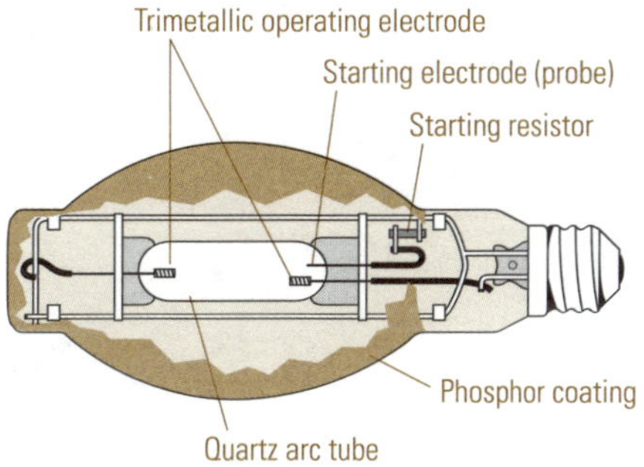

1. Mercury atom (Hg) absorbs impact from electron, elevating its energy level.
2. Mercury atom returns to its normal state, emitting a visible or UV photon.
3. If visible, the photon passes through the phosphor. UV photons strike and energize the phosphor on the lamp wall. Phosphor returns to its normal energy state, emitting visible photon.

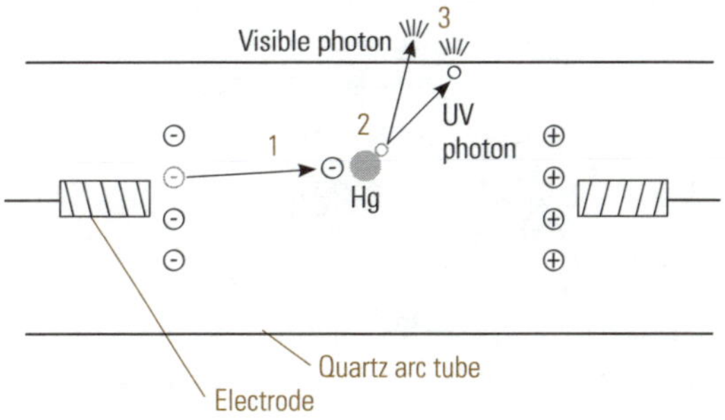

Long life, low initial cost, and color stability are the best features of mercury lamps. Their relatively low efficacy (30 to 65 lumens per watt, excluding ballast losses),[1] however, makes them a poor choice for energy efficiency. Perhaps the most important issue concerning mercury vapor lamps is how to best replace them with other types of HID or fluorescent sources that have better efficacy and color rendering. Even long life has its disadvantages: the light output of mercury vapor lamps drops 40 percent by the end of their rated life (see Figure 9-9) and as much as 60 percent when kept in service far beyond normal rated life, as often occurs without scheduled relamping.

HID TERMINOLOGY

ANSI ballast designation: Alphanumeric code indicating type of lamp served and circuitry employed.

ballast tap: Electrical contacts on ballast supplying voltage to an auxiliary device, such as standby lighting.

CWA: Abbreviation for constant wattage autotransformer, the preferred type of HID ballast.

encased and potted (E&P): Description of ballast designed to operate quietly and with uniform heat distribution.

getter: Flat plate in lamp coated with materials that absorb gaseous impurities; also sometimes called a "flag."

HX: Abbreviation for "high reactance" autotransformer.

maximum distance to lamp: Maximum tolerable length of wiring between lamp and igniter to avoid excessive voltage drop in starting pulse.

normal and high power factor (NPF and HPF): Normal power factor is typically 0.6 or lower, while high power factor is usually 0.9 or higher.

starter/igniter: Electronic device for providing high voltage pulse to ionize gas, initiating flow of current.

starting electrode: Electrode that transmits igniter pulse and is then disconnected.

The most important issue concerning mercury vapor lamps is how to get rid of them.

Most mercury vapor lamps are installed in high-bay spaces such as gymnasiums or outdoor applications such as roadway and security lighting. Mercury lamps make the most sense where relamping access is difficult, accurate color rendering is not required, and existing fixtures prevent easy conversion to other sources.

9.1.1 Self-ballasted mercury vapor lamps

Self-ballasted mercury vapor lamps do not strike an arc; instead, they use a continuous tungsten filament to vaporize mercury. These lamps are available in sizes from 160 to 750 watts, and can be directly substituted for incandescents in the same sockets, a favorite trick of facility managers seeking longer-lived lamps. A large portion of the light emitted from such lamps is really incandescent, however, so their efficacy (10 to 25 lumens per watt) is much lower than that of normal mercury vapor lamps. Like standard mercury lamps, their main advantage is longevity. Since better replacement options are now available (metal halides and large compact fluorescents), self-ballasted MV lamps are not suggested to those seeking energy savings.

9.2 METAL HALIDE LAMPS

Metal halide lamps are similar to mercury lamps, with an important improvement: the addition of iodides of metals such as thallium, indium, and sodium to the arc tube (**Figure 9-4**). These halides form the basis of a regenerative cycle that produces a higher quality and quantity of light than mercury stimulation alone can create. As the lamp reaches operating temperature, the metallic and iodine ions are separated in the electric arc, and the energized metal ions emit light. The gases cool as they reach the outer surface of the arc tube, and the ions recombine and flow by convection back into the arc, where they are again separated.

Together with the emissions from high-pressure mercury, these reactions produce high-quality light at very high efficacies. Were it not for several limitations, metal halide lamps might be considered the ideal light source. Here are some of those limitations:

■ Starting the lamps takes three to five minutes, and restarting after a shutdown or power interruption takes 10 to 20 minutes, as the enclosed gases must cool before the arc can be restruck. Some larger (over 400 watt) lamps are available with instant restrike capability, but they require special ballasts.

■ Metal halide lamps produce relatively high levels of UV radiation that must be controlled with shielding glass in the lamp or fixture (see **3.4.3** for more on UV radiation, and **9.4.5** for products that control ultraviolet output). Protection from possible explosion is also required.

■ The smallest metal halide sources are 32-watt packages producing about 2,500 lumens—nearly as much as a 4-foot fluorescent lamp. This high intensity limits their range of applications and requires the use of sophisticated fixtures. As with other types of lamps, smaller sizes have lower efficacy.

The smallest metal halide sources are 32-watt packages producing about 2,500 lumens—nearly as much as a 4-foot fluorescent lamp.

■ The color of metal halides' light output can shift, sometimes dramatically and randomly. A given batch of lamps can all turn different colors over time. Electronic ballasting and new lamp designs using ceramic arc tubes can limit this undesirable characteristic.

■ The light output of metal halide lamps is sensitive to lamp position. (Mercury-vapor and sodium lamps are not position sensitive.) Three types of lamps are available: BU/BD lamps are intended to be operated in the Base Up or Base Down (vertical) position, HOR or H lamps are designed to operate in the HORizontal position, and UNI or U lamps are designed for UNIversal operation and may be installed in any position.

When universal-burn metal halide lamps are operated in positions other than vertical, the arc bends upward, resulting in nonuniform heating of the arc tube wall. This effect reduces light output by as much as 20 percent, as well as reducing lamp efficiency and lumen maintenance. This operating characteristic should be considered in applications such as building facade and flagpole lighting.

■ Some lamps have asymmetric distribution patterns because of the tubular nature of the source.

Figure 9-4

Metal halide lamp

Metal halide lamps provide excellent color quality and efficacy, but have drawbacks that limit their application.

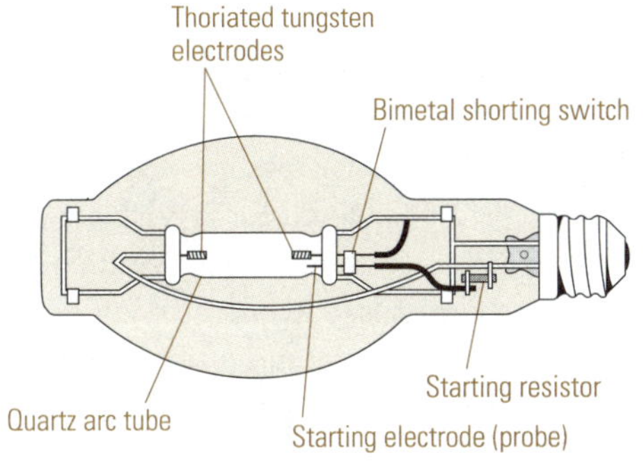

1. Electron impacts metal halide molecule, separating it into its constituent ions.
2. Energized metal ion emits visible photon.
3. Convective cooling at tube wall brings ions together at bottom of tube, where they recombine for another cycle.

(Mercury discharge (Figure 9-3) also occurs in metal halide lamps.)

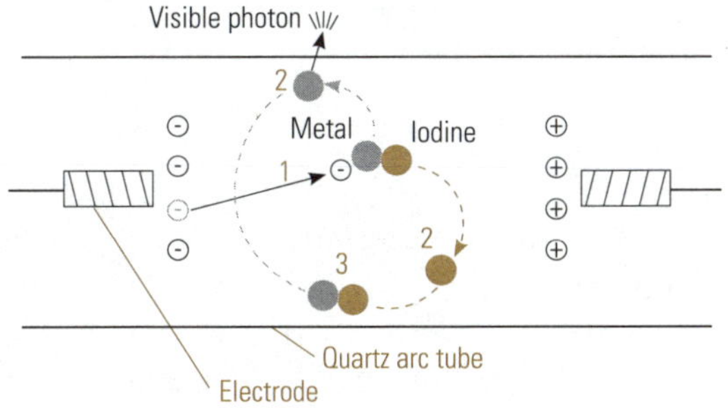

■ Because of their high brightness, metal halide lamps need carefully designed fixtures.

Even with these limitations, metal halide sources remain one of the most effective tools for energy efficient lighting.

Many traditional-style fixtures are available for metal halide lamps. Original Cast Lighting Company, for example, specializes in adapting large incandescent chandeliers to use metal halide lamps up to 1,000 watts, while SPI Lighting offers several bowl-shaped HID fixtures with handsome metal trim.[2] Many of these fixtures include an instant-start/restrike fluorescent lamp (not visible when the metal halide lamp is operating) which ensures at least some light output after a power outage or at startup.

Ceramic metal halide lamps are recent developments that offer improved performance.

Low-output metal halide lamps with reduced power consumption are available from Venture Lighting for use when a screw-in retrofit is needed and a slight drop in illumination is tolerable. Venture's Energy Master™ lamps save 10 to 14 percent of energy by dropping light output by 10 to 12 percent compared to standard metal halide lamps, while maintaining high CRI and rated life.[3]

Venture also offers the Energy Master Plus lamps and matching pulse start ballasts, which use a redesigned arc tube to produce energy savings of up to 20 percent and 50 percent longer lamp life. Other advantages include faster start and restrike times, reduced color shift and improved lumen maintenance. The 150, 200 and 350 watt Energy Master Plus lamps are designed to replace standard 175, 250 and 400 watt metal halide lamps of the older design, with no sacrifice in initial lumen output. These lamps are available in enclosed fixture rated and open fixture rated versions.[4]

Ceramic metal halide (CMH) lamps, sometimes called ceramic discharge metal halide lamps, are recent developments that offer improved performance. In contrast to the glass arc tube used in most metal halide lamps, these lamps use the same polycrystalline alumina (PCA) material used in HPS arc tubes. When operated on matching electronic ballasts, these lamps improve efficacy by 10 to 20 percent, maintain color temperature within 200K, and boost CRI to 80 or more. CMH lamps are available in the PAR 20, PAR 30, and PAR 38 shapes in several beam spreads and ranging from 35 to 150 watts.

The rated lifetime of CMH lamps (6,000 to 7,500 hours) is shorter than that for normal metal halide lamps (10,000 to 20,000 hours), and some CMH lamps have been reported to fail as early as 4,000 hours. One manufacturer, Osram Sylvania, states that the halide salts used to create better color rendering can prove corrosive to the seals at the ends of the ceramic arc tube, and that new seal formulations are being explored to extend the lifetime of the lamps.

Some users mistakenly believe that CMH lamps require a HPS ballast because they use the same arc tube material as HPS lamps. This is not so; CMH lamps should be operated with metal halide ballasts of the same ANSI designation as the lamp.

Specialty metal halide lamps can provide pink, aqua, green, or blue colored light without tinted lenses or filters through selective blending of arc tube chemicals.[5] Applications include special effects for facade and bridge lighting where a seasonal color change may be desirable.

Several manufacturers offer wall sconces using compact metal halide sources such as the 70-watt HQI lamp.[6] Electronic ballasts for these lamps (see Table 9-3) are small enough to fit inside the fixture, allowing easy installation for these high-efficacy alternatives to incandescent wall sconces. Metal halide wall sconces drawing 80 watts have successfully replaced 250-watt halogen fixtures, while providing equal or higher light levels (see Columbia University case study in **9.6**).[7]

Pulse-start systems are a recent development for metal halide lamps. Pulse-start ballasts use a four microsecond pulse of about four kV to ignite the lamp arc. Pulse-start lamps are optimized to work with this voltage pulse, and lack the traditional starting electrode (see Figure 9-4) that normal metal halide systems rely on to ignite the lamp. Pulse-start ballasts can ignite any metal halide lamp, but pulse-start lamps must be connected to pulse-start ballasts, or the lamps will not start. Manufacturers claim that pulse starting systems achieve:

- quicker, more reliable starting (even in cold weather);
- less electrode damage with each start, extending lamp life (with matched ballasts and lamps);
- better lumen maintenance; and
- faster hot restrike times (five to seven minutes, compared to 10 to 20 minutes for normal metal halide systems).

Pulse-start lamps and ballasts are being developed by Osram Sylvania, Venture Lighting, Delta Power Supply, and C.E.W. (marketing the Japanese Iwasaki brand).

9.3 Sodium Lamps

In sodium lamps, a high-frequency, high-voltage pulse ionizes xenon gas in an enclosed tube, which in turn vaporizes a sodium-mercury amalgam (**Figure 9-5**). An electric arc through this vapor causes electrons in the sodium atoms to emit visible light, mostly in the longer wavelengths between yellow and red. The gas mixture is sealed in a translucent polycrystalline aluminum cylinder that transmits 90 percent of the visible light created inside it.[8] Sodium lamps vary widely in their efficacy and color quality; performance is very sensitive to the gas pressure inside the arc cylinder (**Figure 9-6**).[9] The quality of light from some sodium sources may significantly reduce their effective efficacy—see **2.2.6** on "pupil-lumens."

9.3.1 High-pressure sodium lamps

The relatively high pressures in high-pressure sodium (HPS) lamps (approximately 13,300 Pa) create frequent molecular collisions of the

Figure 9-5

High-pressure sodium lamp

Sodium lamps use no phosphors and emit no UV radiation.

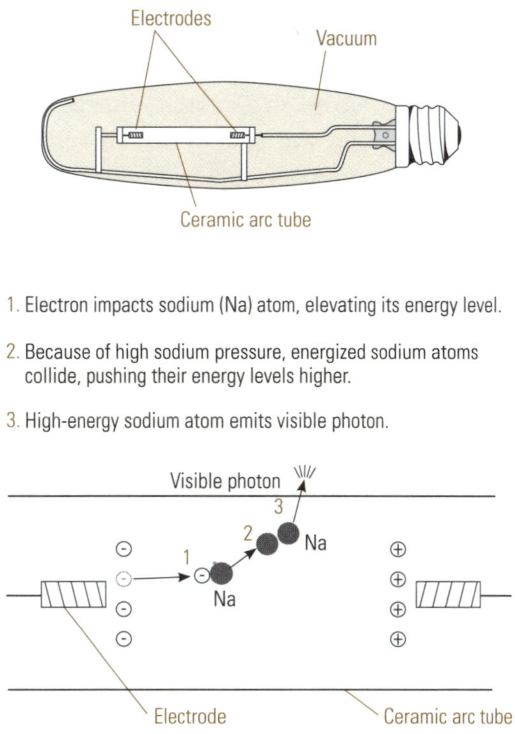

Figure 9-6

Luminous efficacy versus sodium pressure for sodium lamps

Sodium lamp performance is very sensitive to the gas pressure inside the arc cylinder.

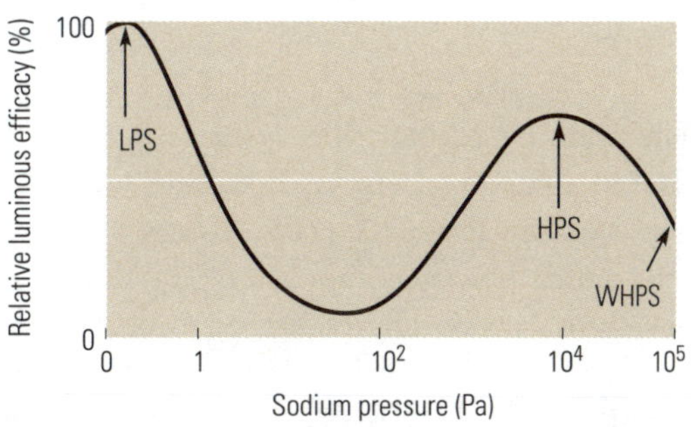

Key:
LPS: Low-pressure sodium
HPS: High-pressure sodium
WHPS: White high-pressure sodium

Source: Philips [9]

discharge gases, resulting in greater emissions of red and green light and higher CRI. (Lowering the pressure eventually narrows output to only two bands of yellow, characteristic of the low-pressure sodium lamp.) Sodium lamps use no phosphors, and their low mercury content creates very little UV output. Some HPS lamps are coated, however, to reduce glare and widen light distribution. All HPS lamps can be operated in any position.

There are three basic grades of HPS lamps with different color rendering indices: the lowest CRI is about 21 (typically used for outdoor lighting), general-purpose indoor units are around 60, and the newer "white" versions boost CRI to 80 or higher. Unfortunately, better color means lower efficacy, and vice versa. **Figure 9-7** plots efficacy versus CRI for several sodium lamps.

Figure 9-7

Color quality and efficacy for sodium lamps

"White" HPS lamps provide much better color rendering than standard sodium lamps, but at much lower efficiency.

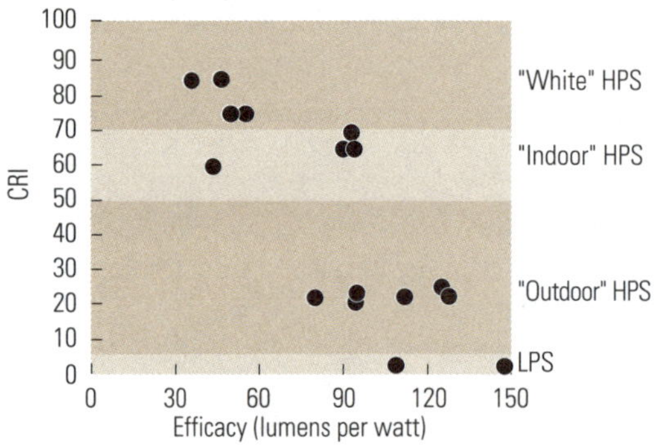

Source: manufacturer data

The primary use of HPS systems is for outdoor security, roadway, and parking lot lighting. Because low-CRI HPS sources are typically used, however, the quality of their yellow light can be unsatisfactory. Modern outdoor lighting systems are increasingly using metal halide sources because of the greater "seeability" of their light (as described in **2.2.6**). In some of these outdoor applications, intelligently applied metal halide lighting may be able to replace HPS lighting with a reduction in installed wattage.

Two new sizes of HPS lamps are now available to bridge the gap between 400 watts and 1,000 watts: a 600-watt, 92,000-lumen lamp rated at 18,000 hours, and a 750-watt, 110,000-lumen lamp, rated at 24,000+ hours. The latter size is being more widely adopted as a standard in this range.

9.3.2 White HPS lamps

The white high-pressure sodium (WHPS) lamps now available are a potentially useful tool for efficient and effective lighting. These lamps use extra-high sodium/mercury pressure in the arc tube to extract CRIs of 80 or higher. As Figure 9-7 shows, however, their efficacy is only 30 to 50 lumens per watt, and they are only available in smaller sizes (35 to 100 watts).

In addition, the halide salts in the arc tube deplete as the lamp ages, decreasing the CRI. For this reason the manufacturers recommend group relamping before end of rated life.

There are two white HPS systems: White Lucalux from GE Lighting, and White SON from Philips Lighting. These systems are available in different wattages and are not compatible with each other. The White SON lamps are available in 35-, 50-, and 100-watt sizes and two base configurations. The White Lucalux lamp comes in one size: a 95-watt T10 envelope with a medium screwbase. Each system uses a large hybrid magnetic ballast with an electronic regulator. The color temperature (CCT) of these lamps is in the "warm" range—from 2,600 to 2,800. One popular application of WHPS systems is for retail merchandise lighting.

9.3.3 Low-pressure sodium lamps

The applications of low-pressure sodium (LPS) lighting are extremely limited by the nearly monochromatic yellow light they produce—two spectral spikes with wavelengths of 589.0 and 589.6 nanometers. One accepted use of these lamps is for streetlighting in areas surrounding observatories, as astronomical equipment can more easily compensate for their "light pollution" than it can with fullerspectrum sources. But since it makes identification of cars, license plates, and other colored items very difficult, many jurisdictions do not allow LPS streetlighting.

The published efficacy of LPS sources appears very high—as much as 180 or more lumens per watt, excluding ballast losses.[10] This would seem to make LPS the most efficient light source available. Recent studies, however, have shown that because of the way eyes register light, the effective illumination power of yellow LPS lamps is much lower than their rated lumen output would indicate. After applying a correction factor of 0.38, the efficacy of a 180-lumens-per-watt lamp declines to 68 "pupil-lumens per watt," where other sources with better color rendering easily surpass it.[11] The effect of this derating is that more lumens (from a brighter lamp) are required of an LPS source to perform many visual tasks than with other sources. Section **2.2.6** discusses this phenomenon in greater detail.

9.4 HID LAMP ISSUES

The complex nature of HID light sources creates several issues that must be considered before specifying or installing them. In particular, conversions from incandescent sources require careful planning to avoid insidious or dramatic failures of HID equipment.

9.4.1 Lamp life

HID lamps have long lifetimes—as much as 40,000 hours (**Figure 9-8**). In general, lower-wattage lamps have shorter lifetimes—sometimes less than 5,000 hours—than those of higher wattage. The average rated life of HID lamps is based on 10 hours of operation per start (unlike fluorescent lamps, which are rated at three hours per start).

Standard HPS lamps normally cycle on and off when they reach end of life. If the lamp is not replaced, the ballast's starter can have its life reduced or it may be destroyed if neglected too long. Double arc-tube lamps can be used to extend the life from 24,000+ hours to 40,000 hours. With these lamps, cycling will not occur until the second arc tube reaches the end of its life. (Osram Sylvania's Lumalux Plus HPS lamps eliminate this cycling by shutting off the current at the end of the lamp's life.)

The effective illumination power of yellow LPS lamps is much lower than their rated lumen output would indicate.

Some MV lamps may continue to burn well past their rated life at a greatly depreciated lumen output. This characteristic may lead to delayed relamping and lighting complaints, even though the system was initially designed and lamped properly. Enforcement of routine group relamping is thus crucial with MV lamps (where they are still used at all). Lumen depreciation over time is different for each lamp type (**Figure 9-**

Figure 9-8

Lifetimes of HID lamps

HID lamps, especially mercury vapor lamps, have some of the longest lifetimes of any light source. Incandescent and fluorescent lifetimes are also shown for comparison.

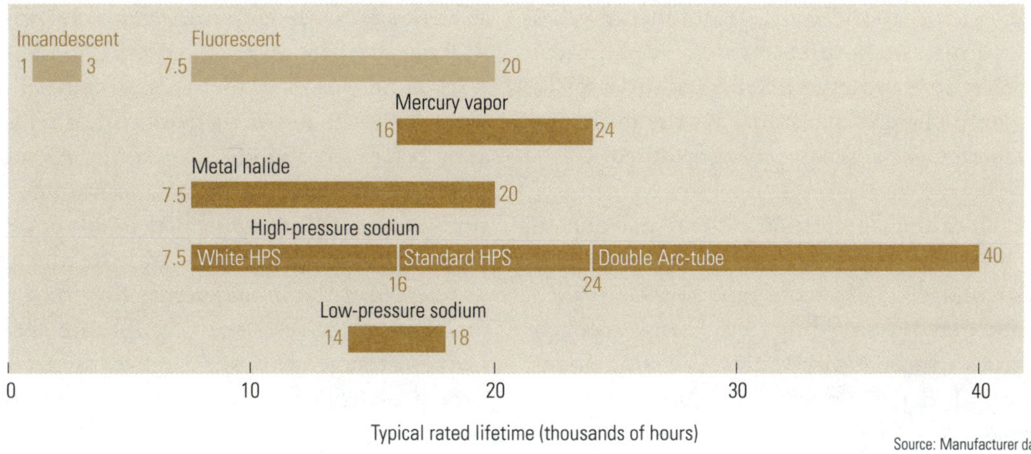

Source: Manufacturer data

Figure 9-9

Lamp lumen depreciation of HID lamps

Like most other lamp types, HID lamps lose light output over their lifetime.

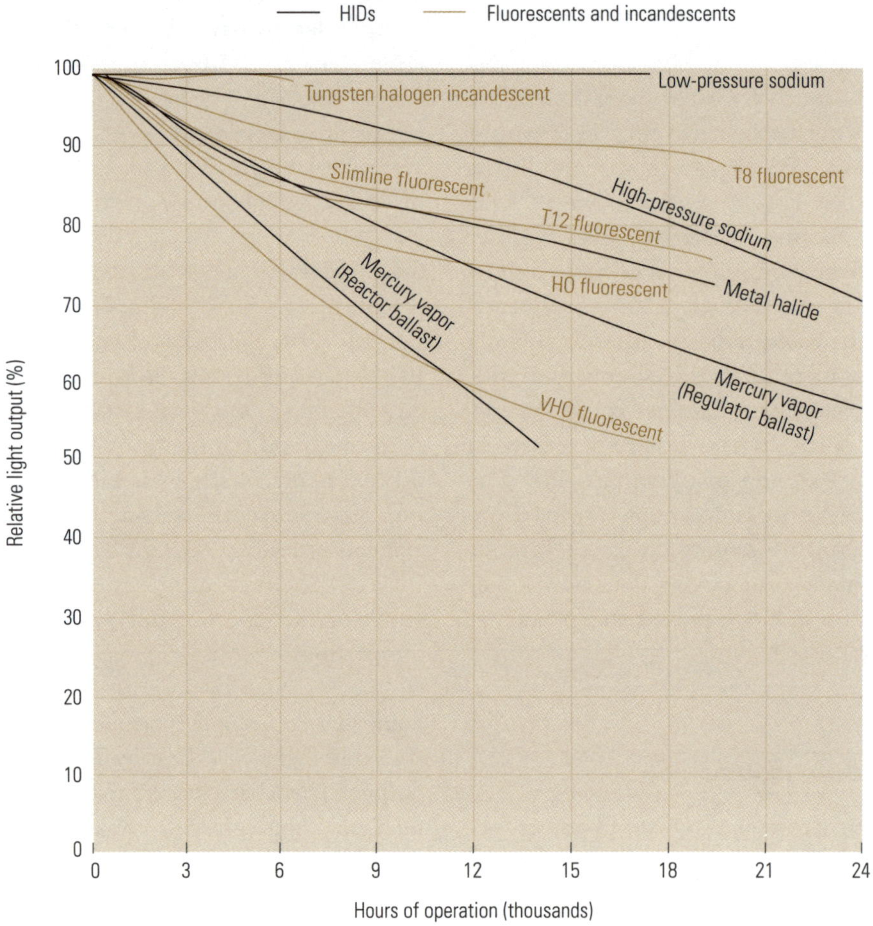

Source: National Lighting Bureau

Because HID sources are typically used in high-ceiling applications, maintenance costs can significantly affect system economics.

9). Group replacement of metal halide lamps at 70 percent of rated life is recommended to minimize problems with color shifting.

Because HID sources are commonly used in high-ceiling applications, maintenance costs can significantly affect system economics. Labor costs can range from $5 to over $100 per lamp change, depending on the difficulty of replacement and equipment required.

In high-ceiling applications, it may make sense to choose an open, lensless fixture, even though this requires use of a more expensive lamp with an integral UV-shielding lens. Lamps in open fixtures can be easily replaced with a lamp clamping device on a pole, greatly reducing change-out time compared to using a ladder to open a lensed fixture. Fixtures are also available with an internal winch that lowers the unit for easy maintenance.

Both LPS and MV lamps increase their wattage as they age, with some MV lamps increasing wattage draw by up to 25 percent. On a 400-watt lamp, for example, this amounts to an extra 100-watt loss if the lamp is not replaced at or before the end of its rated life. As shown in Figure 9-9, such an increase in wattage over time is typically accompanied by a noticeable drop in light output. While drawing much less wattage than the incandescents they may have replaced, such losses may complicate shared-savings arrangements and slightly negate over time a utility's expected demand reduction.

Once again, prompt (or early) group relamping will avoid such situations.

9.4.2 Color shifting

Standard metal halide and HPS lamps change color over time, most noticeably after about 70 percent of the rated lamp life. Spot relamping makes this phenomenon stand out, since lamps of many ages may be grouped near each other, highlighting their differences in color. The following steps can help deal with this problem:

- Group relamp at 70 percent of rated life.
- For critical color applications, use electronic ballasts that control lamp current to maintain nearly constant color.
- Use phosphor-coated MH and MV lamps, which have greater color stability than clear lamps.
- Use ceramic arc tube metal halide lamps, which have improved color stability.

The problem may be further reduced by using color-stable light sources (such as fluorescents) in the same area, or by combining color-stable direct lighting with HID indirect fixtures that reflect light off surfaces not in the immediate view of occupants. Some designers also position multiple lamps so that their light is mixed before reaching the target surface.

9.4.3 Dimming

Dimming options for HID lighting are limited. Color shift (even on new lamps) becomes apparent below about 60 percent of light output for MH and HPS, and below about 40 percent for MV, seriously limiting this capability in areas where color is of concern (but not, perhaps, in warehousing). Some lamp makers may also void their warranties because of the potential for a poorly designed dimming system to stress their lamps. To maintain lamp life, good HID dimming systems always start the lamp at full output and then automatically dim it to the set level. Dimming is discussed in greater detail in **CHAPTER 11**.

As metal halide lamps are dimmed, their output begins to resemble that of mercury vapor lamps.

Despite these problems, step dimming (usually down to 50 percent) with occupancy sensors is becoming popular in warehouse applications and similar applications where occupancy is irregular and major energy savings are attainable. Step-dimming requires switching a capacitor in or out of the lamp circuit. Dimmed MH lamps can double their light output within the first second of being returned to full output.[12]

As metal halide lamps are dimmed, their output begins to resemble that of mercury vapor lamps. Because of this, phosphor-coated metal halide lamps are a better choice for dimming applications, since the phosphor converts the increased UV into visible light. **Figure 9-10** shows the difference in dimming performance between coated and clear MH lamps.

9.4.4 Starting and restarting

As shown in **Figure 9-11**, HID lamps require a warm-up period to fully heat the compounds in the discharge tube. In general, mercury lamps take the most time to start and sodium lamps the least. This characteristic raises special considerations:

- Do occupants need light as soon as they enter the room?
- Will the lights be turned off during the normal workday, and will the delayed restrike time be acceptable?
- What happens during a blackout or brownout?

There are several potential solutions to the delay problem. Instant-restrike lamps (such as fluorescent) can be used in the same space, or an instant-restrike standby light source can be added to each HID fixture (typically a small halogen lamp). Instant-restrike HID lamps are also available, but are limited to large sizes (250 watts and over).

Figure 9-10

Color temperature and CRI shift with dimming

Clear metal halide lamps can experience a shift in color temperature of over 1,000 K and a drop of 35 percent in CRI when dimmed to 50 percent of rated output. Phosphor-coated lamps are much less vulnerable to this color shift.

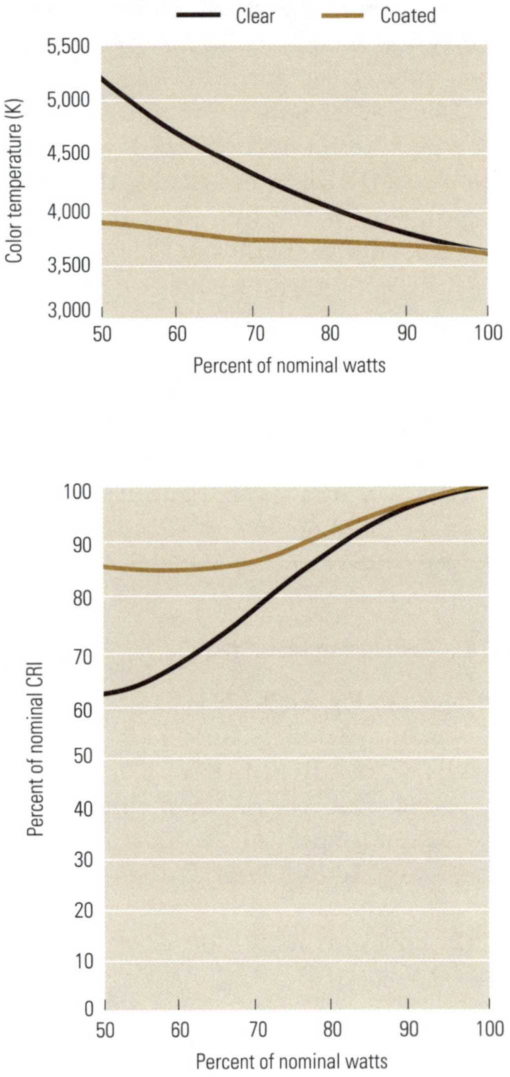

Source: Philips Lighting

Figure 9-11

Starting and restrike times

HID lamps must heat up to full output and cool down before restriking.

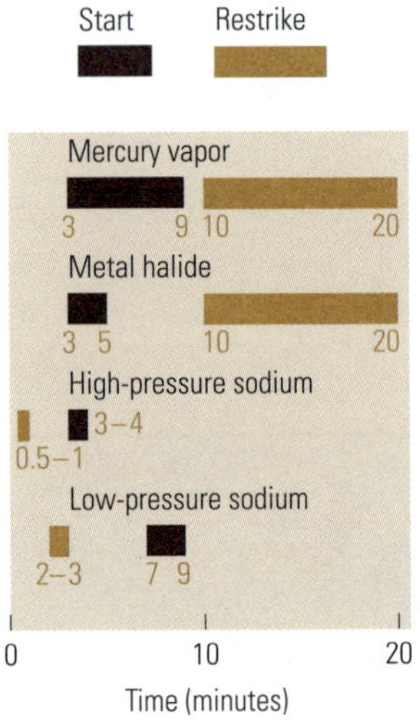

Using instant-restrike lighting can create special ballast requirements. Standby lighting (for use during warmup and restrike periods) usually requires 120-volt supply, but many HID fixtures run on 277-volt power. If standby lighting is needed as part of a fixture retrofit, the new ballasts must have a tap that provides 120-volt power or a separate 120-volt circuit is required. Many HID ballasts do not include this feature.

9.4.5 Safety requirements

Ultraviolet emissions. Metal halide and mercury vapor lamps emit a great deal of UV (mostly in the longer, less biologically harmful, wavelengths) which can gradually damage sensitive fabrics, papers, and artifacts. Upgrading incandescent lighting to MH in a museum, rare books library, or retail area selling sensitive materials (such as silks) may require special filtering to avoid material damage.

Heat-resistant flat glass UV filters are available that cut out almost all UV, but they are not standard on most fixtures. Globes and sconces may present a problem because their glass does not filter much UV, and most flexible UV filters lose their UV-abating capacity at temperatures common to MH and MV fixtures. In such cases, white HPS sources are a better choice because their extremely low UV output requires no filtering. Ultraviolet light and its impacts are discussed in **3.4.3**.

Fixtures with polycarbonate lenses (often found in parking garages and other outdoor installations) will gradually yellow from UV exposure, making the lens ugly and reducing the light output of the fixture. Open-fixture-rated lamps that control UV output are now available for nearly all sizes of metal halide lamps. Venture Lighting offers UV Guard metal halide lamps that incorporate a UV-absorbing quartz shroud around the arc tube to extend the life of fixtures with polycarbonate lenses. UV Guard lamps from 50 to 150 watts are available now, and 200-, 350-, and 400-watt lamps are expected by late 1997.[13]

Explosion protection. Metal halide lamps can "extinguish nonpassively"—in other words, explode—sending out bits of hot glass onto occupants and merchandise. To reduce the chances of such an event, metal halide lamps should be operated in their proper orientations, shut off for 15 minutes at least once a week (the power surge of restarting will cause weak lamps to fail, presumably under controlled circumstances), changed at 70 percent of rated life, and enclosed in fixtures with tempered glass lenses that filter UV and shut off the fixture if opened. Protected lamps are designed to protect against the nonpassive end-of-life explosion by providing either a glass shroud around the arc tube or by automatically shutting off the lamp current if the outer glass envelope is broken.

Mercury vapor lamps do not explode, but may continue to operate when their outer glass bulbs are broken, causing hazardous UV emissions. Such lamps should be contained in fixtures that filter UV. Self-extinguishing mercury lamps open the lamp circuit to shut off the lamp when the outer glass is broken.

High-pressure sodium lamps do not explode or emit high levels of UV under normal operation. It should be noted, however, that using an HID lamp with a wattage much lower than its ballast rating can cause almost any lamp to explode.

Many fixtures designed for HPS and MV lamps are open (that is, have no lens) since neither type can explode. While some MH lamps will fit in these fixtures, they create a hazard unless the fixture is also retrofitted with a tempered glass lens. This important safety step is frequently overlooked or avoided because of its extra expense and the lack of safety lens attachment points on the fixture.

Metal halide fixtures should include a tempered glass lens—an important safety step that is frequently overlooked.

9.4.6 Relamping and inventory control

Stocking and inventory control can be a problem when personnel and vendors do not fully appreciate the sometimes subtle differences between two lamps of the same wattage and type. While the 50-watt Iwasaki Daylux and White SON lamps are both high-CRI HPS lamps, for example, they are not interchangeable: the bases are different and the ballasts they use have different power characteristics. Similarly, a 50-watt Ceramalux and a 50-watt White SON—both HPS and both made by Philips—have different CRIs, use different ballasts, and have different bases. As a result, precise specification of replacement lamps is very important when using HID sources.

Metal halide lamps come in three mounting orientations: horizontal, vertical, and universal. Using horizontal lamps in vertical applications (and vice versa) will reduce light output and lifetime. To minimize inventory and simplify maintenance, it is helpful to use only one type of orientation in a given facility. The universal lamps can be used in any orientation, but they have slightly poorer performance than the specifically oriented lamps.

9.4.7 Lamp and fixture photometrics

While most lamps emit light relatively symmetrically around their own axes, or those of the fixtures containing them, many double-ended HID lamps have highly asymmetric light output. Examination of the output of low-wattage double-ended metal halides, for example, shows that in some fixtures they create a dumbbell-shaped pattern, versus a circular spot for symmetrical lamps. The same problem is apparent in the new White SON PAR lamps. This characteristic can only be seen in advance by comparing the candela distributions in two perpendicular planes, an effort not always made by lighting designers or engineers. **Figure 9-12** shows typical photometrics from a double-ended source.

Installing a fluted glass lens (with the flutes oriented perpendicular to the lamp axis) can make light output more symmetrical. Doing so can spread light too far if the flutes are too large; some experimentation may be needed to attain acceptable results.

9.5 HID BALLASTS

Like other discharge light sources, HID lamps require a ballast to limit current to the electrodes. Other ballast functions include voltage control for starting and restarting, and adjustments in voltage and current to maintain light color and intensity over time. All ballasts suffer internal losses—ranging from 5 to 90 percent of the lamp wattage—that should be included in any calculation of potential savings, especially when switching from nonballasted sources such as incandescents. There are three common types of magnetic HID ballasts, in order of increasing sophistication and cost:

- *Reactor ballasts* are small, cheap, and simple, consisting primarily of an inductor coil. These ballasts have low internal losses, but cause more rapid lumen depreciation than other ballast types. They create high inrush currents when starting, have low power factor (unless corrected by capacitors), and can cause lamps to flicker or shut off if voltage varies. While reducing the first cost of a fixture, they may require upgraded electrical services because of their low power factor.

- *High-reactance autotransformers* are similar to reactors but are capable of boosting line voltage when it is insufficient to start a lamp. They cost more and have higher internal losses than reactor ballasts.

Figure 9-12

Double-ended HID photometrics

Many HID lamp-fixture combinations have very different light distributions in perpendicular planes. These track fixtures use 70-watt metal halide lamps.

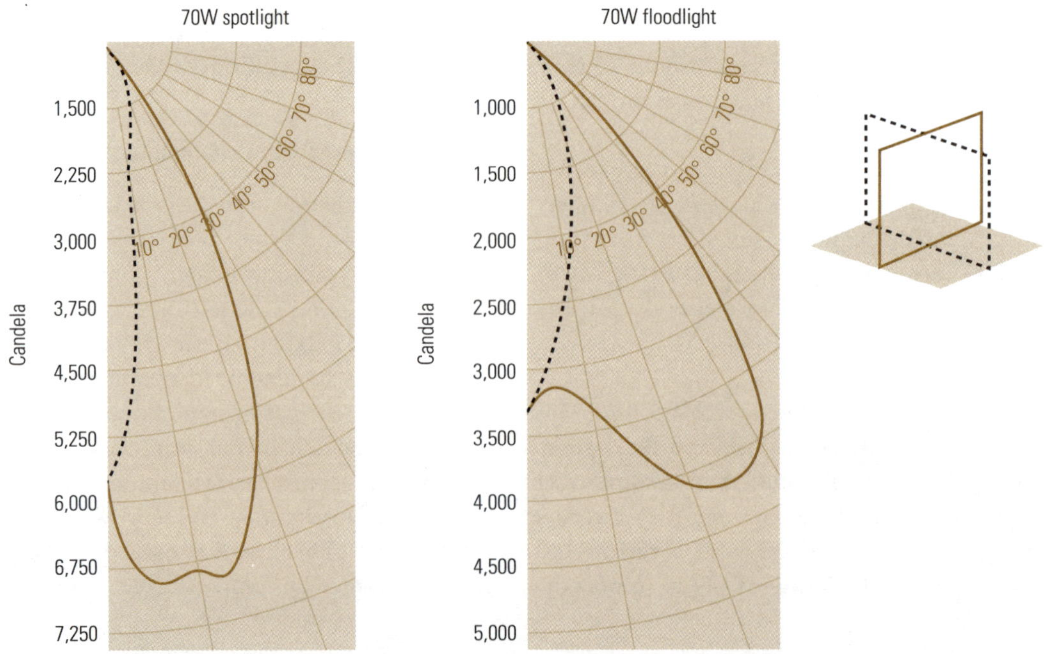

Source: Staff Inc.

■ *Constant-wattage autotransformers,* now the most common type in use, regulate lamp power better than the other ballast types, avoiding flicker and inadvertent shutoffs when line voltage varies, as is common in industrial installations. A capacitor is routinely included in the ballast circuit, boosting power factor to 90 percent or more.

Most HID ballasts use magnetic coils to regulate current. Magnetic ballasts are quite bulky and heavy, and suffer internal losses of 10 percent of lamp wattage or higher—much higher, for low-wattage sources. Total harmonic distortion (THD) is typically below 30 percent, and a noticeable hum is common unless ballasts are encased and well potted. **Figure 9-13** shows ballast losses against lamp wattage for all three types of HID sources, plotted from data in **Table 9-2.**[14]

Electronic ballasts. Electronic ballasts are available for operating metal halide and HPS lamps as large as 400 watts, but mostly for lamps 100 watts or smaller. These ballasts use switching electronics rather than magnetic transformers to control current and voltage to the lamp. Although electronic HID ballasts are usually more expensive than magnetic ballasts, their potential advantages include:

- improved color and light output stability
- longer lamp life and lower lumen depreciation
- the elimination of flicker
- the ability to keep the lamp lighted during line voltage fluctuations
- small size and light weight that allow them to be built into smaller fixtures
- quiet operation
- high power factor and low THD

Figure 9-13

Ballast losses vs lamp wattage

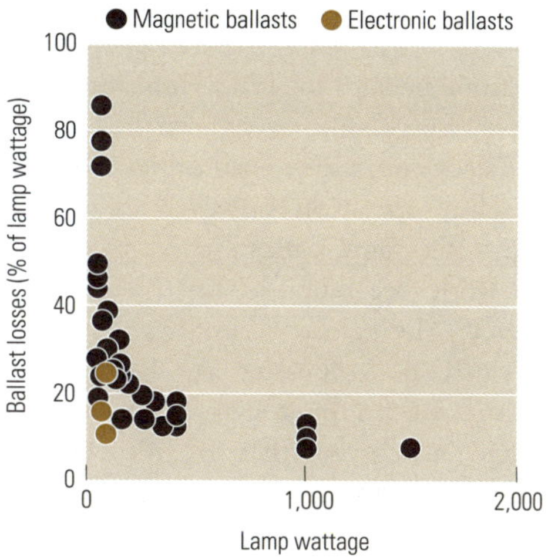

Source: Consolidated Edison Co. [14]

Table 9-2

Typical ballast power consumption for HID lamps

	Lamp wattage	Total wattage	Ballast wattage	% ballast loss
Mercury vapor	40	74	34	85
	50	74	24	48
	75	93	18	24
	100	125	25	25
	175	200	25	14
	250	285	35	14
	400	455	55	14
	1,000	1,000	75	8
Metal halide	32	38	6	19
	50	72	22	44
	70	93	25	36
	100	125	25	25
	150	185	35	23
	175	215	40	23
	250	295	45	18
	360	405	45	13
	400	455	55	14
	1,000	1,080	80	8
	1,500	1,610	110	7
High-pressure sodium	30	53	23	77
	50	64	14	28
	70	95	25	36
	100	130	30	30
	150	188	38	25
	200	245	45	23
	250	295	45	18
	310	365	55	18
	400	465	65	16
	1,000	1,100	100	10
Low-pressure sodium	35	60	25	71
	55	80	25	45
	90	125	35	39
	135	178	43	32
	180	220	40	22

Source: Consolidated Edison Co. [14]

Electronic HID ballasts are more expensive than magnetic ballasts, but are much smaller, lighter, and quieter, and may cost less to install.

Table 9-3 shows the characteristics of electronic ballasts currently on the market for HID lighting.

As Figure 9-13 shows, the internal losses of electronic ballasts for HID lamps are relatively low. Users should *not* expect the same boost in efficacy that electronics achieve in fluorescent lamp operation, however. In fluorescent lamps, electronic ballasts drive the lamps at very high frequencies, thus boosting lamp efficacy. The primary benefits of electronic HID ballasts are smaller size, lighter weight, and the more sophisticated control of lamp voltage and current that provide better color maintenance and lumen output through the lamp's lifetime.

Ballast compatibility. With the exceptions noted below, HID lamps and ballasts should have matching ANSI designations. For example, a 400-watt metal halide lamp that says "M59" on it should be matched with an M59 ballast. Luminaire manufacturers normally label their fixtures with the ANSI designation of the ballast so re-lamping personnel can be sure they are installing the correct lamp.

Many HID lamps will successfully operate on each others' ballasts. As a general rule, lamps of comparable wattage can be used in ballasts made for other types of lamps. Before making any such switch, however, check with the manufacturers of both the ballast and the lamps, or test the new arrangement with a sin-

Table 9-3

Electronic ballasts for HID lamps

Manufacturer	Lamp type	ANSI ballast type	~Dimensions (L x W x H) (inches)	Weight (pounds)	Total wattage (including lamp) (W)	Voltage	Power factor	THD (%)	Approx. cost ($)[b]
Delta Power Supply[c]	70W MH	M98	7x3x2	1.1	73	120–277	0.99	8–12	n/a
	150W MH	M107	7x3x2	2.1	160	120–277	0.99	7–9	n/a
	200W MH	M58	7x3x2	2.2	210	120–277	0.99	7–9	n/a
	350W MH	M131	8x7x5	6.0	362	120–277	0.99	7–9	n/a
	400W MH	M59	8x7x5	4.8	413	120–277	0.96	7–9	n/a
GE Lighting	95W HPS Lucolux	n/a	8.3 x 5.3 x 2.6	4	118	108–132	>0.90	n/a	100
Lights of America	50W HPS std. CRI	S68	2.9 diam. 4.6 length	0.6	~50[a]	120	0.65	120	40
Reliable Ballast Inc.	35W MH	M130	5x4x1.2	0.9	46	120/277	0.99		60–90
	50W MH	M110	5x4x1.2	0.9	58	120/277	0.99	<5 (120V) <20 (277V)	60–90
	70W MH	M85/M98	5x4x1.2	0.9	83	120/277	0.99		60–90
	100W MH	M91/M90	5x4x1.2	0.9	109	120/277	0.99		60–90
WPI Electronics	35W MH	M130	5x4x1.2	1.5	45	120/277	0.95	<29	65
	50W MH	M110	5x4x1.2	1.5	58	120/277	0.95	<29	65
	70W MH	M85/M98	5x4x1.2	1.5	88/82	120/277	0.95	<29	65
	100W MH	M91/M90	5x4x1.4	1.9	116	120/277	0.95	<29	70

Notes: a: But volt-amps are higher because of poor power factor.
b: estimated costs to OEMs/distributors.
c: Delta ballasts are dimmable down to 30 percent of full output. Ballasts listed have high output; standard-output ballasts are also available.
n/a = not available.

Source: Manufacturer data

gle lamp. The retrofit examples in Table 9-4 show several possibilities for direct switches between HID lamps. Major lamp manufacturers also produce HPS lamps that are designed to replace specific mercury vapor lamps without changing ballasts.

This cross-compatibility can potentially make retrofits much more economical by eliminating the need to buy new ballasts. Care is needed when making such substitutions, however. Since a more efficacious lamp of the same approximate wattage must be installed, the result will be to increase light levels while maintaining power consumption. The only way to save energy with this type of retrofit is to light the same area with fewer fixtures; this in turn requires a wider distribution from the remaining fixtures to be effective. Some of the retrofit products discussed in **9.6** are designed specifically to do this.

9.5.1 Ballast location

Special care is needed to locate magnetic HID ballasts when they cannot be enclosed within a fixture. This is not a concern in industrial facilities where esthetics are unimportant, or where fixtures are enclosed in a suspended ceiling, but it can be a serious problem for wall sconces and pendant-mounted globes. Some magnetic ballasts are so large (up to 18 inches long) that excavation into a wall or column is necessary to hide them. Alternatively, they can be remotely wired to the fixture, but there are limits on this option, including the following:

- Power wiring and its enclosure (for example, conduit or wiremold) must be rated for the high-voltage pulse involved in starting the lamp.

- With some ballast-lamp combinations, the distance between the igniter/ballast and lamp is limited to only a few feet; the 150-watt MH HQI lamp, for example, must be within 10 feet of its starter, while igniters for the White SON lamp must be within 3 to 5 feet (depending on the brand of the ballast).[15]

- Any remote enclosure must be NEMA-rated for the heat output and wattage of the ballast, and any decorative enclosure surrounding it should be fire-resistant.

- A remote location must be accessible for ballast maintenance/replacement, usually with a removable metal plate.

9.5.2 Igniters

All high-pressure sodium and lower-wattage (150 watts or smaller) metal halide lamps require a starter, or igniter, to initiate operation. Ballasts that include igniter circuits should also have circuitry that disconnects the igniter when the lamp has failed. Here's why. The igniter sends a pulse into the lamp to heat and ionize the discharge gases, allowing an arc to form. The igniter normally tries to restart the lamp every few seconds until the arc is established. Trying to start a failed lamp will eventually destroy the igniter, and possibly the ballast if the igniter is built into it. While the destruction time varies from a few days to a few weeks, it is possible to damage a ballast just enough so that it starts the next replacement lamp too harshly, burning it out in only a few weeks or months. Group relamping and/or quick replacement of burned-out lamps will reduce this possibility, but specifying a ballast with an automatic starter cutoff will eliminate the problem at a relatively low cost. Electronic HID ballasts normally include cutoff circuitry, and fixtures using the White SON lamp normally employ an electronic starter with automatic cutoff.

At least one lamp manufacturer (Iwasaki, represented in the U.S. by C.E.W. Lighting, Inc.)[16] sells a series of 21-CRI HPS lamps that operate without igniters. While more expensive, they provide the same lumens and fit the same fixtures as standard HPS. This feature makes them especially applicable to highway and parking lot lighting, and any other application where maintenance is costly (such as high-bay locations).

9.6 HID RETROFITS

There are several possibilities for retrofitting to efficient HID sources. The key to HID retrofits is to carefully consider the light distribution of the new fixtures, as the new lamps will usually have much higher light output.

The key to HID retrofits is to carefully consider the light distribution of the new fixtures, as the new lamps will usually have much higher light output.

Incandescent-to-HID retrofits

- Products are available to screw directly into the existing medium edison base. These retrofit "kits" include a new ballast, new reflector, and medium edison base into which a low-wattage HPS lamp can be fitted. Retrofits of this kind will either reduce wattage—such as converting a 500-watt incandescent to a 125-watt HPS—or increase light output—such as converting a 500-watt incandescent to a 400-watt HPS lamp. In the latter case, the number of fixtures can be reduced if the retrofit reflectors do a good job of spreading out the light. Check the power quality of the ballasts in screw-in kits; products have appeared on the market with power factors as low as 0.35 and THD of up to 135 percent.

- A new fixture can be installed, like the one shown[17] in **Figure 9-14**. These fixtures can spread out the much higher light output of the more efficient HID lamp.

HID-to-better-HID retrofits

- HPS lamps are available that can run directly on mercury vapor ballasts of similar wattage. This makes for an easy screw-in retrofit, but again, light distribution must be considered, as the new HPS lamps will put out more light.

- Other same-ballast retrofits are possible with lamps of similar wattage, including upgrades from MV to MH lamps (**Table 9-4**). Check with ballast and lamp manufacturers and conduct small-scale testing before implementing these retrofits on a large scale.

- New fixtures (like the one shown in Figure 9-14) can be installed, complete with new ballasts. Replacing the ballast provides more control over the wattage and light output of the new source. Ballasts-and-lamp-only retrofits are also possible, but may invalidate the UL listing of the old fixture.

Figure 9-14

Wide-spread HID fixture

Fixtures like this one, which sends light out at an angle, can be used to light an area with fewer and brighter sources than fixtures that direct light downward.

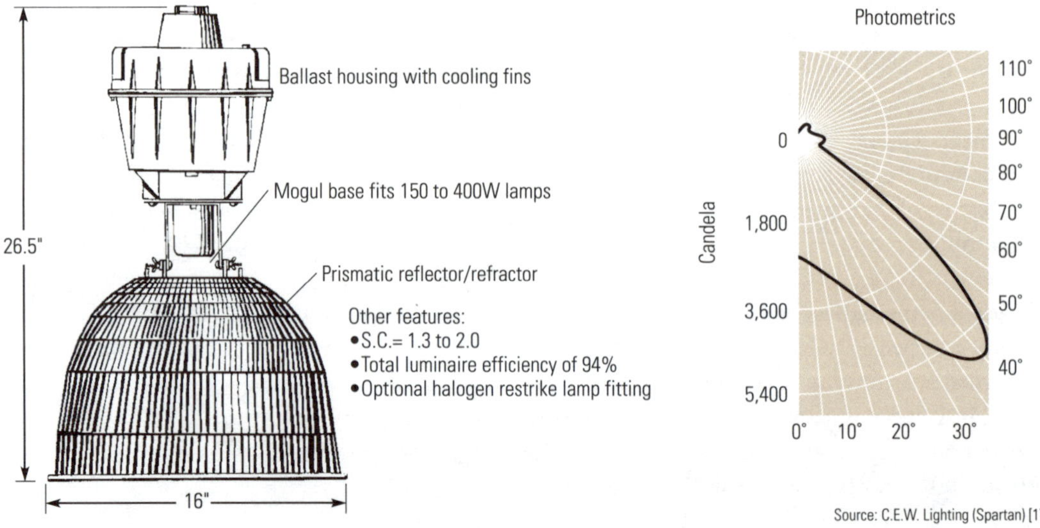

Source: C.E.W. Lighting (Spartan) [17]

Table 9-4

Same-ballast retrofit opportunities within the HID family

All lumens are initial. Lamp enclosures may need to be added to fixtures when high-pressure sodium lamps are retrofitted with metal halides. Note that HPS-to-MH retrofits reduce light output and do not save energy; these retrofits would only be done to improve lighting quality.

Lamp type		Power (W)			Light output (1,000 lm)		
Old	New	Old	New	Savings (%)	Old	New	Increase (%)
Coated MV	HPS	175	150	14	8	13	65
	HPS	250	215	14	12	20	65
	HPS	400	360	10	21	38	81
	HPS	1,000	950	5	60	102	70
	MH	400	325	19	21	28	33
	MH	400	400	0	21	36	71
	MH	1,000	950	5	62	100	61
Clear MV	HPS	175	150	14	8	13	65
	HPS	250	220	12	12	20	67
	HPS	400	360	10	21	38	81
	HPS	1,000	880	12	57	102	79
	MH	400	325	19	21	28	33
	MH	400	400	0	21	36	71
	MH	1,000	950	5	57	100	75
HPS	MH	250	250	0	27	21	-24
	MH	400	400	0	50	36	-28
	MH	250	225	10	25.6	24.8	-3
	MH	400	360	10	45	42.8	-5

Source: GE, Osram Sylvania, Philips

Fluorescent-to-HID retrofits

■ HIDs can replace fluorescents in industrial spaces, especially those that use high-output or VHO lamps in high-bay applications. A complete rewiring with new fixtures and ballasts is required.

HID retrofit issues. Retrofits with low-power-factor ballasts may increase current draw beyond circuit capacity, causing fuses to blow or breakers to open. Suppose, for example, that an incandescent lamp draws 100 watts with a power factor of 1.0, and a 50-watt HPS retrofit kit (with a PF=0.35) is installed in the same socket, increasing light output by 50 percent while roughly halving the wattage. While the incandescent drew 0.83 amps, the retrofit draws 1.19 amps—a 43 percent increase! If 13 such fixtures were ganged on one 15-amp breaker, the HPS retrofit would overload it.

Similarly, a THD of 135 percent on a retrofit ballast could lead to operational problems if many fixtures in one location (such as a convention hall with many incandescent sockets) were to be upgraded using that device. Check THD levels carefully when retrofitting all sockets on one circuit or transformer.

HIDs can replace fluorescents in industrial spaces, especially those that use high-output or VHO lamps in high-bay applications.

Magnetically ballasted retrofits are exceptionally heavy (about four pounds with lamp and reflector). To obtain a UL listing, they may include a chain to attach to the existing fixture's reflector, in case the socket fails and the unit begins to fall out. Before installing, check that the socket is firmly in place and verify that the installer knows the chain *must* be attached. Some contractors ignore this step because there may be no readily accessible point to which the chain can be hooked. Damage or serious injury could result if one of these devices fell out of a fixture in a high ceiling.

CASE STUDY **HID RETROFIT AT COLUMBIA UNIVERSITY**

A group of existing incandescent wall sconces in a cafeteria at Columbia University (**Figure 9-15**) used 250-watt halogen lamps. Mounted on pillars, they operated 24 hours a day, all year, supplemented by pendant-mounted 100-watt incandescent globes. The globes were retrofitted with pairs of 13-watt quad compact fluorescent lamps providing instant-on capability after a power interruption. There was therefore no problem in replacing the halogen fixtures with HID sconces.

Several replacement wall sconces that could accommodate different HID lamp types were available at a cost of about $400 each. **Table 9-5** shows the economics of three options for HID retrofits and how they compare to the halogen baseline. Installation of each fixture cost $50 if all equipment was in the same location as the existing fixture, or $100 if remote ballasting was required. To match the fluorescents, light of about the same color temperature was required, but a slight shift in color was tolerable since the sconce light would be indirect.

Since the MV and White SON lamps (options 1 and 2) required remote ballasting, their installation costs were $100, versus $50 for the HQI fixture (option 3), which was mounted directly on the pillar because it contains the much smaller Powertronic electronic ballast.

In this case, the 70-watt HQI metal halide source slightly beat out the mercury vapor lamp on economics. However, it also provides more light, a higher CRI, and less lumen depreciation, making it a far better choice overall. The new lamps—and their electronic ballasts—have been operating successfully since their installation in 1991.[18]

Figure 9-15

HID retrofit at Columbia University

Source: Columbia University

Installing an HPS lamp in a MV fixture may yield small (about 15 percent) wattage savings, but a large increase in light output (potentially tripling it). If the original light level was acceptable, removing two-thirds of the fixtures to attain greater savings is a tempting option to maintain average light levels. Depending on the light distribution of the existing fixture, this may not work: one would instead find bright spots of light separated by darkened areas. There are at least three ways to reduce this problem:

- Use a coated (instead of clear) HPS lamp; doing so widens light distribution from most fixtures.

- Install a wide-spread retrofit reflector, such as the one shown in Figure 9-14.

- Install a fresnel or other lens that widens light distribution.

Retrofitters finding MH or HPS fixtures already installed should not be deterred from seeking further savings. Many HID fixtures installed before 1980 have poor photometrics, and cheap aluminum reflectors may have oxidized due to lamp heat, greatly reducing their reflectance. New HID fixtures having the same input wattage may provide double the light output of older fixtures, requiring only half as many to be installed during such an upgrade. Once again, care is needed to ensure that proper distribution is maintained when reducing the number of fixtures.

Magnetically ballasted HID track lights will fit into existing tracks, but their weight may exceed the capacity of the track mounting. The Indy Lighting 50-watt White SON track head weighs, for example, 6.5 pounds; mounting eight of these in place of track heads previously holding 150-watt incandescent reflector lamps (at 0.7 pounds each) adds 46.4 pounds to a track never designed for such weight. Older tracks, mounted with lead anchors into plaster, may gradually come loose, creating a serious safety hazard. Electronically ballasted track heads weigh considerably less and may not present such a hazard.

Table 9-5

Columbia University retrofit options

In this case, mercury vapor and metal halide sources had similar paybacks. The latter (option 3) was selected because of its superior light quality and deeper electricity savings.

	Baseline	Option 1	Option 2	Option 3
Lamp data	250W halogen	100W mercury vapor	100W white high-pressure sodium	70W metal halide with electronic ballast
Total power (W)	250	125	130	80
Light output (lm)	5,000	4,600	4,700	5,200
Rated life (h)	2,000	24,000	10,000	6,000
Color temp (K)	3,400	3,700	3,200	3,200
CRI	100	52	80	75
Lamp cost ($)	5	25	40	35
Retrofit costs				
Fixture cost ($)	–	400	400	400
Installation cost ($)	–	100	100	50
Annual costs				
Electricity cost ($/y)	219	109	114	70
Lamp cost ($/yr)	22	9	35	51
Labor cost ($/yr) ($10 per relamping)	44	4	9	15
Total annual cost ($)	285	122	158	136
Economic analysis				
Annual electricity savings (kWh/y)	–	1095	1051	1489
Annual electricity savings ($/y)	–	110	105	149
Total annual savings ($/y)	–	163	127	149
Simple payback (y)	–	**3.07**	**3.93**	**3.02**

Assumptions: Duty = 8,760 h/y
Average electricity cost = 10¢/kWh (including demand changes)

CASE STUDY Central Carolina Bank

The University Square branch of Central Carolina Bank, located in Chapel Hill, North Carolina, chose metal halide lamps as part of its lobby lighting retrofit. The retrofit cut electricity use by almost one-third and provided a much brighter workspace. The bank's lobby—with a 20-foot ceiling and high, wraparound clerestory windows—had been lit by a grid of 28 recessed downlights, each with a 300-watt incandescent lamp (**Figure 9-16**). Illumination at desk level was insufficient: only about 30 footcandles. The retrofit replaced eight of those downlights with surface-mounted metal halide fixtures, each consuming 460 watts including ballast losses. The remaining 20 downlights were disabled, cutting downlighting power from 8.4 kW to 3.7 kW. Other changes included replacing eighteen 90-watt incandescent lamps, located around the perimeter of the lobby space, with compact fluorescent lamps using 16 watts each, cutting 1.6 kW of lighting power to 0.3 kW. In all, the bank reduced its lighting load from over 10 kW to about 4 kW. Illuminance levels after the retrofit (including daylight contributions) were increased to over 90 footcandles. Lighting quality was improved, as well, because the white light produced by the metal halide and compact fluorescent sources blends with daylight much better than the yellowish light produced by the incandescents.[19]

Figure 9-16

Central Carolina Bank lobby retrofit

Source: North Carolina Alternative Energy Center [19]

1 Mark S. Rea, ed., *Lighting Handbook: Reference & Application,* 8th Edition, (New York: Illuminating Engineering Society of North America, 1993), p. 218.

2 Original Cast Lighting Company, 6120 Delmar Boulevard, St. Louis, MO 63112-1204, tel 314-863-1895, fax 314-863-3278; SPI Lighting Inc., 10400 North Enterprise Drive, P.O. Box 635, Mequon, WI 53092, tel 414-242-1420, fax 414-242-6414.

3 Venture Lighting International, Inc., 32000 Aurora Road, Solon, OH, 44139-2814, tel 800-451-2606 or 216-248-3510, fax 800-451-2605, web www.adlt.com/venture.

4 Energy Master Plus data sheet (second quarter 1997), Venture Lighting International, Inc. [3].

5 Venture Lighting Designer Color data sheet (1996), Venture Lighting International, Inc. [3].

6 The five manufacturers we know of are Rambusch, Orgatech, Miroflector, Reggiani, and SPI.

7 Lindsay Audin, 1 Everett Avenue, Ossining, NY 10562, tel/fax 914-762-4939, e-mail energywiz@aol.com.

8 Rea [1], p. 220.

9 Sam Carleton, R.A.J. Keijser, and J.T.C. van Kemenade, "White HPS Lamps with a Color Temperature of 2,700 K," Philips Lighting Company (1990).

10 Philips Lighting Company's 180W SOX180 lamp produces 33,000 initial lumens.

11 James R. Benya, "Bluer Light, Better Sight," *Architectural Record Lighting* (February 1991), p. 63 (corrected values published as errata in subsequent issue).

12 C. Turner, "Metal Halide Dimming," *Lighttimes,* newsletter of the New York chapter of the IES (Summer 1993).

13 David Snider, personal communication (July 17, 1997), Venture Lighting International, Inc [3].

14 Consolidated Edison Company, Commercial and Industrial Program Support Bulletin (May 4, 1993), p. 4.

15 Indy Lighting (maker of White SON fixtures), 12001 Exit Five Parkway, Fishers, IN 46038, tel 317-849-1233, fax 317-576-8006.

16 C.E.W. Lighting, Inc., 14040 Welch Road, Dallas, TX 75244, tel 972-960-1993, fax 972-387-9711.

17 Spartan Lighting Manufacturing Company, 14040 Welch Road, Dallas, TX 75244, tel 972-960-1993, fax 972-387-9711.

18 Lindsay Audin [7].

19 "Lighting Case Study," North Carolina Alternative Energy Corporation (AEC), Better Lighting by Design Program (March 1993), P.O. Box 12699, Research Triangle Park, NC 27709, tel 919-361-8000.

10 Other Lighting Sources

10 OTHER LIGHTING SOURCES

In addition to the conventional illumination sources described in the preceding chapters, a variety of other lighting technologies are available or in development, some with high energy efficiencies. Each source type has unique operating characteristics, advantages, and limitations. Most of these technologies are best suited for special applications rather than for general area lighting.

There are many ways to convert electrical energy into light, and new ideas continue to appear. In this chapter we examine the most useful and encouraging alternatives to the filament and arc discharge lamps that now dominate the industry. The advantages of these sources can include compact size, lack of environmentally damaging materials, high efficacy, long life, and excellent controllability. Perhaps the most important contribution of these other sources is that their variety allows them to elegantly match specific lighting needs that conventional sources must meet with brute force.

10.1 LIGHT EMITTING DIODES

Light emitting diodes (LEDs) use solid-state electronics to create light (**Figure 10-1**). Initially discovered during the rapid development of semiconductors in the 1960s, their use has grown to displace conventional sources in signage, indicator and control panel illumination, and other uses.

HIGHLIGHTS

- Light emitting diodes (LEDs) are dropping in cost, and increasing in efficacy, light output, and color availability. These hardy light sources have established themselves in the niche markets of exit signs, traffic signal lights, and signage. Progress in developing blue LEDs will allow composite units to create white light, further expanding their applications.

- Electroluminescent products may be suitable for signage and other applications not well handled by other sources, if their lifetime can be increased and if production problems can be smoothed out.

- Photoluminescent products ("glow-in-the-dark" materials) have the potential to displace a few powered lighting applications, but low luminance limits their usefulness.

- Radioluminescent materials that glow from nuclear discharge require no electrical power, but have sufficient safety concerns to severely limit their application. Most firms producing tritium units (primarily for exit signs) have diversified into LED products and are phasing out signs containing tritium.

- Radio-frequency-induced fluorescence is a viable option for area and specialty lighting, but energy performance is currently no better than conventional fluorescent systems. The main advantage of RF lighting is long lamp life due to cathode elimination.

- Sulfur lighting (also called microwave lighting) is potentially a great leap forward, especially if combined with advanced lighting distribution technologies, such as fiber optics. Off-the-shelf sulfur lamps and light distribution systems are now available. With the increasing number and visibility of sulfur lighting installations, this could be an important lighting technology in the next century.

Exit signs that use as much as 40 watts of incandescent lighting can be replaced by LED signs using 5 watts or less, even though they have similar efficacy.

LEDs have several limitations, however, including small size, relatively low light output, and narrow color bands defined by the semiconductor materials used in their construction. Their estimated lifetimes range from 5 to 100 years, depending on the type of LED and how "failure" is defined (see box on page 224). The quality of fixture circuitry can also limit the value of LED products. Some exit signs, for example, group several LEDs in series to accommodate the standard voltage of emergency backup batteries. If one LED fails, several go out at once.

Figure 10-1

LED operation

Light emitting diodes (LEDs) use solid-state electronics to creat light. Electrons pass across a junction between two types of materials attached to a semiconductor base. One material (n-type) has excess electrons and the other (p-type) a deficit of electrons, sometimes called "holes." Applying power across this junction causes the electrons to meet the "holes," resulting in the creation of photons at fairly distinct wavelengths (that is, colors) of light. This particular LED is a transparent-substrate AlGaAs unit; other types have different configurations.

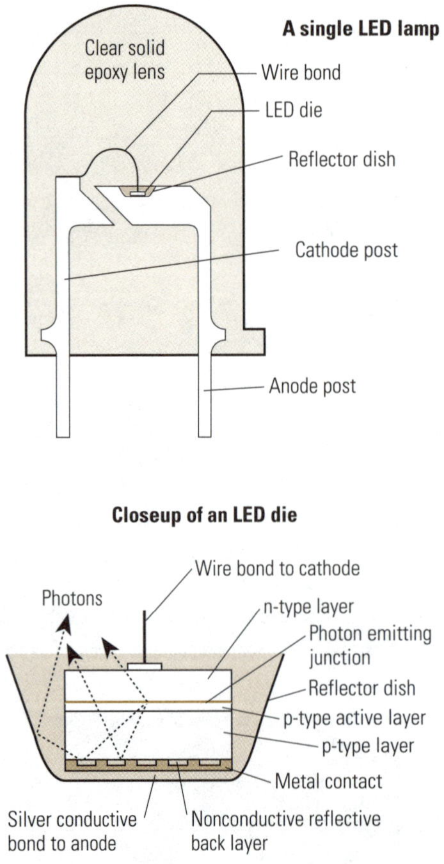

Source: Hewlett-Packard

LEDs are built in two basic packages:

■ The *discrete package*, also called the "through-hole" LED, is the most common (Figure 10-1). Its packaging includes a reflector to gather and focus the light and a plastic dome envelope which determines the beam spread and protects the LED.

■ The *surface mount chip* packaging consists of only a flat plastic lens. Chip LEDs are micro-soldered to a circuit board and, because they have no reflector or dome envelope, provide a wider beamspread at lower intensity than discrete LEDs.

The luminance of LEDs depends on their optical design: the narrower the beam, the higher the intensity. Both types of LEDs can be ordered with a number of lens options, producing various beamspreads and light intensities. LEDs normally operate using 1.6 to 2.0 volts DC at less than 0.1 amp.[1] Luminous intensities vary from about 12 to 15,000 millicandelas,[2] and is proportional to applied current. LEDs can be dimmed, pulsed, or strobed.

Efficacy. The efficacy of LEDs covers a wide range, from 0.01 to nearly 25 lumens per watt (**Figure 10-2**). While this is significantly lower than fluorescents and even some incandescent sources, the small size of LEDs allows very high utilization of their light output, which enables them to do the same task as a more efficacious source with much lower wattage. For example, exit signs that use as much as 40 watts of incandescent lighting can be replaced by LED signs using 5 watts or less, even though they have similar efficacy.

The efficacy, light output, and color selection of LED sources continues to improve with advancements in solid-state physics. The most startling recent news in LEDs came in 1996 when a small Japanese company, Nichia, announced that it had combined indium, gallium, and nitride to create a high-intensity blue

Figure 10-2

History of LED source performance

This graph compares LED output for units with 8° beam spread and 20 mA input current. Performance has improved dramatically since 1990.

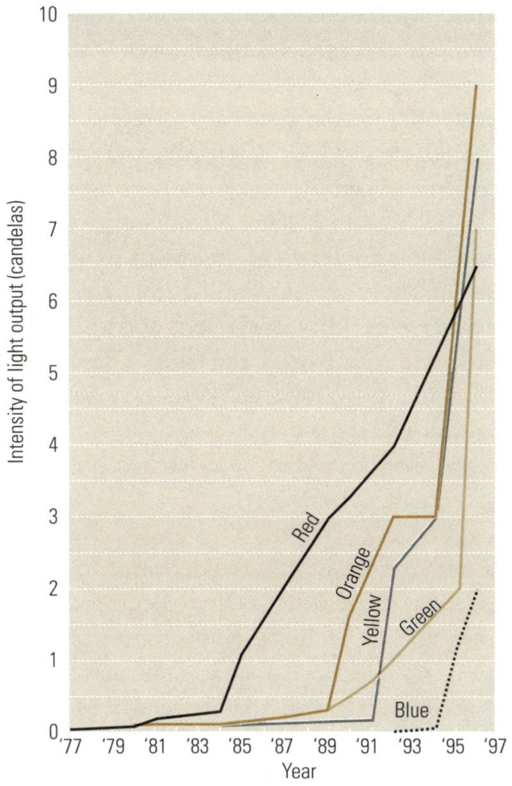

Source: Toshiba

This graph shows the improvement in efficacy of LEDs. The GaP and GaAsP LEDs are green colored, all others are red or amber.

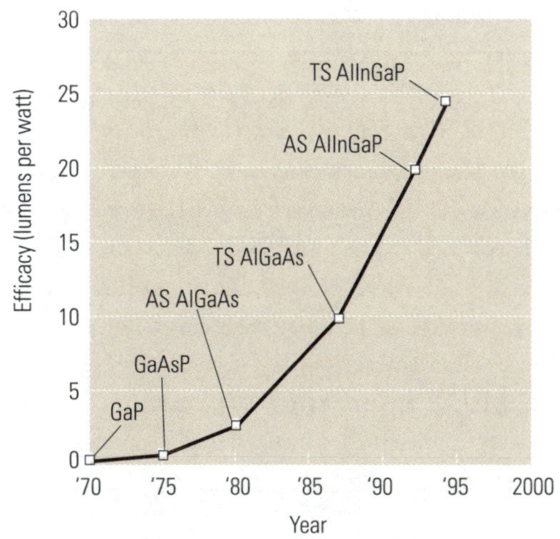

Key:
AS = Absorbent Substrate, TS = Transparent Substrate, Al = Aluminum, As = Arsenide, Ga = Gallium, In = Indium, P = Phosphide

Source: Hewlett Packard

LED.[3] Previous designs of blue LEDs used silicon carbide materials and produced only a feeble light, but the new InGaN units increased light output by two orders of magnitude. In addition, this class of LEDs spans a broad part of the blue-green spectrum range, and even extends into the ultraviolet. Nichia is still the leading producer of InGaN LEDs, but other manufacturers such as Hewlett-Packard are following suit with similar products.

continued on page 226

Luminance

To fully understand the limitations of low-intensity light sources, some background on luminance (surface brightness) is helpful. Standard light sources (incandescent and fluorescent) all create light in a relatively small volume, with surface brightnesses ranging from about 10,000 to 100,000 candelas per square meter, also known as "nits." By comparison, the luminance of a sheet of white paper under about 75 fc of illuminance is about 200 nits. (Divide the number of nits by 3.43 to get footlamberts.) Compare those numbers to the luminances of low-intensity sources:

- LED: 100 to 300 nits (average across sign lettering)
- Photoluminescence: about 0.01 nits (after 15 minutes)
- Electroluminescence: 27 to 125 nits
- Radioluminescence: 1.7 to 7 nits

Only LED and electroluminescent sources can provide brightness comparable to a sheet of illuminated paper. Sources that operate without electric power (photoluminescence and radioluminescence) provide very low luminance levels which are most appropriate when other light sources are off (such as during a power failure) and eyes have adapted to the darkness.

LED Lifetime

One of the more interesting and complicated issues surrounding LED applications is their rated life. If failure is defined as zero light output, then the lifetime of LEDs is very long indeed—based on manufacturers' stress testing, on the order of one hundred thousand hours or more.[4] (The lifetime of electronic components such as LEDs is sometimes expressed in terms of "mean time between failures" (MTBF). This quantity—which may be an impressively long time—is a statistical abstraction that is not very useful for real applications.)[5] Because of their robustness, LEDs are often used in military equipment and other critical applications such as airplane flight decks.

However, for many users the most significant failure mode for LEDs is the "soft" failure of lumen decline. LED users and manufacturers agree that light output of the units changes over time, but how fast and in what direction depend on the type of LED unit and its operating environment. One potential definition of failure—the point at which an LED puts out 50 percent of its initial light output—is an interpretation of the definition for incandescent and fluorescent sources, which holds that the lifetime of a product ends when 50 percent of a batch of samples have ceased to operate. Although loosely based on an existing concept, this definition is arbitrary. A more effective definition of LED failure would be the point at which the light output of the unit is no longer sufficient to perform the required visual task.

For example, if a 5,000 millicandela (mca) unit declines to 2,500 mca over several years, but only 1,000 mca is required for sufficient visibility, the "50 percent rule" is not a useful definition. Unfortunately, required illuminance is poorly defined for some of the key visual tasks that LEDs can fulfill, such as traffic signals. A better understanding of illumination requirements—and long-term LED light output—will eliminate much of the confusion that currently surrounds the issue of LED lifetime.

While much more testing and observation is required, a few general rules can be stated.

Temperature. Like other solid-state devices, the lifetime of LEDs is sensitive to temperature; hotter conditions typically reduce output and lifetime.[6] The critical consideration is the junction temperature, which is a function of the ambient conditions, the heat generated by LED operation, and the heat dissipation ability of the unit. Operating LEDs in high ambient temperatures may require a lower current—and lower light output—to keep the junction temperature at acceptable levels.

Operating current. LEDs last longest when they run at or below their rated current, typically 30 mA or less. Some vendors of new and retrofit products for exit signs have used power supplies that overdrive their LEDs. Overdriving provides higher initial light output but causes rapid lumen depreciation. To avoid this problem, specify exit signs and other LED products that operate at current levels no higher than those recommended by the LED manufacturer.

Voltage and waveform. LEDs are diodes, the electrical equivalent of one-way streets. Because of this, they operate on direct current (DC) rather than alternating current (AC). The LEDs work best if fed a square waveform, as opposed to unfiltered half-wave or full-wave rectified AC, which produce a sine wave shape. More sophisticated producers of LED-equipped devices use circuitry that flattens out the voltage sent to the LED. Some applications may also require circuitry that adjusts current to maintain light output if voltage drops.

Continued on next page

LED Lifetime

Continued from previous page

Manufacturing defects. Quality of construction continually improves as new production processes are developed, but occasional failures due to physical assembly problems can still occur, typically early in operation. For example, too much heat generated during soldering can cause a micro-delamination between the LED die and the epoxy casing. Pre-assembly burn-in can weed out these cases of "infant mortality."

Humidity. To adversely affect LEDs, high humidity must be persistent enough to saturate the epoxy resin surrounding the junction. According to one LED engineer, conditions must be on the order of 85 percent relative humidity for several thousand hours before moisture can penetrate the junction.[7] For outdoor applications, daily or seasonal humidity cycling should be sufficient to dry out the resin.

Time. The largest question concerning LED performance is long-term light output under field conditions. Tests show that light output declines initially, and then either levels off or declines at a much slower rate.[8] Under certain conditions, some types of LEDs increase their light output over time.[9]

Although data on old LED sources is difficult to come by, one firm we contacted has 13-year-old "moving message" LED signs that are working fine, but they do not have data on light output.[10] **Figure 10-3** shows the results of two long-term measurements of LEDs in traffic signals. Two years of field data from an operating intersection in California show a decline in lumen output of about 30 percent, while a signal operated continuously for 15 months at E SOURCE's office showed a decline of nearly 50 percent. (Much of the loss in the E SOURCE test was recoverable when the unit was cycled off briefly.) Both of these signals used earlier-generation AlGaAs LEDs.

LED type. Newer generations of LEDs, particularly the AlInGaP units, reportedly have better lumen maintenance (and therefore greater effective lifetime) than the older types such as AlGaAs. The rapid pace of improvement in LED technology makes it difficult to keep pace; during the time that E SOURCE tested the lumen maintenance of an AlGaAs traffic signal (Figure 10-3), the AlInGaP products entered the market.

Field tests of LED traffic signals have revealed much about the operation of these devices under rugged conditions. For more information, see E SOURCE *Tech Update TU-95-14*, "LED Traffic Lights: Test Results Give Green Light for Additional Installations," (November 1995).

Figure 10-3

LED output declines over time

This chart shows two sets of long-term data for the light output of LED traffic signals. The CalTrans data is from a signal unit that operated in the field for two years, and was measured for light output every six months. The E SOURCE data is from a 12-inch red signal head with 600 AlGaAs LEDs. This signal was indoors at a temperature of about 73°F, and ran nearly continuously (no on-off cycling) for 15 months. Measurements were made with a hand-held illuminance meter centered over the signal face one foot away; control measurements on a fluorescent source made with the same meter stayed within one percent of original output. Although the ultimate drop in output was nearly 50 percent, on-off cycling tests near the end of the 15-month period showed that much of this lumen loss was recoverable from brief "rests."

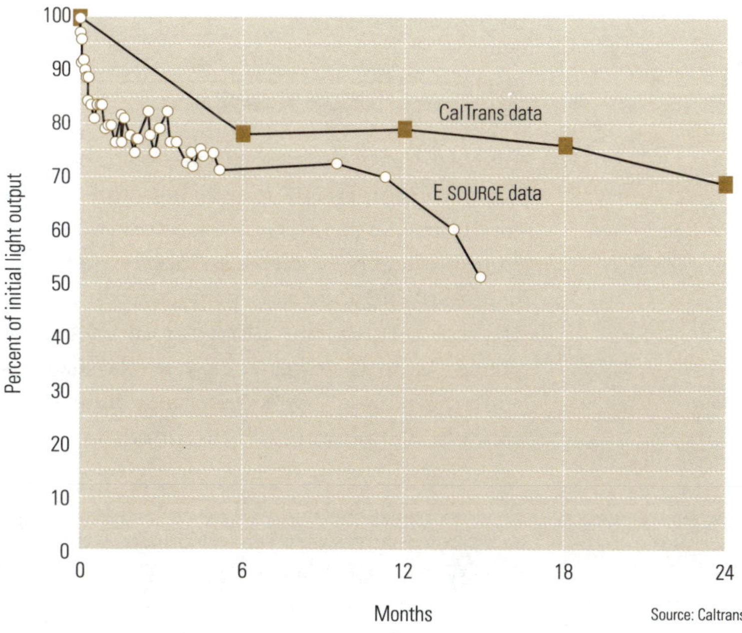

Source: Caltrans, E SOURCE

A strong blue LED provides the missing link in allowing efficient generation of white light from solid-state sources. There are currently three approaches:

- *Combine blue and yellow LEDs.* Blue and yellow are opposite each other in the color spectrum, so a reasonable white can be produced by controlling the intensities of adjacent units.

- *Combine blue, red, and green LEDs.* This is currently the method used in most large high-end outdoor video screens, especially those in Japan. In most designs, the LEDs are bunched in groups of three that appear white when viewed from a distance but resolve into individual colors at close range. Controlling the RGB components separately allows any color to be created.

- *Add phosphor to a blue LED.* By mixing phosphorescent materials into the epoxy lenses of InGaN LEDs, a secondary process creates white light. This method is still experimental but holds promise; the efficacy of first-generation phosphorescent LEDs is close to that of the InGaN LEDs, or about two to five lumens per watt. A significant advantage of this approach is that there is only one pair of power leads to the LED, rather than six as in the RGB designs. However, these units can only create white light.

10.2 Photoluminescence

Glow-in-the-dark materials—known formally as photoluminescent materials—store ambient light energy in chemically treated plastic compounds. No external electric power is required. The absorbed energy is then gradually released as the electrons in the compounds drop to lower energy levels, also re-emitting photons. When ambient lighting is shut off, photoluminescent light output drops exponentially with time (typically by a factor of ten in less than one hour), making it useful for a relatively short period of time, such as during building evacuations. **Figure 10-4** shows typical lumen depreciation curves for some photoluminescent materials.[11] Very low luminance limits application to spaces where the eye will dark-adapt (such as long hallways) during power failures.

Photoluminescent materials are similar to fluorescent lamp phosphors, whose slight photoconversion delay reduces visible flicker between electrical cycles. When fluorescent lamps are shut off in a room sealed off from any other light, a dark-adapted individual can see them glowing (albeit at a very low level) as much as 10 minutes after being shut off, due to the gradual release of stored energy. Application of photoluminescence to signage is relatively new, and is a result of its gradual reduction in cost and increased availability.

The most common application of photoluminescent surfaces for lighting is for illumination of emergency and exit signage. One manufacturer suggests disconnection of power to signs and attachment of a photoluminescent exit sign to the face of the existing sign. A similar treatment is suggested when a tritium-based exit sign has lost most of its luminance. Few building codes, however, specifically accept such installations in lieu of powered light sources.

Figure 10-4

Luminance depreciation of photoluminescent materials

Photoluminescent (glow-in-the-dark) materials experience a steep decline in light output over the first hours of operation, but can still maintain sufficient luminance to be seen by dark-adapted eyes.

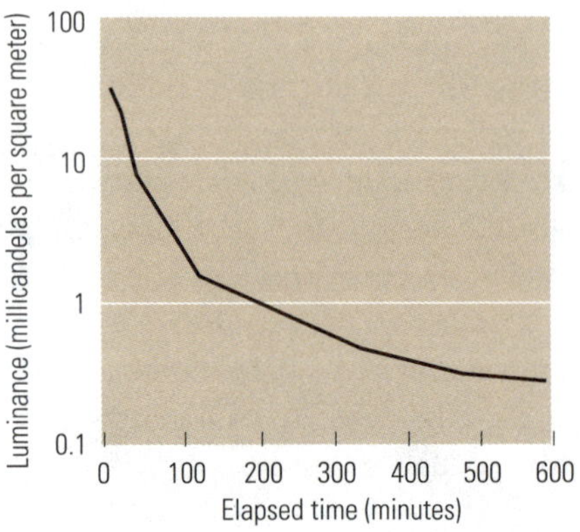

Excitation: 1,000 Lux for 5 minutes

Source: American PermaLight [11]

In some buildings, low-wattage incandescent lamps burn continuously to highlight locations of emergency equipment (such as fire extinguishers) or light switches. If located near a standard light source, a photoluminescent strip performs the same function using no power. The lifetime of such materials is basically unlimited, as long as they remain clean and unpainted. But unless enclosed by glass or plastic, these materials can become dirty, be peeled off by occupants or vandals, or be painted over during renovations.

10.3 Electroluminescence

Electroluminescence (EL) is similar to photoluminescence in that a surface (not a point) becomes luminous due to an input of energy, in this case low-voltage AC power. Built like a large, thin (1.2 mm) capacitor, electroluminescent "lamps" consist of a dielectric containing phosphors, sandwiched between two electrode plates, at least one of which is transparent (**Figure 10-5**). When AC current is applied to the plates, electrons flow in the dielectric, striking the phosphors and making them glow.

Figure 10-5

Construction of an electroluminescent panel

In an EL panel, current flows between two large sheets, causing a phosphor layer to glow.

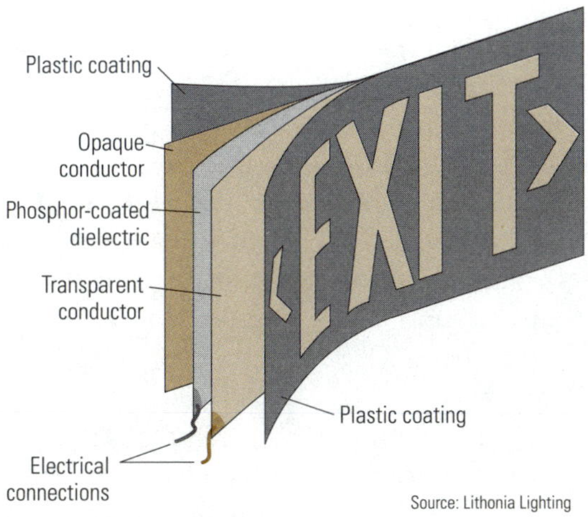

Source: Lithonia Lighting

Electroluminescence dates back to the early 1950s, with the first commercial lighting applications appearing about a decade later. Their use has greatly expanded recently as a source for backlighting liquid crystal diode (LCD) portable computer screens.

Because of low luminance (27 to 125 nits), this source is also limited to applications involving signage. Its uniform luminance characteristics and wafer-like shape make it a candidate for backlighting building directories and similar indicators. The primary use of electroluminescent light is for exit signs, although LEDs now dominate the energy-efficiency portion of this market. For more on exit signs, see E SOURCE *Tech Update TU-94-2*, "LED Exit Signs: Improved Technology Leads the Way to Energy Savings," (April 1994).

The biggest drawback to electroluminescent products is their rapid decay in light output.

The color of light emitted by EL signs depends on the phosphor, and can be green, blue, amber, yellow, or white. Since red signs are produced by placing a filter over a different color EL lamp, they are inherently lower in luminance. Green EL signs have the highest luminance of EL signs.

The biggest drawback to EL products is their rapid decay in light output. Electroluminescent surfaces typically last about 8 years, and replacement is somewhat labor intensive due to the need to unsolder the old surface and solder in the new one, although it should be possible to design modular systems. There are some reports of signs that extend lifetime by boosting initial light output, but these have not been independently confirmed.[12] Possible problems with accessibility and time out-of-service must be considered for any application. **Figure 10-6** shows a typical luminance depreciation curve for an EL source.

One way to deal with lumen degradation is to minimize operating time at full output. For instance, EL exit signs from some manufacturers run at 20 percent output during normal operation but increase to full output when an emergency notification system is activated or when the power goes out (the signs use a battery backup system).

Figure 10-6

Luminance depreciation curve of a typical electroluminescent light source

EL sources experience a steep drop in light output over time.

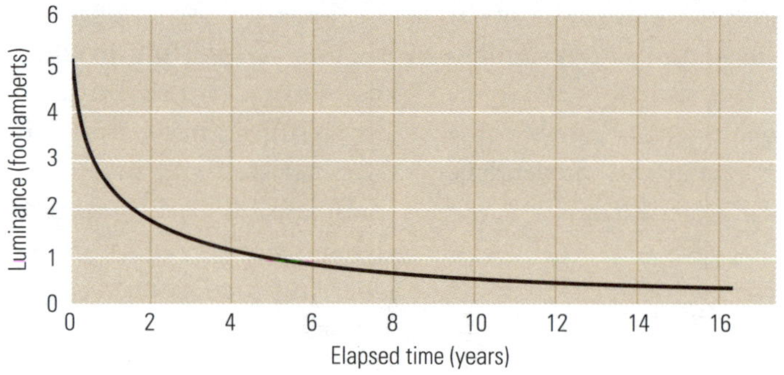

Source: Northstar Technologies

Tritium is especially hazardous because of its potential for rapid uptake into the body. As a form of hydrogen, tritium can easily be included in water molecules.

Lamp failure eventually results from moisture or phosphor degradation. The dielectric material on which the phosphor is imbedded is susceptible to humidity, which typically penetrates the lamp at the wire leads. To alleviate moisture problems, most manufacturers now seal the entire lamp sheet in plastic. The addition of desiccants helps to further control damaging moisture. Similarly, UV filtering materials and color stabilizers are incorporated into the lamp laminates to minimize the material degradation and color shift.

Manufacturers indicate that both the energy consumption and the cost of EL signs may be reduced further by producing signs in which only the letters, rather than the whole panel, are illuminated.

10.4 Radioluminescence

Radioluminescence is based on beta-ray emissions (that is, free electrons) from mildly radioactive isotopes (primarily tritium, an isotope of hydrogen) striking phosphors on the inside of sealed glass tubes and causing them to glow. Since isotopes decay spontaneously, no electric power is needed to energize the electrons, and light output continues as long as decay does.

Radioluminescence is a byproduct of nuclear research, with early applications dating back to the 1940s. Military and scientific applications as equipment indicator lights became possible as the cost of tritium dropped. Tritium was originally produced in large quantities by U.S. government facilities as a component for nuclear bombs, but production in this country ceased when most weapons plants were shut down over the last five years. Tritium-based exit signs became popular during the energy crises of the late 1970s.

Even though radioluminescent products require no electrical energy, E SOURCE cannot recommend them because of the difficulty and/or expense of disposing of the depleted tritium, and the potential hazard from breakage. Radioactive tritium poses no health risk while it is contained in the glass tubing typically used for exit signs (the beta-ray emissions cannot penetrate the glass). If the glass is broken, however, federal regulations call for evacuation of the enclosed spaces near the tritium spill until it is contained and removed.

While beta emissions from tritium are not as damaging as high-energy gamma rays from heavier radioactive elements such as plutonium, tritium is especially hazardous because of its potential for rapid uptake into the body. As a form of hydrogen, tritium can easily be included in water molecules, which can then be absorbed in various biological structures where it is difficult or impossible to retrieve and sequester.

The half-life of tritium is 12.3 years, so the signs will lose half their luminous intensity over this period. Disposing of no-longer-functional but still-radioactive signs can be problematic and expensive, as tritium is classified as a low-level nuclear waste. The federal Low-Level Waste Compact Act was instituted in 1994 to encourage the development of regional disposal sites for low-level radioactive wastes, but court action and inertia on the part of many states has delayed or limited its implementation. As of early 1997, only a few facilities were accepting low-level wastes such as radioluminescent exit signs. A patchwork of agreements and permits exists:[13]

- A facility in Barnwell, South Carolina, accepts wastes from all states except North Carolina.
- An Envirocare plant in Utah accepts wastes from any state.
- A Permafix incinerator facility in Gainesville, Florida, accepts tritium sources, including radioluminescent exit signs.
- A disposal site in Hanford, Washington, accepts wastes from states in the Northwest and Rocky Mountain regions.

For the latest on tritium disposal, contact the state environmental agency in your area.

One company that manufactures radioluminescent exit signs—Evenlite—handles disposal internally, by accepting signs that have reached the end of their lifetimes. The company puts a sticker on each sign that states its creation and recommended replacement dates. Customers returning used signs and purchasing new ones are entitled to free disposal. Without purchasing a replacement sign, the company charges $10 per sign. Evenlite then disassembles the sign and sends the tritium tubes back to their original manufacturer.[14] Another manufacturer will accept used signs for no charge—again, if new replacements are being purchased—but charges $50 per sign otherwise.[15]

10.5 Radio-Frequency (RF) Induced Fluorescence

Targeted high-frequency electromagnetic fields can be used to energize mercury atoms, providing the basis for fluorescent lighting without cathodes. As with electric arc-based fluorescence, the mercury's electrons change energy levels, releasing ultraviolet photons which stimulate phosphors into emitting visible light. RF-induction was initially employed to create ultraviolet light for industrial applications, but came under close study in the mid-1970s as a means to power fluorescent lamps without the life-limiting electrodes, which is where conventional fluorescent lamps typically fail.

The first commercial lighting prototypes were produced by Philips Lighting in Europe in 1990, and the first products became available in 1991. Popular awareness of the concept in this country exploded when Intersource Technologies created a media blitz for its "E-lamp" in mid-1992. As of mid-1997, however, the Intersource product has not appeared on the market.

Philips unveiled its own RF fluorescent light—the QL induction system—in the U.S. in 1992. This system uses a separate ballast and lamp and is available in 55 and 85 watt sizes, with a 150-watt version expected to follow. The rated lifetime of the QL is 60,000 hours. The high cost of this system (about $300) has made it a niche product, used primarily where relamping is costly and/or hazardous. A variety of streetlamp and sconce fixtures are available for the QL system in Europe.

GE Lighting offers the Genura™ RF fluorescent lamp (**Figure 10-7**). Unlike the Philips system, this 23-watt, 1,100-lumen unit contains it own ballast, making it similar to self-ballasted CFL lamps. The Genura is similar in appearance and performance to a 75-watt R30 reflector flood lamp, and screws into a standard medium Edison base. The Genura has a rated lifetime of 20,000 hours, and is available in a 120-volt version in the U.S. and a 240V version in Europe. The end-user cost of the lamp is about $20.

Table 10-1 lists the specifications of available and developing induction sources. While these systems have no electrode to degrade, their efficacy is still limited by the physics of fluorescence and the efficiency of their ballasts. The induction systems claim efficacies of 48 to 70 lumens per watt, making them about as efficient as large compact fluorescent lamps. The primary benefit of this type of lighting is greatly increased lamp life. The Genura unit claims a 20,000-hour life for its product, triple that of the smaller compact fluorescents it could replace, while the QL system offers a 60,000-hour life. Both induction systems have better power quality than many compact fluorescent options, and can be operated in any position. The lack of cathodes means that lifetime is unaffected by number of starts or duty cycling—an attractive property for occupancy-switched applications.

Figure 10-7

GE Genura induction lamp

The General Electric induction lamp has an integral electronic "ballast" and is expected to cost about $20.

Source: General Electric

Applications. Due to their 1,000-lumen output range, applications for the Genura and other induction lamps will likely be limited to replacing incandescent A-lamps and R-lamps up to 75 watts. The QL system has been used in Europe for street lighting, and indoor fixtures (similar to decorative metal halides in size and shape) are in production. The 55-watt version is a possible replacement for incandescents up to about 250 watts, and the combination of 65 lm/W efficacy and high CRI makes it competitive with the 50-watt white SON high pressure sodium and HQI metal halide lamps. The 85-watt variety is a potential alternative to the 100-watt white SON.

Limitations. The cost of currently available RF induction lighting systems is too high to realistically compete with other sources. No significant price drop has occurred between 1994 and 1997, and neither of the manufacturers have indicated that price decreases are likely in the near future. The energy savings of RF induction—compared to conventional fluorescents—are nearly zero, so the units must sell based on their long life and maintenance savings claims, which are difficult to prove for such a new product.

Induction lighting is not immune to lumen depreciation, due to degradation of phosphors and dirt coating the bulb and/or fixture optics. The extended lifetime of the source could make phosphor degradation a significant problem, while the perception that it is maintenance-free may prevent routine cleaning. If lighting engineers overdesign to account for these factors, the system could waste power early in the lamp's life, relative to shorter-lived sources. Intelligent ballasting to increase lumen output late in the lamp's life and a regular maintenance schedule should be able to compensate for such losses, however.

Table 10-1

Characteristics of induction lighting sources

Description	Philips	Philips	General Electric
Model	55W QL system	85W QL system	Genura™ R80/R30
Lamp size	3.3" dia., 5.9" length	4.3" dia., 7.5" length	3.2" dia., 5" length
Ballast size	5.5" x 5.5" x 1.8"	5.5" x 5.5" x 1.9"	Ballast included in lamp
Initial light output (lumens)	3,500	6,000	1,100
Power input to ballast (W)	55	85	23
Initial efficacy (lumens/watt)	64	70	48
Rated life (hours)	60,000	60,000	10,000 +
Power factor	>0.90	>0.90	0.55
THD (%)	<10	<10	130
CRI	>80	>80	82
Color temp	3,000 K or 4,000 K	3,000 K or 4,000 K	3,000 K
Current status	Available in U.S. & Europe	Available in U.S. & Europe	Available in U.S. & Europe
Approximate end-user price	$300	$300	$20
Contact information	Kevin Bonawitz Philips Lighting 200 Franklin Square Drive P.O. Box 6800 Somerset, N.J. 08875 tel 908-563-3596 fax 908-563-3155	Kevin Bonawitz Philips Lighting 200 Franklin Square Drive P.O. Box 6800 Somerset, N.J. 08875 tel 908-563-3596 fax 908-563-3155	Gary Gumz GE Lighting Nela Park Cleveland, OH 44112 tel 216-266-6675 fax 216-266-3433

Source: Manufacturers' data

10.6 SULFUR LIGHTING

Sulfur lighting, also called microwave lighting, is a fundamentally new way to create light that holds great promise for the future. Sulfur lighting uses very high frequency electromagnetic waves (microwaves—2.45 GHz, or 2.45 billion cycles per second) to strip electrons from atoms of sulfur, argon, and other gases, creating a plasma of free electrons and nuclei. The electrons change energy levels, emitting photons in the visible light range (or at other wavelengths, depending on the gas mix). No phosphors or electrodes are used.

Microwave stimulation was first developed as a means to generate high levels of ultraviolet light for industrial processes in the late 1980s. The first commercial lighting product was introduced in 1993, and an off-the shelf unit is now available.

Sulfur lamps are now operating in several installations in the United States, including the National Air and Space Museum and the exterior of the DOE's headquarters (**Figure 10-8**), both in Washington DC. An installation in a Pontiac auto plant in Michigan was also completed in 1996, and several others, including a postal facility, are either planned or under construction as of mid-1997. In Europe, lighting systems serving a shopping center and an office complex are also underway.

Since microwave lighting uses no electrodes or phosphors, lifetime and output are limited primarily by the durability of solid-state power supply components. The lamp is a rotating air-cooled glass sphere about the size of a golf ball, served by a microprocessor-controlled magnetron, much like the energy source in a microwave oven. High efficacies are possible, potentially producing over 100 lm/W of high CRI, high color temperature light in a relatively small package. By adjusting the gas mixture, ultraviolet and infrared emissions can be nearly eliminated, providing high-quality light without harmful components. Unlike fluorescent lamps, there are no hazardous materials, such as mercury or phosphors, which may require special disposal methods. For some users, this feature alone may make the sulfur lamp a potential replacement for conventional sources. Unlike the metal halide lamps it may replace, the microwave light can be fully dimmed, and has quick startup and restrike times (about 10 seconds).

The first commercial lighting product using this technology was the VBL-3400 floodlight, built by Fusion Lighting and marketed by Hutchins International. The unit supplied

Figure 10-8

Sulfur lamp installations

Prototype sulfur lamps light an outdoor walkway at the Forrestal Building and a large room at the National Air and Space Museum, both in Washington DC.

Source: Fusion Lighting [16]

430,000 lumens of 85 CRI, 6,700 K light to illuminate large spaces such as stadiums.

Fusion introduced the second-generation sulfur lamp—the Solar 1000—in early 1996 (**Figure 10-9**, **Table 10-2**). This device is currently sold as part of a package with light distribution systems (see below). The 1,425-watt lamp/fixture is UL-approved and produces a minimum of 135,000 lumens. Many improvements have been made to this technology since it first appeared in 1993; the Solar 1000 is nearly 50 percent more efficient than early prototypes. The system is available with an optional step-dimmer that can provide light output between 100 and 20 percent. Fusion is also developing a system to control groups of up to 150 sulfur lamps from a computer interface.[16]

The key to high efficacy is efficient energy conversion in the magnetron, an electromagnetic waveguide that creates microwaves from electricity. Early sulfur lamp prototypes used magnetrons with efficiencies of about 50 percent, but the Solar 1000 has increased that to about 60 percent. Industrial magnetrons are available with efficiencies up to 80 percent, so further improvements are expected. (With a 100 percent efficient magnetron, the efficiency of the sulfur lamp would be about 170 lumens per watt.)

A well-designed light distribution system is the key to utilizing this extremely bright and compact light source. Fusion offers the Solar 1000 with its own 10-inch light pipe (**Figure 10-10**); similar systems are offered by Cylinear Lighting Co. (Toledo, Ohio) and by 3M, which now makes its own light pipes in addition to supplying many of the reflective and refractive materials used by other light-pipe vendors. Fiberart (St. Petersburg, Florida) is working on a side-emitting fiber optic product designed to replace neon signs. Several innovative fixture styles are also being designed by other manufacturers, including a cruciform light guide where the Solar 1000 sits at the intersection of four rectangular light guide legs.

Many improvements have been made to sulfur lighting technology since it first appeared in 1993.

Figure 10-9

Solar 1000 sulfur lamp

The Solar 1000 produces 130,000 lumens of light with 1375 watts of electrical input energy—nearly 100 lumens per watt. The lamp itself is inside the conical light guide junction; the small device is a dimmer control.

Source: Fusion Lighting [16]

Table 10-2

Characteristics of the Solar 1000 sulfur lamp

Physical characteristics	
Model	Solar 1000
Size	16" x 8.5" x 4.5"
Weight	15 pounds
Lamp diameter	1.5 inches
Light guide junction diameter	10 inches
Light characteristics	
Light output	135,000 lumens
Color temperature	6,000 K
CRI	79
Dimming range	100–20 percent
Strike time	instant, full output in 20 seconds
Restrike time	5 minutes
Electrical characteristics	
Input power	1425 watts
Input voltage	200–300 volts
Input frequency	50–60 Hz
Power factor	0.93
Electrical efficiency	95 lumens per watt

Figure 10-10

Sulfur lamp with light pipe

Most sulfur lamp applications use a light pipe to transmit and distribute the 130,000 lumens created by the golf-ball-sized lamp.
Source: Fusion Lighting

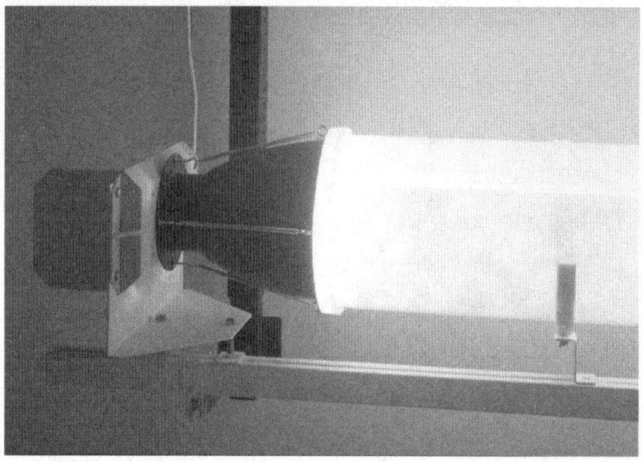

Limitations. Sulfur lighting is still an emerging technology. The primary questions that need to be addressed are the following:

- *Requirement for mechanical rotation.* At present, the sulfur globe must be spun at several hundred revolutions per minute to ensure uniform light output and color in the path of the non-uniform microwave field. This adds significant mechanical complexity that could be avoided if other means are developed, such as varying the intensity of the microwave radiation.

- *RFI and microwave shielding.* While initially a potential safety problem, changes made to meet UL approval have eliminated any hazard to occupants or maintenance personnel. The level of microwave leakage may still be sufficiently high, however, to confuse systems using radar, limiting applicability at airports and in facilities where microwave emitters are used for security systems.

- *Lifetime and economics.* Do to its lack of electrodes and phosphors, the sulfur lamp itself should last nearly indefinitely. The main source of system mortality is the magnetron. Typical magnetron lifetime has been increased from 10,000 hours for early prototypes to the Solar 1000's 15,000 hours, resulting in a five-year replacement warranty. Overall cost effectiveness must be judged against than the HID sources this system may replace.

1 Manufacturer product data, Hewlett Packard Company.

2 Mark S. Rea, ed., *Lighting Handbook: Reference & Application,* 8th Edition (New York: Illuminating Engineering Society of North America, 1993), p. 238.

3 Andrew Pollack, "The Little Light That Could," *The New York Times* (April 29, 1996).

4 Application Note 1005, "Operational Considerations for LED Lamps and Display Devices," Hewlett-Packard, Palo Alto, CA.

5 For example, one manufacturer stress-tests batches of their units to determine failure rates. If no failures occur, they make the conservative assumption that a single unit failed (it is impossible to divide by zero), and calculate MTBF on this basis.

6 Application Note 1005 [4].

7 Dave Evans, personal communication (March 8, 1994), Applications Engineer, Hewlett Packard, San Jose, CA, tel 408-435-6144, fax 408-435-4980.

8 Manufacturer test results, Econolite, 3360 East La Palma, Anaheim, CA 92806, tel 714-630-3700, fax 714-630-6349.

9 Brian Thorbrogger, personal communication (January 5, 1994), Sales Development Engineer, Hewlett-Packard Company, San Jose CA, tel 408-435-4009.

10 Brian Alu, personal communication (March 22, 1994), Data Digital Display, Boulder, CO 80027 tel 303-666-9400.

11 American Permalight, 620 Carob Street, Compton, CA 90220, tel 310-898-3795, fax 310-608-1299.

12 Robert Sardinsky, personal communication (July 5, 1994), Rising Sun Enterprises, 40 Sunset Drive, Unit 1, Basalt, CO 81621, tel 303-927-8051, fax 303-927-3635.

13 Diane D'Arrigo, personal communication (February 22, 1997), Coordinator, Nuclear Information Resource Service, 1424 16th Street NW, Washington, DC 20036, tel 202-328-0002, fax 202-462-2183, e-mail nirsnet@ijc.org, web www.nirs.org.

14 Kim Kvalnes, personal communication (July 5, 1994), Vice President, Evenlite Corporation, 220 VFW Avenue, Graysonville, MD 21638, tel 800-872-0879, fax 410-827-9657.

15 Alice Hutchins, personal communication (July 7, 1994), Self-Powered Lighting, tel 914-592-8230.

16 Fusion Lighting, 7524 Standish Place, Rockville, MD 20855, tel 301-738-0022, fax 301-926-7258.

11 LIGHTING CONTROLS

11 Lighting Controls

Controls are the last step in a lighting efficiency plan. While lighting controls should be considered after high-efficiency sources have been chosen, the controllability of light sources varies widely—usually inversely—with their efficacy. Technological developments continue to provide new control opportunities for fluorescent and HID systems, but cheap, reliable dimming for these sources is still awaited.

Used properly, lighting controls reduce the burn time and/or light output of lighting systems, cutting energy consumption and—usually to a lesser extent—peak demand. Ironically, controls have the most significant impact on demand and consumption in areas served by inefficient lighting—such as incandescent lamps or fluorescent fixtures consuming over 2.5 watts per square foot. As more efficient sources are installed, the amount of potential savings from controls drops. Cost-effectiveness also depends strongly on labor costs and utility rates. The integration of controls into lighting systems may also be complicated by energy code requirements and other considerations such as audible noise and power quality.

Lighting controls pose special challenges for utilities and users for several reasons:

■ Energy, demand, and dollar savings vary widely among installations.

HIGHLIGHTS

■ Elapsed-time switches (also called timer switches) are increasing in sophistication, and can offer many of the benefits of occupancy sensor switches without some of their complexities.

■ Occupancy sensors have great potential for energy savings, but their proper operation depends on equipment choice, location, and training of building staff.

■ The flexibility of dimming can potentially double the energy savings that efficient nondimming systems provide. While fluorescent dimming still doubles (at least) ballast cost, new products continue to appear, and prices continue to fall.

■ New control technologies should be tried out in small-scale demonstrations because of possible interactive failures, such as current-leaking triacs or other incompatibilities.

■ Controls usually save more energy and power off-peak than on-peak. Direct management of peak demand is possible with automatic dimming and centralized control systems.

■ Codes and standards such as ASHRAE 90.1 and California's Title 24 increasingly require some form of automatic lighting control for new construction and renovations. See **APPENDIX C**.

■ The current rating of small control devices such as occupancy sensors continues to climb, while the power draw of efficient lighting systems drops. This trend is making it easier to control more lights with less equipment.

■ It takes more equipment and effort to accurately predict and measure the savings from controls than from lamps and ballasts.

■ In many cases, much of the energy savings occur during off-peak periods where lower rates may apply, yielding smaller-than-expected dollar savings to customers and minimizing the load management benefit to utilities. (Dimming controls—especially those for daylit buildings—can provide substantial peak reductions.)

■ Little information exists on the long-term persistence of savings from controls, and they are vulnerable to misadjustment, vandalism, or disabling—all of which can reduce energy savings over time.

■ Some controls can suffer little-known but avoidable interactions with other building system components, impairing performance, cost-effectiveness, and service life.

■ Human behavior plays a role in determining savings from some lighting controls. One study showed that after sensors were installed in some offices, occupants who previously turned off lights manually let the sensors completely control the lighting. This resulted in more burn hours than expected and

LIGHTING CONTROLS AND PEAK DEMAND

For most industrial, institutional, and commercial facilities, a portion of the electric bill is for the peak demand of the buildings' electric loads, even if it lasts only 15 minutes in one month. Most lighting is typically on during the building peak demand (usually in the early afternoon), which is also usually the utility's peak demand billing period. Peak demand charges for the month may therefore remain about the same even if the lights are turned on only once during the month (assuming all other uses of power are relatively consistent). In such an application, motion sensors will provide savings on energy charges, but only a minimal reduction in the building's peak demand. In areas where demand charges are based on annual peaks, no demand savings will be seen on a bill until 12 months of consistently reduced demand have passed.

In schools and other facilities where space usage routinely drops during the summer, installing sensors may not reduce a utility's summer peak demand. Similarly, where lighting is already controlled by a photocell (such as surrounding an atrium), it may already be off during the utility's peak daytime demand period, limiting the sensor's ability to cut demand.

Furthermore, facilities charged by time-of-use pay less per kWh at night than during the day. For example, if occupancy sensors reduce usage primarily during the overnight period, only the cheaper kWh will be avoided, further reducing the expected dollar savings (**Figure 11-1**).

Figure 11-1

Value of savings from occupancy controls depends on time of use and demand charges

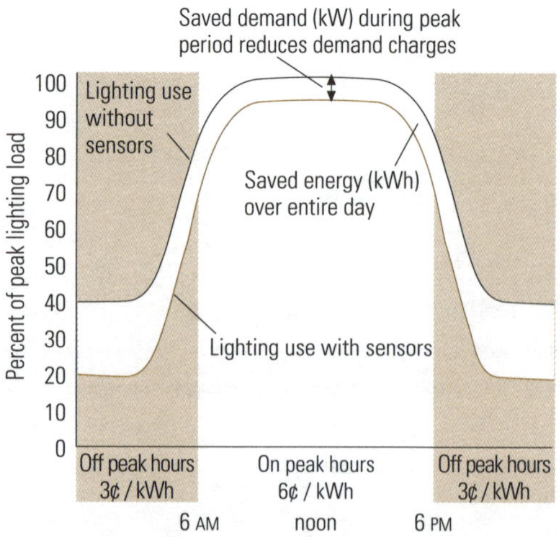

reduced savings by about 30 percent compared to initial estimates.[1]

The greatest impact on both demand *and* consumption occurs in irregularly occupied areas where lights can be shut off or dimmed without creating a problem. Such spaces may include those listed opposite and other areas which are not routinely occupied (by at least one person) during the normal workday.

- individual offices
- secondary school and college classrooms
- warehouses
- storerooms
- lecture halls/auditoriums
- restrooms
- loading docks
- basements
- corridors
- stairwells
- office lounges
- lunchrooms
- conference rooms
- library book stacks

Savings are limited in large spaces where at least one person is often present, such as:

- open-plan offices
- primary school classrooms
- reception areas
- lobbies
- retail spaces
- dormitory and hospital rooms

In all of the above spaces, however, controls can reduce kWh consumption by automatically turning off lights at night.

Goals. In selecting lighting controls, defining *goals first* will simplify choosing *methods later.* Each of the following potential goals should be evaluated when selecting a controls scheme:

- *control*: occupant control of light level and mood

- *light level*: limiting maximum light level and maintaining minimum necessary light level

- *security*: revealing unauthorized intrusion

- *safety*: avoiding accidents in darkened areas

- *notification*: signaling occupants during an emergency, or when closing time is imminent

- *demand management*: shifting peak loads to save money

- *task variation*: adjusting to different light level needs, such as audiovisual presentations.

Interactions. When considering a lighting retrofit, controls should be the last step, following reductions in wattage. There is an approximate correlation between the cost of controls and the installed wattage they serve; major reductions in controls installation cost are possible if wattage has been reduced through other measures. In addition, lighting upgrades typically reduce power density by 50 percent or more, greatly reducing the potential energy and demand savings from controls.

Care is needed when installing controls in an existing building, or when altering a space equipped with lighting controls. Erecting new walls or otherwise subdividing a space may create situations where one person inadvertently controls another's lighting. Electrical contractors must take such changes into account by rewiring controls appropriately as part of their work.

Some buildings are equipped with control systems that communicate with relays through powerline-carrier (PLC) technology that is vulnerable to harmonics created by electronic ballasts and some dimmers. Make sure PLC-based control systems are compatible with any proposed control equipment.

11.1 ON-OFF CONTROLS

The heart of lighting control is turning off unnecessary lights. A wide variety of simple and complex technologies are devoted to this function, providing the designer or retrofitter with many options.

Automatic switching options—elapsed-time, clock, energy management (EMS), occupancy, and photocell systems—may require a separate relay for making and breaking the electrical connection to the lighting load. The relay is an additional piece of equipment that receives a signal from the controlling device. Many vendors offer a single relay that can control up to 20 amps, the typical allowable lighting load per individual circuit breaker. A typical 20-amp relay costs about $20 to $25, and should be included in the economic calculations for control systems, as well as conduit wiring and NEMA-rated enclosures.[2] Larger lighting contactors that can simultaneously make/break connections to several 20-amp circuits cost proportionally more—typically $25 to $50 per 20-amp circuit.[3]

11.1.1 Manual switching

In commercial buildings, it makes sense to group-switch banks of lights based on their daylight exposure.

Toggle switches need little explanation; it is their absence that confounds lighting control. Some buildings erected on a low budget switch whole building floors at once, while others have no switches, and must be controlled at the circuit breaker. This practice, in addition to being electrically unsound, will soon be illegal. The National Energy Act of 1992 mandates ASHRAE 90.1 for new construction, which calls for at least one lighting control (such as a switch) in each space and at each separate task. Adding standard switches in each space also simplifies installation of other options (such as wall-mounted timers or motion sensors) in the future.

In commercial buildings, it makes sense to **group-switch** banks of lights based on their daylight exposure (**Figure 11-2**). In typical open-plan office spaces, workers in perimeter cubicles can exercise manual daylight control if turning off their lights doesn't affect the workers in the interior. This simple concept is lacking in many existing commercial buildings, which gang entire floors on single switch relays, or group lights according to compass directions rather than daylight potential.

Another simple idea for open-plan office space is to include **pull-chain switches** for individual light fixtures that are group-switched. This gives occupants direct control of the lights above their cubicles, and can be especially popular for workers using video display terminals. In some cases, workers are desperate enough for this control that they pry their way into the fixture to remove some or all of the lamps.

Unfortunately, many pull-chain switches may not be reliable for long-term operation because of their flimsy construction. This concept also relies on a light fixture pattern that corresponds closely with the cubicle pattern below; in most cases the two are completely unrelated. With a careful layout and a high-quality product, however, pull-chain switching can be an inexpensive and practical control scheme.

Figure 11-2

Switching plan based on daylight potential

Separately switching perimeter spaces in open-plan offices allows manual daylighting control. This floor is shown with only the core zone fixtures lit.

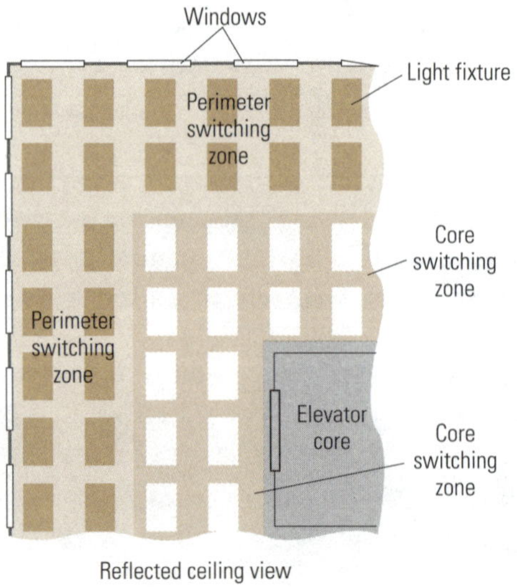

Fixture patterning in alternating rows or checkerboard configurations (**Figure 11-3**) can also provide several light levels, but may create unacceptable variations due to the distance between fixtures that remain lit when others are off. If such a strategy is being considered, fixtures should have a high (about 2.0) spacing-to-mounting height (S/MH) ratio to avoid creating dark spots. For example, a space with a 10-foot ceiling and 8 feet between fixtures would work well with fixture patterning, while a space with an 8-foot ceiling and fixtures 12 feet apart would end up with significant dark areas. Special lenses may be used to increase the S/MH ratio of existing fixtures (see **7.3.2**). Patterning of compact fluorescent (CFL) downlights—which usually have a high S/MH ratio—is another option.

Some track lighting systems control groups of track heads by multiple flips of a single switch, such as the "Smart Start" track system from Capri Lighting.[4] In this system, a relay package interprets the switching pattern from the user, turning on the first, second, or both groups of track heads. Other track systems are available with multiple circuits controlled by multiple wall switches, such as the Halo two-circuit and four-circuit tracks.[5]

Figure 11-3

Fixture patterning

Fixture patterning allows varying light levels, but may create dark spots. This floor is shown with half of its fixtures lit.

Lamp patterning involves wiring the lamps in one fixture on separate circuits. Patterning may be much more cost-effective than fluorescent dimming systems, which are still relatively expensive. A typical arrangement is with 3-lamp fixtures (**Figure 11-4**), where the two outboard lamps are served by one ballast and the single center lamp is served by another, allowing three distinct light levels controllable from two wall switches. Lamp patterning provides better light distribution than fixture patterning, because dark spots are eliminated. However, the wiring is slightly more complicated, and additional ballasts may be required. Both lamp and fixture patterning are options for new construction and major retrofits, where the fixtures are being replaced or rewired.

11.1.2 Elapsed-time switches

These units—also called timer switches—typically fit into or over a standard wall switch box and allow occupants to turn lights on for a defined period, set either by the occupant or by the installer. Lights go off at the end of that period, unless the time cycle has been restarted by the occupant (or manually turned off sooner). Time intervals range from

Figure 11-4

Lamp patterning

Separately switching the individual lamps in a fixture can be an inexpensive and effective means of controlling light levels. This floor is shown with two-thirds of the lamps lit.

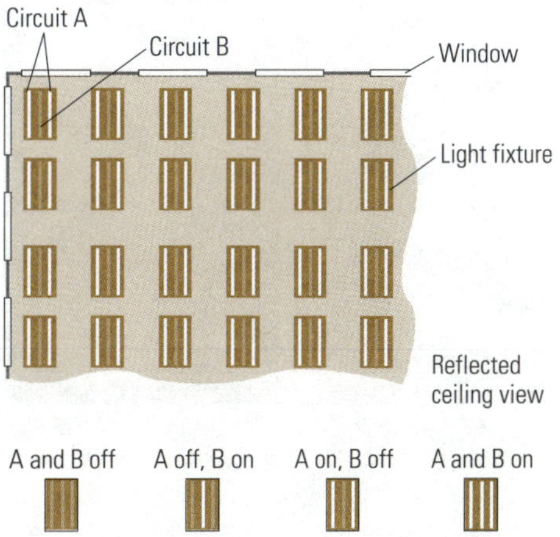

10 minutes to 12 hours. Since they require no consideration of a room's shape, size, or tasks, elapsed-time switches are a much simpler retrofit than occupancy sensors, and are less prone to user maladjustment. Due to their low cost ($15 to $40), these switches offer an economical way to cut unnecessary lighting usage, especially when fixed-duration events occur within a space.

Elapsed-time switches offer an economical way to cut unnecessary lighting usage.

There are two basic types of elapsed-time switches: mechanical and electronic. The mechanical units are little more than spring-wound kitchen timers connected to a small relay. Maximum time settings of 20 minutes are common, though timers with periods up to 12 hours or more are also available. These simple devices can suffer mechanical failures relatively quickly when used in high traffic areas such as school libraries or public spaces.[6] Spring-wound and electronic elapsed-time switches are shown in **Figure 11-5**.

Like the spring-wound type, electronic elapsed time switches provide a defined maximum "on" time, after which lights go off automatically. The user does not, however, set the interval: this is done by the installer using a hidden screw terminal or DIP switches. Since they look like conventional toggle switches, occupants are unaware of the electronic device's existence (when set properly), greatly reducing vandalism and theft. These timers also provide an easy and economical way to comply with energy codes calling for automatic lighting controls.

Earlier generations of electronic timer switches could only control up to 300 watts, and some did not work well with electronic ballasts because of problems in handling inrush current (see **7.2.4** for more on this issue). New designs of timer switches use zero-crossover circuitry and can control up to eight amps (800 or more watts) of incandescent or electronically ballasted fluorescent lamps. A relay can be added between the switch and the lights to increase the wattage capacity of the unit. Electronic timer switches cost about $30 to $40 each in quantity.

In 1995, the Watt Stopper and Paragon each introduced solid state switches with advanced capabilities. The Paragon ET unit looks like a standard switch, can be set for any time period between one minute and 18 hours, and throws the toggle back to its original position at the end of the time cycle. Versions are available that will blink lights once before shutting them off to warn occupants of impending darkness. The Watt Stopper TS-200 uses DIP switches to set times between five minutes and 12 hours, and offers the same blinking option. In addition, the TS-200 can be set to beep every five seconds for a minute before darkness, and includes a backlit LCD screen showing time remaining before lights are turned off. Both of these products are discussed in greater detail in E SOURCE *Product Profile PP-95-3*, "New Electronic Timers Simplify Lighting Control" (June 1995).

Choosing a short time interval is tempting for energy savings, but can create functional problems. Librarians returning books to shelves, for example, may need to remain in an aisle longer than the 10 to 20 minutes that browsers typically require. In one application, short intervals resulted in user frustration, disconnection of the switches, and 24-hour operation of stack lighting.[7]

Figure 11-5

Elapsed-time switches for lighting control

Elapsed-time switches are an inexpensive means of controlling lights in irregularly used spaces.

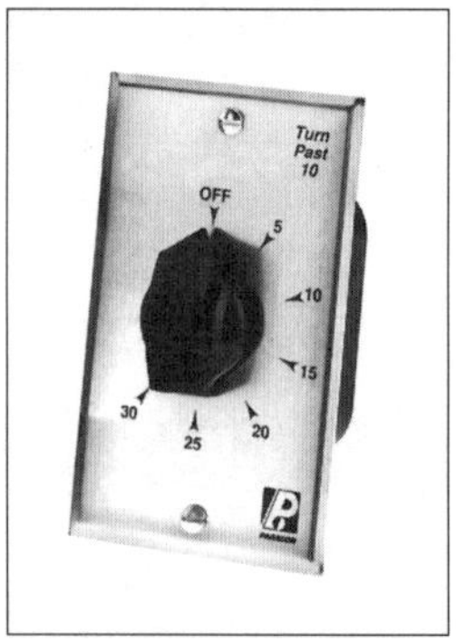

Spring-wound switch

Paragon ET electronic toggle switch

The Watt Stopper's TS-200 electronic pushbutton switch

Source: Paragon Electric Company, The Watt Stopper

Early solid-state switches used triacs (also known as reversing thyristors), which made them incompatible with electronic ballasts because of current leakage, as discussed below. Triacs have been all but eliminated from the current generation of timer switches.[8]

11.1.3 Clock switches

Where occupancy follows a well-defined pattern (such as in merchandising), clock switches may be the most cost-effective option. Clock switches control lights by turning them on and off at prearranged times, regardless of occupancy; they are typically located in electric closets which house lighting power panels (**Figure 11-6**). These devices cost relatively little to install, are generally invulnerable to vandalism or user maladjustment, and can control large loads with single sets of contactors. Clock switch equipment can be mechanical devices employing motors, springs, and relays, or sophisticated electronic systems that handle several time schedules simultaneously. Typical unit prices are shown in **Table 11-1**.

Mechanical clock switches may require manual correction for daylight saving time, or after a power failure (unless equipped with battery backup, which can triple their price). Spring-wound mechanical backup is also available for maintaining operation during power outages. Some mechanical clock timers may also require a neutral wire at the light switch, which may be unavailable, or an expensive retrofit. (Without the neutral wire, the device has no means of obtaining power for its own operation.) Electronic clock switches, on the other hand, routinely include battery backup and can be easily programmed to adjust for the shift to daylight savings time, as well as complete scheduling for holidays.

As with other electronic switching devices, those that use triacs can "trickle" a small amount of power to lamps even when they are shut off. Even without triacs, some clock switches and sensors trickle a small amount of current to lamps when in the "off" position as a continuity check. Two-pin (preheat) compact fluorescent lamps will burn out very quickly if subjected to trickled current because it makes the lamp's starter glow, which then discharges to the cathode, causing it to sputter without starting the lamp. Trickling may also affect the operation of certain electronic ballasts: a capacitor in the ballast circuit is charged by the trickled power, and then discharges into the lamp, causing the cathode to sputter, reducing lamp life. To avoid these problems, specify units without triacs.

Electronic clock switches routinely include battery backup and can be easily programmed to adjust for the shift to daylight savings time.

Table 11-1

Typical prices of clock switches for lighting control

24-hour mechanical	$40
7-day mechanical	$85
7-day mechanical with battery backup	$265
1-channel 365-day electronic	$160 + relay
4-channel 365-day electronic	$480 + relay

Source: Manufacturer quotes

Figure 11-6

Mechanical and electronic clock switches

Clock switches are inexpensive and perform simple schedule control of lighting equipment.

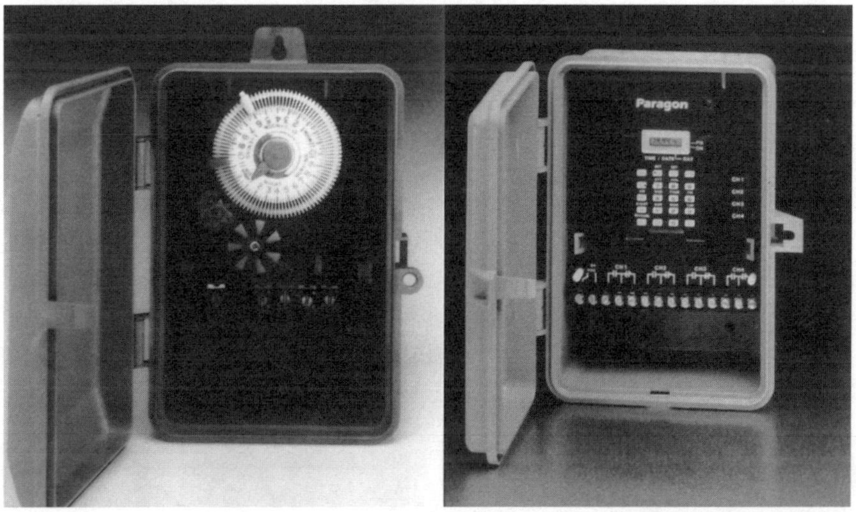

Source: Paragon Electric Company

Unless a space has been designed to use daylight, daylight control with photocells is limited to fixtures within about 12 feet of perimeter windows.

11.1.4 EMS controls

Energy management systems (EMS) perform the same function as clock switches but with additional sophistication and features. A common EMS function is a "sweep" system that automatically cycles lights one floor at a time, signaling occupants that lights will shortly be shut off at the end of the workday. With this type of system, occupants can then override the shutdown in their area by touching a local switch or phoning the EMS panel operator.

Remote relays are typically used with an EMS to control entire circuits at the breaker panel. Such systems were designed to work with magnetic ballasts, which (unlike electronic ballasts) do not create a large inrush current during start-up. Caution is suggested when retrofitting such a system with electronic ballasts, which may open circuit breakers when they are confronted by the large inrush current from many electronic ballasts turning on simultaneously.

If an EMS is going to be used just for time scheduling, two factors are key in determining if the EMS is more cost-effective than simple local controls:

- the typical per-point cost for the EMS, which may range from $200 to $1,500, depending on the type of system, its expandability, and the cost of installation labor, and

- the maximum average lighting wattage that can be controlled by a point.

For example, a $100 local controller can handle several hundred watts for about 30¢ per watt. To yield comparable economics, a $600/point EMS would need to control at least 2,000 watts ($600 ÷ 2,000 = 30¢) of lighting at each point. If it handles more wattage per point, then it is more cost-effective than local controls.

The next question is whether *after a lighting upgrade* it is possible to shut off (or dim) 2,000 watts at one time without compromising the flexible use of a space. At a typical one-watt-per-square-foot post-upgrade power density, this example would involve controlling a 2,000-square-foot area with one EMS point. Are there spaces equal to or greater in size that can be controlled remotely without interfering with their operation? If so, consider EMS control; if not, use local controls (such as sensors, timers and photocells).

11.1.5 Photocell controls

On-off photocell controls sense natural light indoors or outdoors and turn lights on when light levels fall below a defined point. Originally used for on-off control of outdoor and atrium lighting, photocells are slowly becoming more popular as a means to sense natural daylight and reduce or shut off electric lighting in perimeter spaces.

Unless a space has been architecturally designed to use daylight, daylight control with photocells is limited to fixtures within about 12 feet of perimeter windows. Shutting off fixtures beyond that point may result in relative darkness for some occupants. Photocells are usually connected to a relay or dimming system that controls a bank or pattern of light fixtures. Typical single-circuit photocell controls cost about $50.[9]

Wireless pagers (see **11.1.7**) can now fulfill some outdoor lighting control instead of photocells and/or timers.

11.1.6 Occupancy controls

Occupancy sensors can be a cost-effective energy saving tool if applied properly. Sensor installations sometimes yield smaller-than-expected savings, however, due to factors such as improper product selection or installation, unanticipated interactions with other building components, or failure to adequately consider utility time-of-use rates and demand charges. Users and utility DSM programs can maximize the performance and cost-effectiveness of sensor installations by considering these issues.

Because of the popularity of occupancy sensors and their wide range of application issues, we examine this technology in detail here. Many of the installation issues and economic considerations discussed here also apply to other control technologies.

Four types of signals. Most occupancy sensors turn lights on or off by detecting heat or a shift in the frequency of reflected ultrasonic sound waves. Units that use audible sound or microwaves are less common. **Table 11-2** lists manufacturers of infrared and ultrasonic sensors.

Passive infrared (PIR) sensors are the most common type, and are able to "see" heat emitted by occupants. Triggering occurs when a change in infrared levels is detected, such as when a warm object moves in or out of view of one of the sensor's "eyes." PIR sensors are quite resistant to false triggering. They are best used within a 15-foot range, for two principal reasons:

- There are potential "dead" spots between their wedge-shaped sensory patterns that get wider with distance (see **Figure 11-7**).

- Being "passive"—they do not send out any signal—they depend on the intensity of the heat output of the moving part of the subject, which drops with the square of the distance.

Ultrasonic (US) sensors emit a high frequency (over 20,000 cycles per second) sound, above the human and animal audibility ranges, and listen for a change in frequency of the reflected sound. By emitting a signal, instead of only receiving it, they are able to cover larger areas than PIR sensors and are noticeably more sensitive. US sensors are also more prone to false triggering: air motion, due to a person running by a doorway, or the on-off cycling of an HVAC system, may trigger a poorly located or adjusted sensor.

Table 11-2

Manufacturers of occupancy sensors

● Passive infrared (PIR) ● Ultrasonic (US) ● PIR/US combination ● PIR/Audible

PIR	**BRK Electronics** A Division of Pittway Corp. 780 McClure Road Aurora, IL 60504-2495 tel 708-851-7330
PIR	**Bryant Electric Co.** Division of Westinghouse Electric Co. 185 Plains Road Milford, CT 06460-2465 tel 203-876-3600 fax 203-876-3675
US	**Earlwood Technologies** 1055 W. Germantown Pike Norristown, PA 19403 tel 215-539-4210 800-233-5193 fax 215-539-5921
PIR	**Heath Co.** 455 Riverview Drive P.O. Box 1228 Benton Harbor, MI 49023-1288 tel 616-925-2986 fax 616-925-5990
PIR	**Honeywell, Inc.** 1885 Douglas Drive N. Golden Valley, MN 55422-4386 tel 612-542-7500 800-345-6770, ext. 758
PIR	**Hubbell, Inc.** Wiring Device Division P.O. Box 3999 Bridgeport, CT 06605-0933 tel 203-337-3100 fax 203-368-6245
PIR	**Leviton Manufacturing Corp.** 59-25 Little Neck Parkway Little Neck, NY 11362-2591 tel 718-229-4040 fax 718-631-6598
PIR	**Lightolier Controls** 100 Lighting Way Secaucus, NJ 07096-1508 tel 201-864-3000 800-526-2731
PIR	**Lithonia Control Systems** One Lithonia Way Decatur, GA 30035 tel 404-987-4400 fax 404-987-1002
PIR, US	**MyTech Corp.** 706 Brentwood Street Austin, TX 78752 tel 800-888-8006 fax 512-450-1215
US	**Novitas, Inc.** 1657 Euclid Street Santa Monica, CA 90404 tel 310-452-7890 fax 310-452-6168
PIR	**Pace Technologies** P.O. Box 30394 Walnut Creek, CA 94598 tel 415-934-6274 fax 415-934-6778
PIR	**Pass & Seymour Legrand** P.O. Box 4822 Syracuse, NY 13221 tel 800-223-4185 fax 315-468-6305
PIR	**RAB Electric Manufacturing, Inc.** 170 Ludlow Avenue Northvale, NJ 07647-0970 tel 201-784-8600 fax 201-784-0077
PIR, PIR/Audible	**Sensor Switch, Inc.** 10 Capital Drive Wallingford, CT 06492 tel 203-265-2842 fax 203-269-9621
PIR, PIR/US	**Solidyne, Inc.** 1202 Carnegie Street Rolling Meadows, IL 60008 tel 708-394-3333 fax 708-394-8083
PIR	**TORK** One Grove Street Mount Vernon, NY 10550 tel 914-664-3542 fax 914-664-5052
PIR, US, PIR/US	**Unenco, Inc.** 2555 Nicholson Street San Leandro, CA 94577 tel 800-227-0452 fax 415-895-5753
PIR	**Visonic, Ltd.** 20 Northwood Drive Bloomfield, CT 06002 tel 203-243-0833
PIR, US, PIR/US	**The Watt Stopper** 2800 De La Cruz Boulevard Santa Clara, CA 95050 tel 408-988-5331 800-879-8585 fax 408-988-5373

PIR and US sensors dominate the market. Several manufacturers offer both types, and some sell sensors using both technologies simultaneously. Such "dual technology" sensors combine the sensitivity of ultrasonics with the passive infrared units' resistance to false triggering.

Audible sound sensors listen for noise made by people and machines, and assume that such sound is a result of occupant activity. They may react to nearby vibration (such as a slammed door) or to street noise, and require relatively high sound levels (higher than a typical quiet office). They are best applied in an industrial facility or warehouse where machinery, not people, frequently breaks the silence.

Microwave sensors are similar to ultrasonics in that they emit a signal—in this case, a radio signal—and measure a change in frequency when that signal is reflected. At present, only one company[10] makes microwave sensors and very little data exists on field experience or installed cost.

Figure 11-7

Representative sensor coverage diagram

Ultrasonic sensors can detect motion at any point within the contour lines. Infrared sensors "see" only in the wedge-shaped zones and they don't generally "see" as far as ultrasonic units. Some sensors "see" further straight ahead than to the side. The ranges shown here are representative; some sensors may be more or less sensitive.

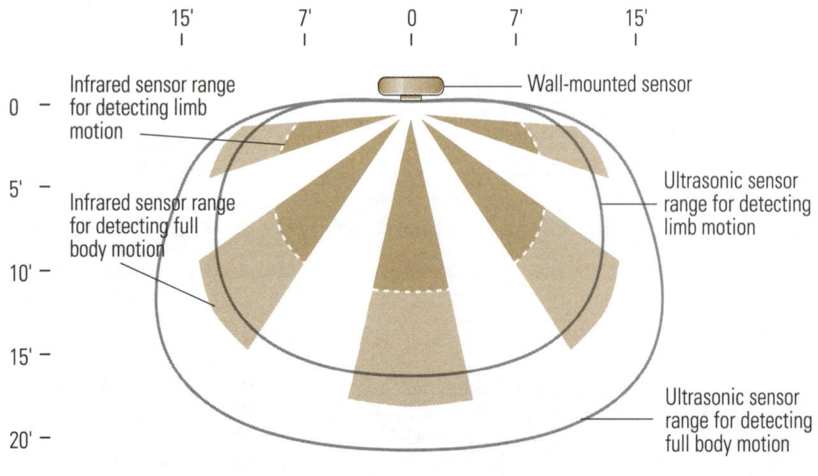

Two mounting configurations. "Wall-mounted" sensors are the most common type, and typically replace an existing wall switch. Most have these characteristics:

- They typically cost between $30 and $90, uninstalled.
- Their sensor pattern spreads primarily to the left and right (60° to 90° in each direction) in the horizontal plane, but only minimally in the vertical plane.
- An off-auto switch is available to either manually shut off the lights when a space is occupied, or allow the sensor to control lighting; some also have a manual override to keep lights on when a room is empty, ensuring that light is available even if a sensor fails.
- Time delay and sensitivity controls are accessible without removing the sensor from the switch box.
- No external relay is needed; most wall-mounted sensors can handle about 700 watts of fluorescent lighting.
- A light-emitting diode (LED) on the casing shows when the sensor "sees" motion and/or initiates the time delay period.

"Ceiling-mounted" sensors are typically used in larger rooms because they can cover greater areas, and usually have these characteristics:

- Most cost between $50 and $120, uninstalled, with relay and low-voltage power supply.
- A separate low-voltage power supply and relay (typically able to handle over 2,000 watts of lighting) powers the sensors and controls the lighting; the sensor triggers the relay to send power to the lights.
- Sensory patterns are available to suit the shape and size of the room; a 360° sensor

"sees" in all directions, while a "one-way" sensor may see only 90° to either side of the sensor in the horizontal plane, and/or 90° in the vertical plane.

■ Time delay and sensitivity controls are mounted on the sensor, making them inaccessible to occupants.

■ Several ceiling-mounted sensors may be used together in one room, or may be combined with a wall sensor, in order to cover an oddly-shaped or large room (over 800 ft^2).

Quantifying potential savings. Consolidated Edison found that occupancy sensors installed under its DSM programs yielded average reductions in energy and demand of about 30 percent.[11] Any given installation can range widely from this average, however. Table 11-3 illustrates the typical range of savings from sensor installations in various types of spaces. Except for open-plan offices and classrooms, the savings often range by a factor of two or three. This uncertainty makes it difficult to predict the performance and cost-effectiveness of any given installation. Several options are available, however, for making more precise estimates or measurements:

■ *Lighting loggers* count lighting hours, record the time-of-use and/or duration, and in some models, correlate it with sensed occupancy. (For more on loggers, see E SOURCE *Tech Update TU-94-4*, "Market Survey: Lighting Loggers and Occupancy Loggers," April 1994.) These battery-operated devices can be placed inconspicuously in rooms and retrieved for later analysis in relation to room schedules or measured occupancy to estimate the savings potential available through occupancy sensor control. If one goal is to reduce peak demand, the most sophisticated "loggers" are more useful, since they will report when lights are on during the peak demand period.

Table 11-3

Typical range of savings from occupancy sensors

Savings range by a factor of two or three in most applications, with the exception of open plan offices and classrooms. Actual savings may differ.

Type of room	Energy savings
Private office	13–50%
Open plan office	20–28%
Classroom	40–46%
Conference room	22–65%
Rest room	30–90%
Corridors	30–80%
Storage area/closet	45–80%

Source: Advanced Lighting Guidelines, U.S. EPA

■ *School schedules* may indicate when classrooms and lecture halls are in use, while office and cleaning hours are well defined for open plan offices, lunch rooms, rest rooms and the corridors that serve them.

■ *Recording ammeters* may be connected at lighting breaker panels to determine when banks of rooms on a common electric feeder are using their lights. Savings can be determined by comparing the ammeter's data to work and cleaning schedules.

■ *Random surveys* (through observation of a building's exterior at night) may reveal rooms where lights have been inadvertently left on, as may routine interrogation of custodial and security personnel.

■ *Existing timers*, scheduling controllers, and EMS operating plans will indicate current usage.

A commercial facilities study performed by San Diego Gas & Electric Company found that measured burn hours in halls and lobbies were 50 to 72 percent greater, respectively, than estimates based on operation schedules. In private areas and conference rooms, however, burn hours were 29 to 46 percent less than estimates.[12] Incorrect estimates could result in significant payback errors for sensors, which are commonly specified for smaller spaces where they can be easily installed at existing switch locations.

The seasonal variations in operation of a facility also need to be considered to avoid incorrectly extrapolating one month's data to a full year.

Consolidated Edison found that occupancy sensors installed under its DSM programs yielded average reductions in energy and demand of about 30 percent.

If occupants complain about sensor operation, a common response is to raise sensitivity to the maximum, which may cause more frequent false triggering.

Sustainability of savings. Unlike an invisible electronic ballast that, once installed, will continue to draw fewer watts than its predecessor, sensors are quite visible and can be improperly adjusted, stolen or vandalized, or fooled into perceiving human motion when a space is unoccupied. Manual override switches on some sensors can cause lights to remain on even when spaces are unoccupied.

We know of no data on the long term persistence of savings from sensors. Anecdotal information indicates, however, that continued proper operation depends upon precautions taken by users in equipment choice, location, and training of building staff.

At one university in New York City, dozens of wall-mounted sensors were stolen because theft-resistant mounting screws were not used, and other sensors were vandalized.[13] In a large downtown New York hotel, cleaning crews mistook sensors for eavesdropping devices and disabled them. On the other hand, inclusion of building personnel in the sensor planning and installation process at a Newport Beach, California office complex was successful in attaining cooperation and acceptance.[14]

If occupants complain about sensor operation, a common response is to raise sensitivity to the maximum, which may cause more frequent false triggering. If such action still fails to resolve the problem, a manual switch may be re-installed, eliminating any savings. Proper design, installation, and follow-up adjustment are therefore essential parts of any sensor program. Building maintenance personnel must be trained to keep sensors operational, rather than disconnecting them when problems occur.

Sensor interactions with other systems. Sensors interact both physically and economically with other systems in a facility, and require some thought during the planning stage of a lighting upgrade to avoid later complications.

Check the maximum allowable wattage of sensors rather than claims of square-foot cov-

CASE STUDY **OCCUPANCY SENSOR INSTALLATION**

A promotional video from one sensor manufacturer states that the installation of occupancy sensors on a 10-kW lighting circuit in a major New York City building reduced lighting energy consumption by 56 percent. This is a particularly high but plausible savings level, because prior to the sensor installation, service personnel turned on all lights in the building before occupants arrived in the morning. However, because the reduction in peak lighting demand was only 20 percent, the dollar savings—taking into account the cost of power at the time of savings and the actual demand charges involved—are about 38 percent (**Figure 11-8**). Sensor manufacturers tend to stress the energy savings from their products in their promotional activities and do not typically place as much emphasis on demand and dollar reductions, so users should be careful to evaluate savings projections in the context of local utility rate structures.

Figure 11-8

Savings from a sensor installation in New York City

When demand changes are a significant portion of the total electric bill—as in this case—the dollar savings from sensor installation may be considerably less than the energy savings, if much of the energy savings occur during off peak periods, when energy rates are lower and demand charges do not apply.

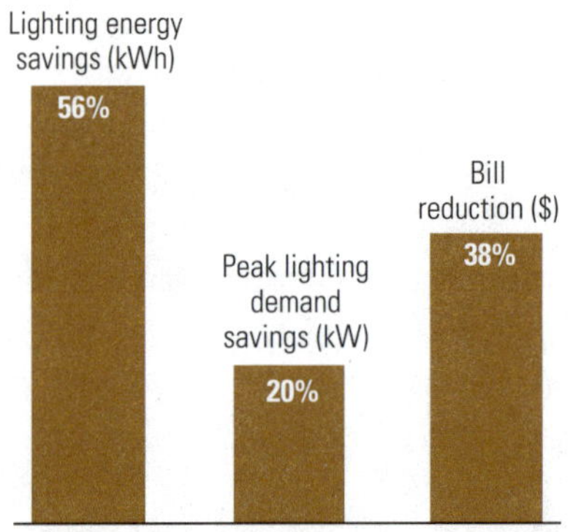

erage. For example, a wall sensor advertised as covering 2,700 square feet may have a peak capacity of only 760 watts—but even the most efficient lighting would require over 1,000 watts to properly illuminate an office of that size. To fully cover such a space would require a second sensor and/or a pilot-duty relay, significantly increasing the total installed cost.

Most comprehensive lighting upgrades sharply reduce lighting wattage, making it possible that a room's (or circuit's) wattage may be less than the minimum needed for the sensor to work. Wall-mounted sensors require at least 40 to 60 watts, but a single-lamp T8 fixture equipped with an electronic ballast draws only 31 watts. While the cost-effectiveness of controlling such a small wattage may be open to question, it may be required under some codes (see sidebar).

Potential incompatibility with electronic ballasts. Many sensors are designed to handle the relatively mild starting power surge (also called inrush current) needed by magnetic ballasts, but are challenged by the significantly larger (though much shorter) surge demanded by electronic ballasts. As a result, some models will suffer shortened lives when controlling their rated wattage of electronic ballasts. See **7.2.4** for more on inrush currents and how to deal with them.

Most vendors have upgraded at least some of their sensors to make them compatible with the starting surges from electronic ballasts. Other sensor manufacturers are likely to follow suit. Because sensor products are changing regularly and because some vulnerable early models remain in distributors' stock, users of electronic ballasts should verify compatibility with particular sensor models by contacting manufacturers directly.

Not only can some ballasts damage sensors, some sensors can damage ballasts or lamps. Many wall-mounted sensors utilize a "triac" (also called a reversing thyristor), which is a semiconductor that allows the sensor to "steal" for its own operation a portion of the power supplied to a fixture's ballast. By chopping off a section of the power sine wave, a triac is able to control the timing of the sensor's cycle while working with other components like a solid-state relay to control power supplied to the ballast. (Ceiling-mounted sensors utilize separate mechanical relays to control power to lights, so there is no triac-type disturbance in the power they supply.) While magnetic ballasts do not appear to notice this slightly altered waveform, some electronic ballasts may experience it in two unfortunate ways:

- as a spike, leading to ballast shutdown to avoid damage, whereupon the lights may go off or flicker;

- as insufficient power to maintain the lamp arc, leading to frequent, rapid, and typically invisible restarting of the lamp, shortening its life.

Compact fluorescent lamp life is especially vulnerable to cycles shorter than the standard three hours.

Some electronic ballast manufacturers, such as Magnetek, state that use of their equipment on a triac-equipped circuit will void the ballast warranty.[15] Advance Transformer Co. strongly recommends that its ballasts not be used with triac-equipped sensors.[16] Some sensors isolate their triacs from the circuitry supplying power to the ballast by employing a tiny mechanical relay that provides a clean sine wave source of power while increasing the wattage controllable by the sensor. Specifying use of such a mechanical relay can avoid triac-related problems.

Unless literature accompanying a sensor specifically addresses the issue of potential damage to either the sensor or the ballast, users should request clarification from vendors regarding the compatibility of any particular ballast/sensor combination. Users should also check with the electronic ballast manufacturer to determine if the ballast's use with a given sensor will affect the ballast warranty.

Potential incompatibility with area-wide dimming. Some buildings have area-wide dimming systems that change the waveform of the power to the fixture circuits, resulting in an across-the-board power reduction. Some sensors will work acceptably well for a time and then fail because their solid-state circuitry may be stressed by the corrupted waveform.

Impacts on lamp operation and life. The lifetime of fluorescent lamps may be significantly shortened by rapid on-off cycling. While rated lamp life is given in hours, actual life is affected by the number of times the lamp is started. Standard 4-foot T12 and T8 lamps can typically be started about 7,000 times during a 20,000-hour life before they burn out, and more starts per month could mean shorter lamp life, more frequent burnouts, more lamp purchases, and more lamp maintenance. Field experience indicates that compact fluorescent lamp life is especially vulnerable to cycles shorter than the standard three hours used in developing rated life.[17]

Some have argued that sensors may increase the interval between relampings by stretching out a shorter lamp life over more hours when the lamps are kept off. This is only true when the sensor keeps the lamps off for relatively long intervals between relatively few on-cycles. For example, sensor control may extend the relamping interval in a storeroom that is visited briefly twice a day. In a frequently visited copy room, on the other hand, sensor control is likely to lead to more frequent relamping because of rapid cycling.

While the electricity saved by automatically shutting off the lamps is almost always worth more dollars than the added cost of relamping (including labor), the increased lighting system maintenance may interfere with other maintenance duties, require overtime or hiring of additional help, and confound efforts toward group relamping, which assumes relatively consistent lamp usage throughout a space.

Setting sensors to their maximum cycle time (over 20 minutes) minimizes the reduction in lamp life while sacrificing some energy savings. Each time someone makes a motion within this time period, the sensor's timing cycle is restarted, and the lights remain on. Only after the space has been continuously unoccupied for more than 20 minutes will the lights go off.

High-intensity discharge (HID) lamps (such as metal halide or high-pressure sodium) take several minutes to warm up, during which time they provide relatively little light. Some types may take even longer to restrike immediately after being shut off. Due to these delays, standard sensors should not be used with such lamps. While limited at this time to higher wattage HID lamps, control systems (such as those by Holophane) exist that avoid this problem by dropping light output (and energy use) to a low level during unoccupied periods while never fully shutting off the lamps.[18] When signaled by a sensor designed for this type of control, the warmed-up lamp rapidly reaches full brightness.

Codes and Controls

New York State's Energy Conservation Construction Code mandates some form of automatic control, which could also include timers or an energy management system, in all new construction covered by the code, and in all upgrades that affect 50 percent or more of an existing building's lighting controls, for spaces not in 24-hour use.[19]

The National Energy Policy Act of 1992 mandates ASHRAE 90.1—which encourages sensors—as its standard for acceptable lighting design in new facilities. ASHRAE 90.1 allows discounting of installed lighting wattage in a new facility to meet its power density limits if controlled by automatic devices such as timers and sensors.[20] California's Title 24 has similar provisions. Under the National Energy Policy Act, all state energy codes should include 90.1's procedures, or else demonstrate how an alternative is comparable.

False triggering. The thermal "signature" seen in a room by an infrared sensor can be altered by other, nonhuman, heat sources. Some sensors can, for example, be fooled by curtains moving in front of sunny windows, objects hanging between sensors and light sources (such as mobiles and balloons), pets, or heating equipment that cycles on and off (such as an oven). A few sensors screen out these sources by requiring multiple signals to be triggered, which also slightly delays triggering of the lights.

While ultrasonic sensors are generally more sensitive than infrared units to small body motions, they are also much more prone to false triggering. They can be confused by random air motion, or movement by anything within the sound field, including a computer printer or air from a diffuser. They are also sensitive to changes in air density, making false triggering easier at certain times of the year in spaces where temperatures vary due to climate (e.g., a warehouse or stairwell). The biggest source of false triggering, however, is probably the incorrect positioning of the sensor, so that its field of view extends beyond the controlled space.

Interaction with emergency, safety, communications, and security systems. While all ultrasonic sensors operate immediately after receiving power, most older and some modern infrared sensors do not have "instant-on" capability. Those without instant-on circuits take several minutes to warm up, during which time the lights remain off. This can be inconvenient in rooms with noninstant-on ceiling-mounted infrared sensors when the lights are turned off manually at a wall switch (during a slide presentation, for instance). That will cut off power to the sensor (if it is wired in series with the switch) which will then need several minutes to warm up and turn on the lights after the wall switch is turned back on. This problem can be avoided at some effort and expense by installing extra wiring to ensure that the ceiling-mounted sensor receives continuous power, even when the light switch on the wall is turned off. This is not an issue with wall-mounted infrared sensors because they are wired to receive power even when the sensor has been manually turned off.

Special care is needed when using sensors to control any lighting needed for emergencies. Delayed operation after a power failure (if sensors are not "instant-on") could create a serious hazard during an evacuation. Moreover, emergency lighting often remains on continuously to keep battery-backup ballasts continually charged; doing so avoids running dedicated power wiring. Controlling such lighting with a sensor will rapidly wear out these rechargeable batteries, which are not designed to handle numerous discharge cycles.

One vendor's sensor may be capable of "seeing" full body motion up to 30 feet away, but a different brand may be unable to discern the same motion at half that distance.

Limits in existing wiring and circuitry. Some sensors require grounding to avoid false triggering. If the existing building wiring does not include a ground, or if it is otherwise not fully functional, such sensors will turn on lights each time there is a slight electrical disturbance (such as startup of nearby motors). Others require a neutral wire, which may not exist in buildings where codes do not require them.

In large open offices, several switches may control the lighting, but having part of the room go dark when it's unoccupied may be disconcerting to employees staying late after others have left. If users do not want this to happen, sensors and lighting circuits need to be wired to one high-capacity master relay, further increasing installed cost.

Lack of uniform sensor standards. While most vendors provide general technical data on their products, no commonly accepted sensitivity standard exists for motion sensors at this time. For example, one vendor's sensor may be capable of "seeing" full body motion up to 30 feet away, but a different brand may be unable to discern the same motion at half that distance.

Barring adoption of uniform standards by manufacturers in the near future, interested users should try out samples and speak to other users to ensure the sensors will meet their needs. More broadly, users and utilities

There are other ways to achieve—at a lower cost—much of the savings gained from sensors.

should consider adopting specifications for occupancy sensors that will help avoid the potential problems cited above and help ensure the trouble-free, cost-effective operation of sensors in appropriate applications.

Alternatives to occupancy sensors. Except where required by new energy codes, lights need not be turned off automatically; humans can do it, too. In some cases, training custodial or security personnel to shut off lights as part of their jobs may be more cost-effective than sensors, even if such people are not as consistent as sensors. Field surveys of two schools performed by the Lighting Research Center, for example, found that conscientious direction of school personnel captured most of the savings normally associated with sensors.[21] There are other ways to achieve—at a lower cost—much of the savings gained from sensors. For example, if lighting circuits are easily accessed and large areas can be turned off at one time, wiring into an existing energy management system may be cheaper (**11.1.4**). Library stacks, storage rooms, and open-plan offices can be controlled by timers or elapsed-time switches (**11.1.2**), which provide a defined "on" period after initiation, regardless of occupancy.

Recent sensor developments. Improvements in sensor design, circuitry, and range are occurring very quickly. For example, ultrasonic "boosters" greatly increase the capability of one wall-mounted sensor to cover very large or oddly shaped rooms, without additional

A BASIC SENSOR SPECIFICATION

A good sensor will have:

- An easily accessible, well-labeled, and adjustable "on" period between sensed motions, preferably as short as 15 seconds (or shorter with a "cheater key" for easy testing) and as long as 20 minutes to ensure no false "offs" (such as when a computer operator's back is to the sensor).

- An easily accessible, well-labeled, and adjustable sensitivity control that provides a wide range of sensitivity to account for sensor placement and possible sources of false triggering (for example, a nearby air diffuser).

- Protection from false triggering due to radio-frequency interference or poor grounding.

- Circuitry that fails in the "on" position so failure will avoid a hazardous condition, and provide light while the failed sensor is being removed.

- No means to override it to "on" without a special key or tool.

- Clear and accurate diagrams showing shape and size of the sensed zones (and preferably a scaled template to allow easy design layout with scaled floor plans) for full body motion, and separately for hand/arm motion.

- Warranty covering replacement for at least three years after installation.

- No troublesome circuitry that limits its applicability with other basic equipment, such as electronic ballasts; such assurances should be written into the warranty.

- Sufficient sensitivity to be triggered by a minor arm motion, such as reaching for a telephone, anywhere in the room.

Many sensors on the market today meet these criteria. Those with triac-equipped circuits and those lacking "instant-on" circuitry may be inappropriate for certain applications, as discussed above.

wiring. When connected to a heavy-duty relay, a single sensor could effectively cover several thousand square feet, needing only hand/arm motion to trigger it.

In other developments, SensorSwitch now offers infrared sensors that work in combination with audible sound to avoid false offs; some new infrared sensors are equipped with double "eye" sets to minimize "blind spots; and many of MyTech's ultrasonic sensors are now equipped with a "smart" chip that learns the shape of a room after several on-off cycles, making it more sensitive to movement within a space.

Multiple-step dimming systems are now entering the market. Instead of cycling lamps off and on, sensors would be designed to drop light levels down to a field-adjustable minimum point when a space is unoccupied. Energy savings could be increased by setting the sensor period much lower than the typical 20 minutes, since lamp life will be unaffected by frequent cycling. (Lamps should still be manually or automatically turned off at night, however.) Another potential development: occupancy-switched outlets or power bars that can control task lights or computer equipment in individual offices.

11.1.7 Wireless pager controls

Radio control is also an option for certain types of lighting equipment—especially outdoor systems. Profile Systems offers the Profile 2000, a pager-driven lighting controller designed to replace photocell and timer controllers in outdoor applications such as billboards, parking lot luminaires, and highway lighting. The system consists of a radio receiver/relay box mounted between the power line and the load and a digital pager network that covers most of the urban United States. According to the manufacturer, the system avoids the maintenance and reliability problems that can occur with photocells and timers. (Although photocells can fail because of environmental factors—such as bird droppings—they can deliver good reliability with them when they are installed properly and protected adequately.)[22] The company claims lighting system cost reductions of 20 to 40 percent based on energy and maintenance savings.[23]

To use the system, a user purchases the receiver/relay boxes for about $400 each and furnishes the vendor with on and off instructions, which can be defined relative to daily sunrise/sunset times or as absolute times (for example, turn on at 30 minutes before sunset, turn off at 3:00 AM). The pager network then programs the boxes in the field. Four times each day, pager signals transmit the day's sunrise/sunset times and synchronize the clocks on the network, correcting for any seasonal time changes, clock drift, or power interruptions. The system offers additional features for billboard operators—for example, the lights on the billboard can be programmed to turn off exactly when the advertising contract ends. About 10,000 of these pager controllers are currently installed in 38 states, mostly for billboard lighting. The company is expanding into other applications such as parking lots, parks, and recreational facilities.

Profile now offers a two-way unit that costs $800 and can relay the lighting system's status (burned out lamps, etc.) back to the company. This system includes a bar-code reader that field workers can use to inform the company that repairs have been completed or a new billboard has been posted. Profile has installed about 1,500 of these two-way systems.

11.2 Dimming Controls

Most power reducers cut both light output and wattage to about the same degree, despite occasional claims to the contrary.

Dimming controls reduce the output of light sources. Compared to on-off controls, dimming adds sophistication, potentially deepens energy savings, better aligns lighting service with human needs, and can extend lamp life. Unfortunately, dimming also adds complexity and expense to lighting systems, and can shorten lamp life under some conditions. Dimming systems should therefore be considered carefully and compared to simpler systems that may also produce the desired lighting effect and/or energy savings.

Lighting designers draw a distinction between "energy dimming," which includes products that allow dimming as low as 10 to 20 percent of full light output, and "architectural dimming," which drops light levels as low as one percent. If the goal is energy savings, products that meet the former definition may be preferable, as they usually cost less than full-range dimming products.

If dimming is used for visual effect, remember that the human eye involuntarily compensates for reductions in light levels. As discussed in **2.2.5**, for example, a perceived 50 percent reduction in illuminance requires a 75 percent reduction in measured levels. This "inverse square law of perception" creates a unique obstacle to some dimming systems: unless they can reduce actual light levels by at least 90 percent (typical during audiovisual presentations), they will be perceived as providing too much light at their lowest setting. No HID system can dim this low, for example, without extinguishing the lamps. Many dimming electronic ballasts drop only to about 20 percent. The bottom line: you may need deeper dimming than you think for applications such as audiovisual presentations.

11.2.1 Power reducers

Power reducers attach to fluorescent lamps or ballasts and cut power (and light level) by a defined percentage, typically 30 or 50 percent. Once installed, this reduction cannot be changed except by removing the power reducers. They became popular in the 1970s as a one-shot method to reduce light output and wattage in overlit spaces. Most power reducers cut both light output and wattage to about the same degree, despite occasional claims to the contrary. In some cases, merely lowering the temperature of a fixture by cutting wattage does lead to greater light output, but this is an artifact of fluorescent lamp technology, not a direct effect of the power reducer. Power reducers are also discussed in **7.2.8.**

Power reducers are typically wired in series with existing T12 magnetic ballasts (they cannot be used with T8 lamps or electronic ballasts) and can cause problems when used with reduced-wattage ballasts or lamps. Some power reducers slip between lamps and their sockets, eliminating any wiring. Because they also reduce power to lamp cathodes, lamp life can potentially decrease due to cathode sputtering.

These devices are most appropriate for retrofits because no wiring is required beyond the fixtures in which they are installed. For new construction, other options such as low-ballast-factor ballasts, fewer lamps, or smaller lamps generally provide higher efficacy than power reducers. The advent of low-cost electronic ballasts, some with reduced wattage capacity, has made most power reducing technology outdated because the same result may now be achieved at a lower wattage, usually at a similar installation cost. Power reducers may, however, prove useful in such limited applications as reducing light output and wattage in built-in task lighting, which is often grossly overlit and expensive to retrofit.

The Ultrawatt is a power reducer that works on an entire 20-amp electrical circuit. This device is installed near the electrical panel to cut the power to fluorescent or HID lighting by about 25 percent. Tests show that the power savings are achieved, but fluorescent systems lose about 18 percent and HID systems lose about 27 to 35 percent of light output when operated with the Ultrawatt. For more on this product, see E SOURCE *Product Profile PP-97-2*, "The Ultrawatt Lighting Controller," (May 1997).

11.2.2 Step-dimming controls

Step-dimming products provide discrete reductions in light output, and include a variety of innovative products, many relatively inexpensive. The familiar three-way incandescent lamps are perhaps the most popular step-dimming products currently available.

Two step-dimming ballasts are currently available on the U.S. market, both made by Magnetek (**Table 11-4**); others (EBT's Varitron ballast and an earlier system by Motorola) have been discontinued. The S50 is a two-step ballast that provides 50 or 100 percent output, and the S30 is a three-step ballast that provides 30, 60, or 100 percent output. The light level is controlled by an additional wall switch that requires an extra set of line-voltage wires. To preserve lamp life, the ballasts start lamps at full power before dimming, and maintain or increase cathode heating during reduced-output operation.

Magnetek's step-dimming ballasts cost about $15 more than standard electronic ballasts, so they provide a relatively low-cost way to accommodate new code requirements for lower levels of lighting. Since these systems require the installation of an additional switch leg, however, retrofitting such ballasts into existing rooms may be difficult. The light level switch may also be controlled by an occupancy sensor, photocell, timer or clock switch.

Step-dimming products occupy an intermediate position in the array of energy-saving ballast options, offering more light control and energy savings than non-dimming ballasts, but costing less than the more versatile continuous-dimming ballasts. One of the advantages of step-dimming products is that occupancy systems can dim fluorescent lamps rather than turn them off, which can reduce cycling and extend lamp life. These units also offer a viable way to reduce lighting levels

One of the advantages of step-dimming products is that occupancy systems can dim, rather than turn off, fluorescent lights during unoccupied periods, thereby reducing the cycling of lamps and extending their lives.

Table 11-4

Magnetek's step-dimming fluorescent ballasts

These products were introduced in 1996 (the previous version, the LLSS, was introduced in 1989) and both are currently shipping.

Product name	S50 Light Level Switching Electronic Ballast	S30 Light Level Switching Electronic Ballast
Product types	two-level	three-level
Number/types of lamp(s)	2-F32T8	2-F32T8
Dimming method	current reduction with full cathode heating	current reduction with full cathode heating
Light levels	50%, 100%	30%, 60%, 100%
Ballast factor at full output	0.88	0.88
Ballast factor at lowest output	0.48	0.27
THD at full output	<10%	<10%
THD at lowest output	<15%	<15%
Power factor at full output	>0.99	>0.99
Power factor at lowest output	>0.95	>0.95
Approximate end user price	$35 to $45	$35 to $45
Warranty	5 years, $10 labor allowance	5 years, $10 labor allowance

Contact information
Magnetek Lighting Products Group
26 Century Blvd.
P.O. Box 290159
Nashville, TN 37214
tel 800-BALLAST

during cleaning hours, and to shed peak demand in common areas such as corridors.

Although the market penetration of step-dimming products is low (estimated at well under one percent), if current products gain acceptance and build a record of reliability, this class of energy-saving options could increase in importance and market share as new construction designs attempt to accommodate the more stringent energy codes required by the 1992 Energy Policy Act.

HID step-dimming. Several HID fixture manufacturers (**Table 11-5**) provide bi-level ballasts for their industrial equipment between 150 and 1,500 watts for metal halide and HPS lamps. As discussed in **9.4.3**, there is some color shift when lamp wattage is cut by 50 percent, but the result is acceptable in most industrial applications (such as warehousing) where bi-level lighting could save energy. The reduction to lower light output is typically triggered by a photocell or occupancy sensor. By dimming instead of switching off, this arrangement avoids the long restrike times of HID sources.

11.2.3 Continuous dimming controls

Incandescent dimming. Early incandescent dimming systems used rheostats (variable resistors) that reduced current to lamps by consuming some of the power that would have gone to them. Doing so reduced light levels but saved little or no energy, since the rheostats dissipated their consumed power as heat, occasionally requiring ventilated enclosures and fans. By the late 1950s, thyristors were developed that reduced light output and power input by chopping off parts of the voltage sine wave, effectively turning the lamp off and on at a rate beyond human perception.

Modern thyristor dimmers thus reduce power consumption and avoid visible flicker. This process also causes the filament to run cooler, reducing color temperature and making lamps (and spaces) appear more yellow. Wattage does not drop linearly with light output, resulting in reduced efficacy at

Table 11-5

HID step-dimming products

Vendor	Power range	Maximum precentage reduction in light output	Notes
Davis Controls Corp. 5400 Newport Drive, #8, Rolling Meadows, IL 60008 tel 708-253-4585 fax 708-253-5314	Up to 16 amps	~25	Minimum setting is adjustable
Day-Brite Benjamin P.O. Box 1687, Tupelo, MS 38803 tel 601-842-7212 fax 601-841-5501	MH: 175–1,500W HPS: 150–1,000W	16–36 14–17	Thomas & Day-Brite fixtures only
GE Lighting Nela Park, Cleveland, OH 44112 tel 216-266-2653 fax 216-266-2780	MH: 250–1,000W HPS: 150–750W	30 20	GE fixtures only
Holophane Co. 214 Oakwood Avenue, Newark, OH 43055 tel 614-345-9631 fax 614-349-4426	MH: 250–400W HPS: 250–400W	30 20–40	Holophane fixtures only
Magnetek 26 Century Blvd., P.O. Box 290159, Nashville, TN 37214 tel 800-BALLAST, fax 615-316-5188, faxback info line 708-445-4444	MH: 175–400 W HPS 175–400 W	about 40 percent	Available in fixtures through OEMs
Ruud Lighting, Inc. 9201 Washington Avenue, Racine, WI 53406 tel 800-236-7000 fax 800-236-7500	MH: 175–400W HPS: 150–400W	26–33 13–27	Ruud fixtures only
WideLite P.O. Box 606, San Marcos, TX 78667 tel 512-392-5821 fax 512-353-5822	MH: 175–1,500W HPS: 150–1,000W	12–26 depending on lamp type	CWA ballasts only

dimmed conditions. Because of the reduction in RMS voltage, lamp life is usually increased greatly in standard lamps, but may be reduced in halogen lamps, which should be burned at full power occasionally to redeposit tungsten on the filament. The rapid cycling of dimmed incandescent lamps may create an audible noise, heard as a slight high-pitched hiss in quiet locations. Increased flicker may become obvious on low-voltage incandescent systems using electronic transformers when light output is reduced below 50 percent.

Fluorescent dimming. Inexpensive, reliable continuous dimming of fluorescent lights is the still-awaited "holy grail" of energy efficient lighting. Dimming can be achieved for nearly all fluorescent systems, whether magnetic or electronic, rapid- or instant-start, and "dimmable" or "nondimmable." There are different methods for achieving this end (see box, next page), and products continue to emerge, although dimming products still represent only a fraction of the fluorescent ballast market. In this section, we review the types of dimming fluorescent products available and discuss the key considerations in selecting a product. E SOURCE *Tech Update TU-95-12*, "Fluorescent Dimming for Energy Management" (September 1995) provides a model-by-model survey of products available as of late 1995.

Two types of electronic ballasts are available for dimming standard fluorescent lamps. Most dimming ballasts use separate low-voltage control wires which direct the ballast to supply full output, minimum output, or something in between. Several ballasts introduced since 1994, however, work with standard incandescent-type dimmers, eliminating the need to run additional wiring.

Low-voltage-controlled fluorescent ballasts. As of mid-1997, there were nearly two dozen brands of dimmable electronic ballasts on the U.S. market that use separate control wires carrying either a 0–10-volt DC or a pulse-width modulated (PWM) control signal. Most of these ballasts are rapid-start units that dim the lamps by limiting current flow (see box, next page). Although low-voltage systems require additional control wiring, they offer the advantages of a relatively standard control interface, an expanding product base, and flexibility in control schemes. For example, they can be easily connected to and controlled by energy management systems, which offer 0–10V output channels (see **Figure 11-9**).

Figure 11-9

Schematic of dimming ballast installation

The typical arrangement for fluorescent dimming requires use of a controller to supply a low-voltage signal to the dimming ballast, which then reduces arc current.

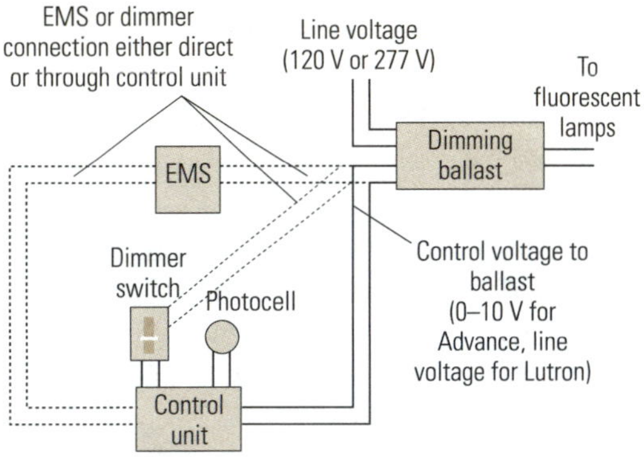

Source: Advance, Lutron

Over the past decade, many dimming ballasts have entered and left the market in the same year, or been announced but never shipped. Perhaps more than any other lighting technology, dimming fluorescent ballasts have yet to establish a solid record of reliable performance and product stability. (Of the five brands of dimming systems installed in E SOURCE's offices as demonstration units, two are no longer on the market.)

The dimming ballasts currently on the market with the longest performance history and installed base are the Advance Mark VII and the Lutron Hi-Lume. Newer models are available from a wide variety of vendors, including Magnetek, Motorola, Electronic Lighting Inc., Energy Savings Inc., DWA, Eclipse Technologies Inc., and Conik Technologies Inc. While differences exist between products (such as minimum light output level, type and quantity of lamps controlled, and controller compatibility), the dimming ballast for four-foot

T8 lamps is slowly approaching commodity status. Prices are dropping, although dimming ballasts still cost at least twice as much as non-dimming products.

Power line-controlled dimming ballasts. Several new products can dim fluorescent lamps with standard or specialized thyristor wall dimmers ("incandescent" dimmers) on the line-voltage switch leg—no extra wires are necessary. These ballasts manage to take the chopped voltage waveform delivered by the thyristor and use it to control the lamp arc. (A dimmer connected to a standard ballast would simply make the lamps flicker or shut off as it chopped the voltage.)

Most of the six brands available at the beginning of 1997 (the Advance Mark X, the B+L Nuance, Energy Savings Inc.'s Dim-L, Stocker & Yale's Dimwatt, and models from Tek-Tron and Solium) boast reasonably good technical specifications: low THD, high power factor,

How to Dim Fluorescents

There are three primary ways to dim fluorescent and other discharge lights (**Figure 11-10**):

Voltage reduction. Light output reductions of up to 50 percent are possible by reducing the voltage across the lamp's cathodes, but the minimum cathode fall voltage must be exceeded at some point in the wave cycle, or the lamp's discharge arc will die. Voltage dimming can be accomplished with a ballast that adjusts voltage, or with a separate voltage controller upstream of the ballast. The primary limitation of this technique is its minimal dimming capability. Voltage reductions upstream of the ballast may be negated by ballasts that automatically adjust for voltage drops by increasing current. For example, with standard magnetic ballasts, a 10 percent drop in voltage to the ballast results in only a five percent reduction in light output.

Current reduction. Just as low-ballast-factor ballasts provide normal voltage with lower current to reduce light output, dimming systems can reduce the output current to the lamp ends. This is the most common method of dimming fluorescent systems. At low output, it is important to maintain or even increase cathode heating to avoid low-temperature sputtering of the cathode's emissive coating.

Waveform modification. Electronic ballasts drive lamps with voltage waves on the order of 20,000 times (or more) per second. By clipping or eliminating some of the output waves (or parts of each wave), the total power to the lamp is reduced, while maintaining a sufficiently high driving frequency that the lamp discharge "coasts" through the power interruption. This type of dimming is particularly susceptible to harmonic distortion, because its rapid on-off switching can create reflected voltage waves.

Figure 11-10

How fluorescent lamp dimming works

There are three primary ways to dim fluorescent and other discharge lamps.

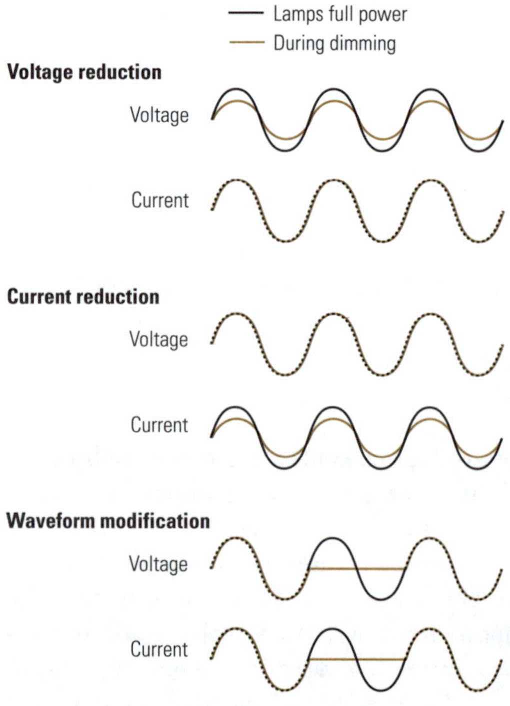

high maximum light output, and three- to five-year warrantees. As of mid-1997, only one maker (Advance) offers both 120 and 277 volt models. Pricing is competitive with low-voltage-controlled dimming ballasts.

Not all of these products can use all dimmers—there are over a dozen manufacturers of incandescent dimmers, not all of which have been tested with the new ballasts. Advance recommends a short list of high-quality thyristors for use with its product. B+L claims its ballast will work with any off-the-shelf dimmer; E SOURCE has tested this product with a $4.95 hardware store dimmer with good results. If ballasts and dimmers are mismatched, symptoms may include the inability to reach full light output, hum or other noise at the dimmer, and the inability to start the lamp except at a high setting.

Another dimmer-related issue is power rating: although a typical dimmer can handle up to 2000 watts of incandescent lamps, most manufacturers suggest that only 500 watts of these new dimming fluorescent ballasts be gathered onto a single dimmer. Finally, not all models use the same procedure to dim lamps: some start lamps at full output and the automatically dim them, while others start them at low output, which may adversely affect lamp life.

Dimming ballast checklist. When choosing either type of dimming fluorescent ballast, the key considerations to review are:

- *Dimming level.* what is the minimum dimming level needed in the space? While most products can dim to 20 percent of full output—a range sometime referred to as "energy dimming"—only a few products can dim down to five percent or less. These "architectural dimming" ballasts typically cost significantly more.

- *Lamps served.* Dimming ballasts are not available in as wide an array of lamp configurations as nondimming ballasts.

- *Technical specifications.* Independent lab reports should be reviewed to verify the ballast's dimming range and its power input, THD, power factor, crest factor, at various dimming levels.

- *Control system compatibility.* The ballast, controller, and any additional devices such as photosensors or occupancy sensors must all be able to work together—not an easy feat when they may all come from different manufacturers. A small-scale test will help sort out these issues before specifying products for large installations.

- *Availability.* Many systems or products are "introduced" before they are truly available. New products are often available in only one configuration (i.e., for two four-foot T8 lamps).

- *Reliability.* Some manufacturers cite failure rates, although these are not tabulated by any official means. If requested, manufacturers or sales reps may also supply references to other purchasers to gauge their experience with the product.

- *Warranty.* A good warranty will replace failed ballasts for at least 3 years, and provide a labor allowance of at least $10 per failed unit.

Panel-level dimming. A third dimming option is to use an external power controller and standard ballasts. One example of this approach is the Ultrawatt (see **11.2.1**), which provides about a 25 percent reduction in power input and a similar reduction in light output. Another is the Peschel lighting control system, which uses an industrial-grade autotransformer to reduce circuit voltage well upstream of the ballasts. This arrangement—which is perhaps better described as a light *trimming* system than a *dimming* system—is best applied to large banks of lights that are switched simultaneously, such as in high-bay retail stores, supermarkets, and large open-plan offices. Dimming is limited to 40 percent reductions in lighting output—substantially less in many cases—making the system suitable only for energy management purposes rather than full-range light control.

A typical application for the Peschel system is for spaces that have been upgraded to high-frequency ballasts but are still overlit. The system can be used with nearly any ballast that does *not* include active input circuitry to adjust for off-voltage conditions (such as the Advance Mark V), even if they are not designed to dim. Four types of voltage controllers provide options such as EMS interfacing, on-board programming and scheduling of light levels, load shedding, photocell/daylight control, lumen maintenance, and continuous or step dimming control. Lights are always started at full power and then dimmed; Peschel claims that lamp life is not decreased.

The Peschel system is designed primarily to control HID lights; with fluorescents, the wide array of ballast manufacturers and the small-scale control typically required make this centralized approach inflexible compared to individual dimming ballasts. The system is not recommended for use in dimming anything less than a complete 20A lighting circuit, which can include as many as 50 T8 lamps. The best results occur with 277-volt ballasts. A unit to control a single 20-amp lighting circuit costs about $1,400, while large units that can control voltage to as much as 250 kW of lighting cost about $130 per kW.[24]

Applications. When combined with light sensors, dimming control can correct for fixture dirt buildup and lamp lumen depreciation (commonly called "lumen maintenance") or take advantage of incoming daylight. Using daylighting for dimming control requires careful attention to issues such as the linearity of photosensor response, calibration ability, sensitivity to different types of light (for example, outdoor versus electric), and sensor view angles. Daylight dimming applications and case studies are also discussed in **CHAPTER 5**.

Since they can be installed in almost any fixture, dimmable electronic fluorescent ballasts offer a major opportunity for displacing incandescent dimming systems. It is common, for example, to find conference rooms equipped with fluorescent fixtures for maximum light levels and incandescent fixtures on dimmers to accommodate visual presentations that require a much lower ambient light level. In this type of situation, dimmable ballasts can replace existing fluorescent ballasts and the incandescent fixtures can be removed, capped off, or retrofitted with compact fluorescent devices (once the dimmers have been removed).

A recent study from the Florida Solar Energy Center provides an example of the energy savings available from dimming fluorescent systems.[25] Two identical offices with south-facing windows were measured, one with a conventional nondimming lighting system (magnetic ballasts, white painted troffer, T12 lamps, and prismatic lens), and the other with an energy efficient dimming system (dimming electronic ballasts, open paracube troffer, specular reflector, and T8 lamps).

Figure 11-11 shows the power consumption and average lighting levels in the two spaces for a typical sunny day. The more efficient fixture drops the maximum lighting power from 157 watts to 114 watts, a 27 percent reduction. But during daylight hours, the dimming controls cut power down to about 70 watts, effectively doubling the power reduction to a total of 55 percent compared to the conventional system. As would be expected, light levels were lower with the dimming system, but were always well above the 50-footcandle design level—indicating that even deeper dimming could have been used.

Dimming of compact fluorescent lamps (CFLs) is advancing more slowly than for linear fluorescent lamps. The advent of amalgam-based four-pin lamps and electronic CFL ballasts has, however, led to advancements in the number of lamps and wattages that can be dimmed, the variety of dimmable lamp shapes, and the types of dimming control systems. For further information on these options, see **8.8**.

Some general rules apply to fluorescent dimming systems. Maintaining a warm cathode greatly improves lamp life (see Figure 7-3 on fluorescent operation). Starting lamps at full power helps assure a warm cathode for the high stress of arc initiation, while maintaining cathode heat at dimmed conditions prevents low-temperature sputtering. Instant-start lamps or ballasts do not provide cathode heat, and so may reduce lamp life. For the same reason, cathode-cutout lamps and ballasts do not make sense for dimming systems. In very cold environments, full-power fluorescents are a more reliable choice than dimming fluorescents.

Unlike incandescent and HID lamps, dimming fluorescent lamps does not change their color temperature. Deeply dimmed fluorescents can alter the appearance of illuminated objects, however, especially if the lamps have high CRI and high color temperature. This effect is shown in Figure 2-13, and occurs because the eye's night-vision rods play a larger role at low light levels. The rods do not sense color, so colors appear grayer. Light sources that shift toward low color temperatures (i.e. incandescent) when dimmed provide some correction for this effect by making objects appear more colorful.

Continuous dimming of high-intensity discharge (HID) lamps. Several electronic ballasts have been offered for dimming certain HID lamps, but their track record has been sketchy, often resulting in product withdrawal. Limited dimming of HID lamps using magnetic ballasts can be accomplished with the voltage reduction technologies described above. HID dimming is limited because of color shifting, reduced CRI, increased flicker, impact on lamp life, and inadvertent lamp shutdown during line voltage variations. As shown in Figure 9-10, the color of HID lamps shifts considerably during dimming, limiting them in any aesthetic applications. This problem is mitigated somewhat for coated metal halide and mercury lamps, due to the color constan-

Figure 11-11

Power and lighting levels for dimming vs nondimming systems

In this test, more efficient lighting equipment cut connected load by 27% (from 157 to 114W). This test kept lights on 24 hours per day. Dimming cut average workday power consumption even further, for a 55% savings (157 to 70W).

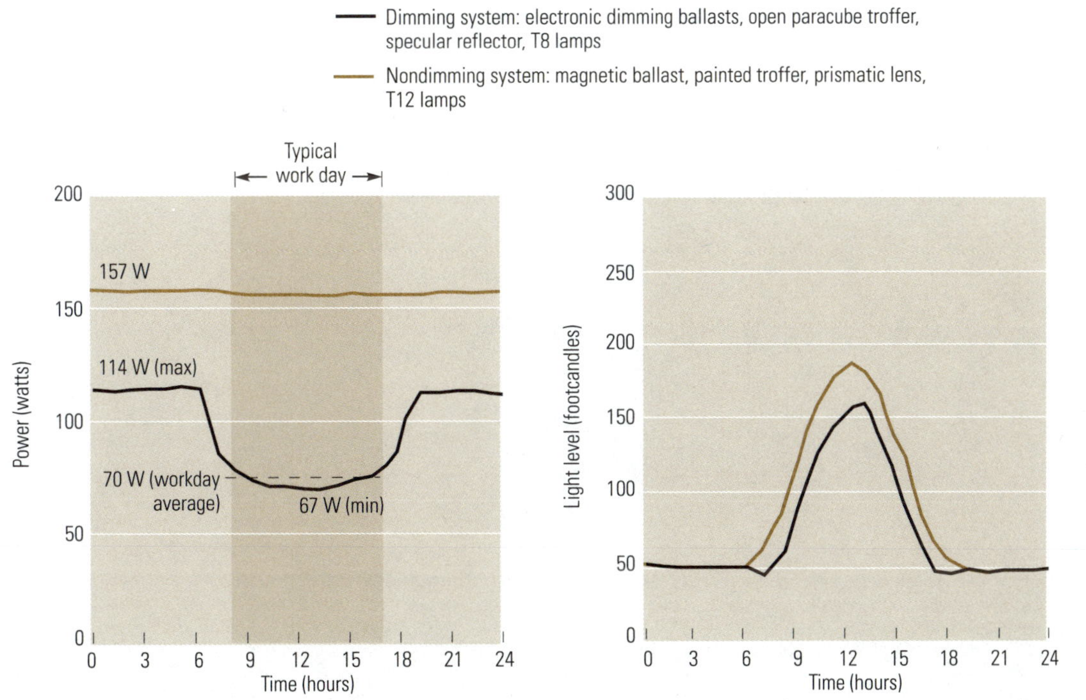

Source: Florida Solar Energy Center [25]

cy of their phosphors. Dimming may also reduce HID gas pressure, which makes flicker more visible, especially in high pressure sodium lamps.

A promising new line of electronic dimming ballasts for metal halide lamps is available from Delta Power Supply. Ballasts are available for 70W, 150W, 200W, 350W, and 400W lamps. These ballasts offer continuous dimming down to 30 percent of full output, have high power factor and low THD, and auto-sense incoming voltage between 120 and 277 volts.[26] See Table 9-3 for specifications for these ballasts.

1 Scott Pigg, Mark Eilers, and John Reed, "Behavioral Aspects of Lighting and Occupancy Sensors in Private Offices: A Case Study of a University Office Building," *Proceedings 1996 ACEEE Summer Study on Energy Efficiency in Buildings*, American Council for an Energy Efficient Economy, 1001 Connecticut Avenue NW, Suite 801, Washington, DC 20036, tel 202-429-8873, fax 202-429-2248.

2 Undiscounted list price for the Watt Stopper Power Pack relay is $25.

3 List price for a 12-pole contactor is $701, with a discounted wholesale price of $421. Grainger, Inc., 1994 General Catalog #385.

4 Capri Lighting, 6430 East Slauson Avenue, Los Angeles, CA 90040, tel 213-726-1800, fax 800-444-1886.

5 Halo Lighting, 400 Busse Road, Elk Grove Village, IL 60007, tel 708-956-8400, fax 708-956-1537.

6 Lindsay Audin, personal experience, Uris Business School, Columbia University, mid-1980s.

7 This occurred in Columbia University's Uris Business School.

8 Paul Malkowski, personal communication (April 20 1994), Paragon Electric Company, 606 Parkway Boulevard, Box 28, Two Rivers, WI 54241, tel 414-793-1161, fax 414-793-3736.

9 List price for the Watt Stopper LightSaver® controller.

10 Orion Alpha Sensors, manufactured by Rigel Systems, 2993 Copper Road, Santa Clara, CA 95051, tel 408-737-0591, fax 408-737-0592.

11 Frank Napoli, personal communication (June 1993), Consolidated Edison, tel 212-460-2761.

12 Leslie Owashi, Dean Schiffman, and Andrew Sickels, "Lighting Hours of Operation: Building Type versus Space Use Characteristics for the Commercial Sector," *Proceedings 1994 ACEEE Summer Study on Energy Efficiency in Buildings*, American Council for an Energy Efficient Economy, 1001 Connecticut Avenue NW, Suite 801, Washington, DC 20036, tel 202-429-8873, fax 202-429-2248.

13 Howard Kessler, personal communication (May 1993), principal, Kessler Associates, New York, NY.

14 Steven Wiegmann, personal communication (July 7, 1993), CB Commercial Real Estate Group, Newport Beach, CA.

15 Jim Sekinger, personal communication (1992), EBT, 23600 Telo Avenue, Torrance, CA 90505, tel 310-784-2000, fax 310-534-8214. John Burgoyne, personal communication (August 1993), Burgoyne Sales, Edison, NJ (Magnetek rep for New York City area), tel 908-225-0450.

16 Mark Kurczak, personal communication (August 1993), Advance Transformer Company, 10275 West Higgins Road, Rosemont, IL 60018, tel 708-390-5281, fax 708-390-5109.

17 Personal experience of Lindsay Audin at Columbia University (1990–1993).

18 Holophane AERS dimming system brochure (1992), Holophane Company Inc., 214 Oakland Avenue, Newark, OH 43055, tel 614-345-9631.

19 New York State Energy Conservation Construction Code (March 1, 1991), New York State Energy Office, Albany, NY 12202.

20 "Energy Efficient Design of New Buildings Except Low-Rise Residential Buildings," ASHRAE.IES 90.1-1989 (1989), American Society of Heating, Refrigerating and Air-Conditioning Engineers, Inc., 1791 Tullie Circle NE, Atlanta, GA 30329, tel 404-636-8400.

21 "Occupancy Sensors," Specifier Report (October 1992), Lighting Research Center, Rensselaer Polytechnic Institute, Troy, NY 12180-3590 tel 518-276-8716, fax 518-276-2999.

22 Personal experience of Lindsay Audin at Columbia University, where over a dozen outdoor lighting circuits are controlled with photocells that work very reliably and need only occasional cleaning.

23 Product information from Susan Newman, Marketing Communications Manager, Profile Systems, 1000 East 80th Place, Suite 509 North, Merrillville IN 46410, tel 219-757-3575, fax 219-757-3541. For inquiries, contact Carl Triemstra, Director of Sales and Marketing.

24 End-user cost based on quantity pricing.

25 Danny Parker, Ted Stedman, and Jeff Sonne, "Side-by-Side Testing of Commercial Office Lighting Systems," Florida Solar Energy Center, 300 State Road 401, Cape Canaveral FL 32920, Paper FSEC-PF-265-93 (1993).

26 Jack Kennedy, personal communication (September 18, 1997), Director of Sales, Delta Power Supply, 607 Redna Terrace, Cincinnati, OH 45215, tel 513-771-3835, fax 513-771-1221.

12 LIGHTING MAINTENANCE

Lighting maintenance is essential for persistent energy savings. Maintenance starts with proper design and specification, but must be enforced after installation by on-site personnel, when the job is out of the designers' hands. Education of electrical maintenance, purchasing, and inventory personnel—through written maintenance procedures—is essential to avoid the degradation of an energy efficient design. Alternatively, maintenance can be contracted to a knowledgeable lighting management company.

Maintaining light quality, output, and energy savings requires three actions:

■ *designing lighting systems* with components that minimize light loss over time, are easy to maintain, and using the fewest types of lamps;

■ *training personnel* in proper maintenance techniques, including cleaning and relamping; and

■ *controlling purchasing and inventory* to ensure that only the right replacement components are available.

Figure 12-1 shows the relative costs of energy, labor, and lamp replacement for a facility with high energy and labor costs; energy clearly dominates the other contributors to total cost.

Most lighting maintenance is handled by in-house personnel at large facilities, some of whom are thorough and sophisticated in their knowledge and methods. As a general rule, however, many lighting upgrades degrade significantly after only a few years of operation, due to the lack of enforcement and training of on-site personnel. Energy service companies (ESCOs) have learned to handle this problem by including lighting maintenance during the terms of their contracts.

Lamp costs may increase significantly after a lighting upgrade; it is important to budget for these higher purchase prices. For example, replacing a long-life halogen lamp with a compact metal halide lamp may cut wattage by nearly 70 percent, increase lamp life by only 40 percent, and raise lamp cost by over 100 percent. In cost-effective upgrades, of course, any increase in lamp cost is overwhelmed by the energy savings, but this fact is of little interest to the person handling a limited lamp budget.

HIGHLIGHTS

■ Proper design and maintenance can reduce oversizing design margins, cutting the connected load of lighting systems.

■ Group relamping will help to maintain lighting quality, and usually saves money as well.

■ Fluorescent lamp disposal remains an open-ended environmental problem in most states, but new low-mercury lamps from the major manufacturers offer an easy way to meet codes while allowing disposal of lamps as regular trash (as long as it is not incinerated). See **CHAPTER 3** for more on lamp and ballast disposal issues.

■ Ceiling fixtures should be cleaned at every relamping, and lenses and lamp sockets should be replaced at every group reballasting.

Figure 12-1

Relative costs of energy, labor, and lamps

Energy charges dominate the total operating cost of most lamps.

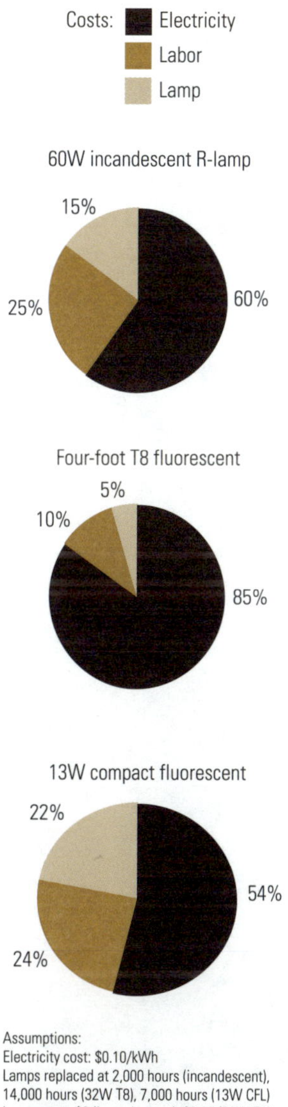

Assumptions:
Electricity cost: $0.10/kWh
Lamps replaced at 2,000 hours (incandescent), 14,000 hours (32W T8), 7,000 hours (13W CFL)
Lamp costs: $3 (incandescent), $2.25 (32W T8), $4.50 (13W CFL)
Labor cost: $5 per relamping

Poor lighting maintenance can easily degrade the visual quality in a space, which can reduce productivity and discourage future upgrades by setting a bad example. While this can be annoying or inconvenient, neither of these issues directly affect the dollar value of energy savings. There are three primary ways that energy dollars can be *directly* lost through poor maintenance:

- compact fluorescent lamps are replaced with higher-wattage incandescent lamps;
- failure to keep lighting controls operative increases burn time and/or power draw; and
- failure to relamp promptly encourages occupants to install their own (usually incandescent) task lighting.

TERMINOLOGY

ballast factor (BF): the percentage of a lamp's rated lumen output provided by a given ballast.

diethylhexyl phthalate (DEHP): solvent used in dielectric fluids in ballasts having wet capacitors, which were common between 1979 and 1984. Like PCB ballasts, DEHP ballasts cannot be discarded in sanitary landfills.

group relamping: complete replacement of a group of lamps on a defined schedule, usually at about 70 to 80 percent of rated life, in order to minimize burnouts.

igniter: fixture component designed to initiate an arc in HPS lamps (and some metal halide lamps) through a high-voltage pulse; also called a "starter."

lamp lumen depreciation (LLD): a factor expressing the remaining percent of rated lumen output after a given burn period (for example, an LLD of 0.75 indicates that lumen output has dropped 25 percent due to phosphor depreciation and/or lamp dirt buildup).

luminaire dirt depreciation (LDD): a factor expressing the remaining percent of a fixture's rated lumen output after a given period (for example, an LDD = 0.9 indicates that 10 percent of light output has been lost due to dirt buildup).

maintenance category: a fixture classification (ranging from I to VI) indicating how rapidly the fixture will depreciate due to dirt buildup over time.

nonrecoverable light loss factor: one of several factors causing light loss that cannot be corrected by normal lamp or fixture maintenance (such as low building voltage, poor ballast factor, or unusual space temperatures).

Occupational Safety and Health Administration (OSHA): federal agency enforcing standards on worker exposure to hazardous chemicals, such as lamp mercury.

polychlorinated biphenyls (PCB): compounds used in magnetic-ballast capacitors up to 1978, now considered a hazardous material requiring special disposal.

rated life: the point at which 50 percent of a group of lamps have burned out, after being restarted at given constant intervals.

recoverable light loss factor: one of several factors causing light loss that can be corrected by maintenance (such as lamp burnouts, dirt buildup, or lamp aging).

RCRA (The Resource Conservation and Recovery Act): the enabling legislation for federal control of hazardous wastes such as mercury and polychlorinated biphenyls (PCBs).

spot relamping: lamp replacement on an as-needed basis, only after burnout has occurred.

These failure modes can be addressed by an aggressive maintenance program using the suggestions in this chapter. One key is to focus final responsibility for savings on one person, such as an energy or utilities manager, rather than diffusing it among several individuals with other agendas (such as relampers and purchasing/inventory personnel). Incentive systems such as bonuses or awards will also help maintenance personnel stay on their toes.

12.1 Designing for Maintenance

Good maintenance starts long before installation of the first lamp: it begins with the *choice* of lamp. To account for lighting system degradation, designers typically oversize initial light output by 35 percent or more to "maintain" minimum target illumination levels at the time lamps are changed and fixtures cleaned (see **3.1**). Proper design and maintenance—using high-quality lighting components that degrade less or slower—can cut that excess nearly in half, saving nearly 15 percent in connected load. Minimizing that 35 percent safety factor involves choosing lamps with *maximum rated life*, to reduce the burnout rate, and *minimum lumen depreciation*, to ensure high output. Maintenance costs can be further minimized by limiting the number of different types of lamps that must be stocked.

A maintenance-conscious lighting design should consider the following options:

- Standardizing on a few lamp types and one lamp color: limiting choices to only 2-foot, 4-foot, and U-shaped T-8 lamps, for example, provides sufficient fixture choices to handle most lighting work; choosing one lamp color temperature and CRI eliminates the potential to mix colors in one space or fixture.

The exact composition of phosphor blends in fluorescent lamps varies between manufacturers. In the past, settling on one brand of lamp could help ensure color consistency, but the initiation of common "rare earth" phosphor standards now makes brand changes less obvious. All T8 lamps (and some newer T12s) are now rated with a code indicating color rendering index (CRI) range and color temperature that is consistent among manufacturers. A RE735 lamp, for example, has a CRI of about 75 and a 3,500 K color temperature, regardless of its origin. While very slight differences in color may still exist between different manufactures, these variations are small enough to allow mixing brands without being noticeable to the average occupant.

A maintenance-conscious lighting design should standardize on a few lamp types and one lamp color.

- Using lamps with a rated life of 20,000 hours (or higher) whenever possible: this means minimizing standard compact fluorescents (CFLs) having a 10,000-hour rated life in favor of fixtures accommodating larger

Impacts of the 1992 EPACT Legislation on Lighting Maintenance

With the exception of 34-watt "energy saving" lamps, EPACT has effectively eliminated most T12 cool-white lamps. Although stocks of the standard F40T12 still exist, conversion to T8 lamps now makes sense simply to ensure long-term lamp availability, in addition to the energy benefits. (Expanded production of T8 lamps has been sufficient to avoid any supply problems.) Major lamp vendors have also increased production of higher-efficacy incandescent lamps (either halogen- or krypton-based) to replace the R-lamps and PAR-lamps that fail EPACT standards. While EPACT mandated ASHRAE/IES Standard 90.1 as a national minimum lighting standard, funding for its enforcement has not been provided. As a result, many states have not upgraded their building codes to comply with 90.1's requirements for new buildings.

*A maintainable lighting design does not **offer** options to save. Rather it **takes away** options to waste.*

SEVEN WAYS TO MAINTAIN ENERGY SAVINGS

1. During system design, minimize the variety of lamps used.

2. Use theft-resistant compact fluorescent retrofit kits.

3. Choose fixtures that use lamp-specific sockets.

4. Convert lighting upgrade specifications into purchasing specifications.

5. Regularly train lighting maintenance personnel.

6. Budget and train for group relamping.

7. Focus responsibility for lighting system maintenance on one person instead of diffusing it among several.

Figure 12-2

Mortality curves for incandescent and fluorescent lamps

At the rated life of a lamp, only half of the lamps out of a large group have failed. Note also that many lamps will normally fail before their rated life; this fact is often overlooked.

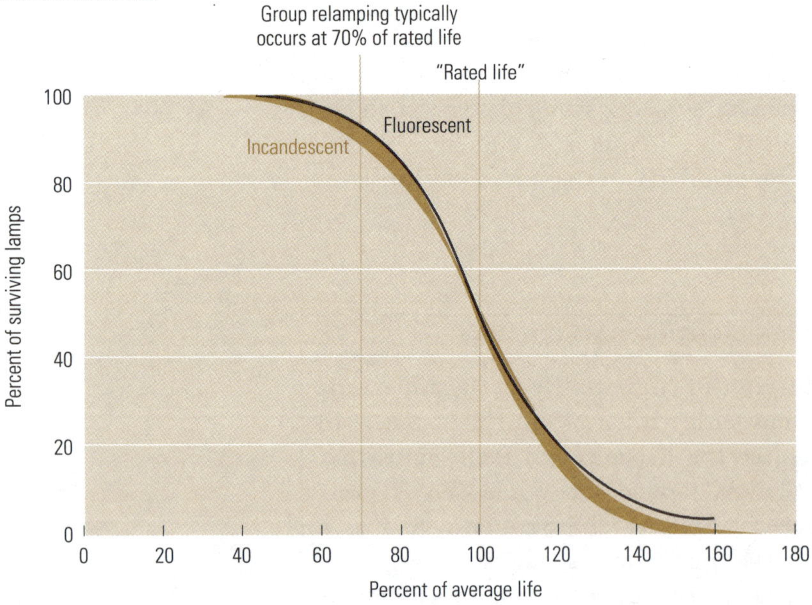

Source: IES

CFLs or low-wattage (16- or 24-watt) U-lamps, all of which have 20,000 hour lives.

■ If CFLs are essential to achieve design goals or maintain the character of a space, hardwired fixtures are preferable to retrofit kits as insurance against improper relamping with incandescents.

■ If no fluorescent lamp offers the necessary "warm" coloration, "white" HPS is preferable to halogen because no incandescent can be screwed into such a fixture.

12.1.1 Lamp lifetime

Although lamps appear to have simple rated lifetimes, the behavior of a large group of lamps must be described in statistical terms. By definition, the rated life of any group of lamps has passed when 50 percent of them have burned out. As shown in **Figure 12-2**, lamps expire on a nonlinear distribution curve, with most surviving until about 70 percent of rated life, after which the burnout rate increases rapidly. For this reason, group relamping is usually recommended at 70 percent of the lamps' rated life. Analyses of spot versus group relamping should consider that over the rated life, only *half* of the lamps will burn out—not all of them, as is commonly assumed.

There are at least two ways to improve the field lifetime of lamps: use incandescent lamps rated at higher-than-line voltage (130 volts), and lengthen the average burn cycles for fluorescents beyond the standard rating cycle of 3 hours per start. Both of these methods reduce the stress on the lamps, postponing the most common failure mode. Adding a hidden preset dimmer to 24-hour incandescent lighting (for example, in elevators) is a common way to greatly stretch lamp life where light levels are not crucial. Figure 7-12 shows typical life extensions that can be gained by increasing fluorescent burn cycle. For example, continuously-burning fluorescent lamps can last 35,000 hours or more. Applications with predictable scheduling can take advantage of this phenomenon by relamping at 70 percent of the *extended* life of the lamps.

12.1.2 Lumen depreciation

Lighting designs that use lamps with slow lumen depreciation can use fewer installed watts compared to faster-depreciating lamps. An example is shown in **Table 12-1**, based on the lumen depreciation curves in **Figure 12-3**. If 70 percent group relamping is used, the T8 lamps require only a 9 percent oversizing factor, compared to 22 percent for T12 Cool Whites. To take advantage of this effect, ballasts with slightly lower output can be used, or the number of fixtures can be reduced. This savings is *in addition to* the inherent improvements in system efficacy that come with a switch between lamp types.

12.2 MAINTENANCE CONSIDERATIONS FOR SPECIFIC LIGHTING TECHNOLOGIES

A variety of technical considerations can impede or improve the maintenance of different lighting sources.

12.2.1 Incandescent lamps

Lighting upgrades that replace standard incandescents with Edison-base halogen or halogen-infrared lamps are the most difficult to maintain, unless all incandescents have been changed in this manner and any remaining incandescent stock is removed from the premises. Otherwise, relampers will use standard incandescents if halogen stocks run out, degrading the installation and eliminating any savings.

Table 12-1

Comparison of design margins for lamps with different lumen maintenance profiles

Lamps with high-quality phosphors require less oversizing to account for lumen depreciation.

	Four-foot T8 Philips TL80	Four-foot T12 Philips Cool White
Rated lifetime (hours) at 3 hours/start	20,000	20,000
Initial lumens after 100 hours burn-in	3,050	3,050
Design lumens after 40% of rated life (8,000 hours)	2,850	2,650
Relamping lumens after 70% of rated life (14,000 hours)	2,800	2,500
Design margin (%) to provide same light level at relamping	109	122

Halogen lamps—especially those lacking a lens—can explode if spattered by water or spotted with dirt, due to the temperature differentials that can occur at the points of contact with such materials. Relampers should handle the lamps with clean gloves (or at least with clean hands), and reinstall protective glass covers after each relamping. Clean gloves can be stored with the lamp inventory to make them readily available.

12.2.2 Fluorescent lamps

The physical interchangeability of T12 and T8 lamps can easily create maintenance problems, since they require entirely different ballasts. Relampers or maintenance staff may seek to "finish up" existing T12 stocks in a new T8 installation, so those stocks should be removed *before* this mistake is made. Failure to do so will result in a shoddy appearance and a major drop

Figure 12-3

Lumen depreciation curves for high-CRI T8 and Cool White T12 lamps

The slower degradation of lamps with high-quality phosphors can translate to lower installed wattage.

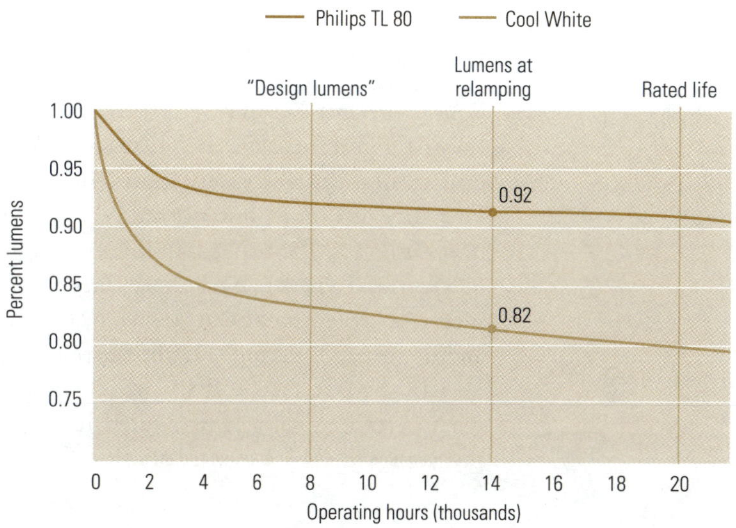

Source: Philips Lighting

in light levels. Any T12 lamps running on T8 ballasts will stand out easily due to their lower output and different color, and will burn out quickly as their cathodes sputter.

T8s can also mistakenly be installed in T12-ballasted fixtures. In this case, two results are possible: the lamps will put out greater light for a brief period before burnout (the cathodes will be rapidly burned out by the T12 ballast's higher current), or the lamps will fail to start (because of the mismatch between the starting requirements of T8 lamps and T12 ballasts).

When ballasts eventually require replacement, fluorescent lamp sockets should also be replaced.

Some ballasts with impedance-sensing intelligence—such as the Advance Mark V—are capable of supplying T12 or T8 lamps without degrading the operation of either. While these ballasts give building managers maximum flexibility, they allow relampers to "mix and match" high-CRI T8s with cool white T12s. Even after EPACT (see sidebar on page 275) takes effect, 34-watt T12 lamps with low CRI will be available, so the possibility of poor performance will remain as long as T12-compatible fixtures are installed in the building.

Proper handling of fluorescent and HID lamps (all of which contain some mercury) impacts not only our environment but also the workers exposed to handling them; Section **3.3** includes a complete discussion of lamp and ballast disposal.

12.2.3 Fluorescent ballasts

Ballast lifetime is typically 10 to 20 years for linear fluorescents, and is not usually considered in illumination calculations. Ballast performance also does not normally deteriorate over time. On the other hand, a wide variety of similar-looking ballasts exists, and an incorrect replacement can reduce light output and/or increase wattage. Light output is limited by ballast factor (BF), which indicates what percentage of a lamp's rated lumen output is provided by a given ballast.

Ballast factor for two different standard ballasts powering the same type of lamp can vary from 0.75 to 0.92, so up to 18 percent of light output can be lost if the wrong replacement ballast is installed. Where high- or low-power ballasts are used (with BFs ranging between 0.75 and 1.35), improper replacement could result in even larger changes in light and wattage level. Even ballasts with the same BF may not use the same wattage; a standard two-lamp magnetic T12 ballast, for example, may draw 92 watts while an electronic version draws only 73 watts, but light output from lamps powered by each will be the same.

Caution is therefore suggested when considering fluorescent ballasts with nonstandard ballast factors. It is best to avoid mixing different ballast factors in the same room, and good to avoid mixing high- and low-BF equipment in the same building. In most cases, a lighting design can be acceptably configured using either standard and high, or standard and low, ballast factors. Difference between members of each of these pairs will not be noticeable, but shifting between high and low BF would be quite obvious: light levels could change by over 60 percent. In-house personnel should not be expected to stock or identify the many different types of ballasts that could be used in an uncontrolled design. Ballast factor and wattage data are available from independent ballast test reports and from manufacturers' data sheets.

When ballasts eventually require replacement, fluorescent lamp sockets should also be replaced. During that normally long interval, it is likely that sockets have sufficiently oxidized to create resistance, shortening lamp life and making starting more difficult.

It also makes sense to standardize on either rapid- or instant-start ballasts, because of their different effects on lamp life and different wiring configurations.

Most lighting upgrades have already removed ballasts containing PCBs and DEHP (that is, those made prior to 1984); issues related to their disposal are covered in **3.3.2**.

12.2.4 Compact fluorescent lamps

The biggest maintenance problem with screw-in compact fluorescent lamps is their premature removal, either by relampers who need to replace those that failed (and didn't stock any replacements) or by thieves aware of their value. Fortunately, theft-resistant options are now available that limit "disappearances." Most require a special tool to install and remove the kits, while one model actually chews its way into the socket, making removal quite difficult. Access to installation tools should be restricted to maintenance supervisors to keep them out of the wrong hands.

Relamping some magnetically ballasted modular screw-in CFLs can be a problem when the ballast suffers a short circuit (a typical failure mode). A relamper may instead believe only the lamp is burned out, and replace it with a new lamp. The damaged ballast then immediately burns out the new lamp (costing $4 to $12 each), creating an unacceptable—and unnecessary—expense. The only way to avoid making this error is to check the ballast's lamp socket using a voltmeter that has leads soldered into a CFL base. A high reading (near line voltage) indicates the ballast has a short circuit; a zero reading shows that connection to the socket has failed. A low reading (typically about 15 volts) indicates the ballast is OK. Using such a device requires care because it bypasses the normally narrow access of a CFL socket and creates a shock hazard. An enterprising inventor would perform a great service by developing a self-contained ballast checker to simplify this process.

Because CFL screw-in ballasts are not yet a stock item in most electrical supply houses, users should keep their own supply so that no incentive is created to "temporarily" replace such a device with an incandescent lamp.

12.2.5 High-intensity discharge lamps and ballasts

HID lamps present their own special maintenance problems. The Philips White SON, for example, has a special polarized bipin socket that eliminates using the wrong type of HPS lamp. While large and small notches show the relamper the right way to insert the lamp, these notches are too similar in size to make incorrect insertion impossible. Sockets are often not visible when set deeply into a fixture, and it is never safe to "feel around" an open socket to find the larger notch. The lamp will fit and start when installed backwards, but then suffer a short life—an expensive proposition at $20 to $30 per lamp. Relampers need to be shown the proper way to install the lamps, and fixtures need to be marked near their sockets to highlight correct orientation.

Many metal halide lamps require proper orientation for maximum life. While compact metal halide fixtures make incorrect orientation impossible, some recessed metal halide fixtures (often seen in supermarkets and conference centers) can accommodate any lamp with a mogul base, regardless of its designed orientation (horizontal or vertical). Universal-orientation metal halide lamps avoid this problem, but they have lower efficacy. Mixing differently oriented halide fixtures in one building should be avoided by designers and specifiers, and enforced on contractors who may combine them when stock in one or the other runs out during installation work.

Metal halide lamps can also explode, usually after operating beyond their rated lives. As a result, many fixtures are equipped with glass covers designed to contain such an event. When a cover is lifted or removed, it activates a switch that shuts off the lamp. Some fixtures are now available, however, without such covers (they impede relamping, and special metal halide lamps with their own enclosures make them unnecessary). In most cases, however, there is no barrier to using a standard metal halide lamp in such an open fixture, should that be the only lamp available to untrained personnel. Chances of a lamp explosion can also be reduced by cycling the lamps off for 15 minutes a week and relamping them before the end of their rated lives.

Mercury vapor lamps sometimes seem immortal, never burning out.

Mercury vapor lamps have a unique characteristic: they sometimes seem immortal, never burning out. Unfortunately, they instead may lose over half their rated light output, leading to complaints or encouraging addition of unnecessary incandescent lighting. Group relamping at regular intervals is essential when using mercury vapor sources.

HID lamps also shift in color as they age, making group relamping essential where color constancy is important. While the color change may not noticeably affect the look of a space, the differences can be distracting when looking at a fixture or a white ceiling above an indirect HID fixture.

As discussed in **9.5.2**, high-pressure sodium lamps, and some smaller metal halide lamps, are started by igniters (also called "starters"), which may be an integral part of the ballast. Delayed lamp replacement can damage an igniter, which may then shorten the life of the next lamp installed. Some magnetic and nearly all electronic HID ballasts are available with automatic starter cutoff when a lamp fails to start in a short period of time, usually less than 3 minutes. While initially more expensive, such equipment can quickly recoup its extra cost in saved labor and HID lamp costs, which often exceed $25 per lamp.

Standard HPS lamps will cycle on and off just prior to failure, signaling relampers to replace them. Testing for damaged igniters is possible with a Philips Igniter Tester (available from Philips lamp vendors), which screws into the lamp's socket. Iwasaki provides HPS lamps with built-in igniters that avoid this problem, but the fixture's existing igniter must be removed first.

Most metal halide ballasts last 6 to 12 years, but experience with some early versions of the Osram Powertronic ballasts showed a failure rate exceeding 20 percent in the first two years of operation.[1] This ballast supplies power to Osram's 70-watt HQI metal halide lamp. While the ballast is warranted for three years, and Osram has honored all failures with free replacements, users should date such ballasts when they are installed and maintain proper purchasing records. As often occurs with new electronic devices, there may be a "break-in" period for this technology, and users should be prepared by stocking a few extras since this item is not warehoused by many ballast vendors.

12.3 TRAINING

Without proper training, even motivated maintenance personnel are unlikely to keep a good lighting system operating properly. Knowledge of the equipment, an understanding of the procedures, and an appreciation for the relative costs of energy, labor, and lamps, are essential parts of instruction. As shown in Figure 12-1, the greatest cost is energy, but this fact is not obvious to those concerned with buying or installing lamps. Training is the key to creating that awareness. A written lighting maintenance policy incorporating the items mentioned in this chapter will help ensure that energy-saving procedures remain in effect and are brought to the attention of new employees. Where on-site personnel are unable to properly handle lighting maintenance, use of an outside firm should be considered to minimize labor, inventory, and component costs.

Annual retraining is also necessary to account for lost memory, changes in equipment, and staff turnover. It is especially important immediately after a lighting upgrade. Compact fluorescent lamp/ballast pin compatibility, the differences between standard HPS and "white" HPS, and metal halide safety are essential topics to avoid equipment damage, design degradation, and injuries. Such training should include hands-on work with new devices and a quiz to indicate which employees may need further technical education.

A variety of training and certification programs are offered by national organizations for the design and maintenance of efficient lighting systems (**Table 12-2**). Many utilities also sponsor local training programs.

In small or very large facilities, use of a lighting management company may simplify maintenance, and, when employee benefits are counted in, may actually be cheaper. Small facilities (those under 100,000 square feet) rarely have the staff to handle a group relamping, while large facilities (over 500,000 square feet) may have sufficient work to keep an outside crew busy on a weekly or even daily basis. Using an outside firm familiar with new lighting technologies can eliminate the need to stock many lamps, train employees, and deal with union personnel, while offering quick correction when a written specification is not followed.

12.4 Cleaning

Cleaning is important for light fixtures to maintain reflectance and transmissivity. Here we examine the impacts and methods of cleaning the most common fixture type: the direct ceiling downlight with reflectors and lenses or louvers.

12.4.1 Lenses

All lenses get dirty and discolor with age, and the timing and choice of a replacement can alter light output by over 20 percent. Failure to keep lenses clean can cut output by another 10 percent. A new clear acrylic lens transmits over 92 percent of a fixture's light, as shown in **Figure 12-4**, while an opalescent (also called "milky white") diffuser may allow less than 50 percent of light to pass through it.[2]

A good rule of thumb is to *clean* lenses at every relamping, and *replace* them whenever ballasts are changed out, since by that time they will have suffered significant irreversible loss of transmittance. Cleaning can be done with mild dishwashing detergent; both sides of the lens should be well rinsed and allowed to air-dry. Dry wiping should be avoided due to its potential to create static electricity, which attracts dust. While wiping lenses with a moist cloth may remove much of the loose dirt, elimination of the remainder could require heroic efforts whose labor cost exceeds that of a new lens. Like parabolic louvers (see page 282), some lenses can be effectively cleaned with ultrasonic equipment.

While some city and state fire codes prohibit the use of polystyrene because it will burn rapidly during a fire, its use should also be avoided for other reasons. Unlike acrylic or polycarbonate plastics, polystyrene becomes noticeably yellow when subjected to ultraviolet emissions from standard fluorescent lamps. This discoloration reduces light output and creates a dingy cast on all surfaces. Purchasing specifications should not allow its substitution for the other two types of plastics when buying replacement lenses.

Table 12-2
Training and certification programs for efficient lighting systems

Organization	Program	Contact information
AEE (Association of Energy Engineers)	Certified Lighting Efficiency Professional program: a one-day class and three-hour test oriented toward design and specification of energy efficient lighting systems	**Association of Energy Engineers** Lesha Thomas 4025 Pleasantdale Road, Suite 420 Atlanta, GA 30340 tel 770-447-5083 (x 218) fax 770-446-3969
EPA (U.S. Environmental Protection Agency)	Lighting Upgrade Workshops: Two-day program focused on energy efficient lighting, auditing and implementing lighting upgrades, maintenance and disposal, and an orientation to the Green Lights program and software tools. Those that pass a 90-minute test are identified as EPA "Surveyor Allies"	**Green Lights Hotline** tel 202-775-6650 fax 202-775-6680
IES (Illuminating Engineering Society)	Certificate of Technical Knowledge: a three-hour test that focuses on the basics and theory of lighting and lighting design	**Illuminating Engineering Society** 120 Wall Street, 17th Floor New York, NY 10005-4001 tel 212-248-5000 fax 212-248-5017
NALMCO (The interNational Association of Lighting Management Companies)	Certified Lighting Management Contractor program: Home-study course and one-day test focused on lighting systems specification and management	**NALMCO** Jennifer Busch 34C Washington Road Princeton Junction, NJ 08550 tel 609-799-5501 fax 609-799-7032
NCQLP (National Council on the Qualification of Lighting Professionals)	Supported by many lighting organizations and the US EPA, this group is developing a unified certification program involving a test and background check. Its first test is scheduled for late 1997.	**NCQLP** Betty Cole, Certification Manager 4401 East-West Highway, Suite 305 Bethesda, MD 20814 tel 301-654-1212 fax 301-654-4273

Figure 12-4

Light loss through fixture lenses

Old and dirty lenses can absorb as much as one third of the light that tries to pass through them. In this case, a brief cleaning only reduced losses from 29% to 19% (relative to a new K-12 lens).

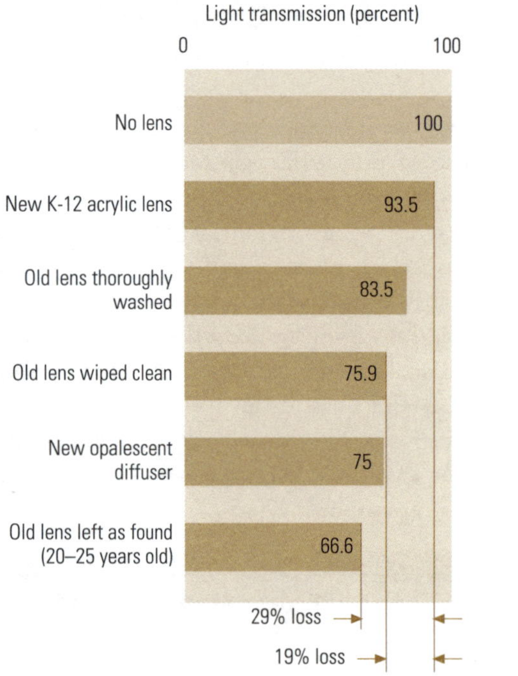

Source: Data from measurements taken at Columbia University, on various lenses taken from McVickar School of Social Work, NYC, 1993.

The parabolic cube louver is difficult to keep clean.

12.4.2 Parabolic louvers

The parabolic cube louver is a popular design option for controlling glare, but is difficult to keep clean. Its shape (like that of all louvers) exposes a large area to airborne dirt, and its many cube-shaped cells require a great deal of manual labor to be wiped clean. Professional lighting management companies instead use an ultrasonic bath and special cleansers to remove the dirt. Sending ultrasonic sound through water causes cavitation (that is, the formation of bubbles) at the louver's surfaces, scrubbing them in the process. While this process works well with metal louvers, a specular plastic louver sample should be tested first to ensure that its coating is not damaged by this action. Several special cleaning solutions (such as Parabolix 100) have been marketed to handle such louvers without the need for an ultrasonic bath.[3]

12.4.3 Fixtures

Because reflective surfaces in fixtures are crucial to maintaining light output, they should be cleaned at each relamping, and possibly more often in dirty or industrial spaces. As discussed above, different types of fixtures are more vulnerable to dirt buildup, and a good designer will consider the use of gasketing between lenses and fixture bodies to seal out dirt where it is a problem.

Washing white-painted reflective surfaces can sometimes restore them to near-new reflectance. Dishwashing detergent (or a mild solution of trisodium phosphate (TSP) cleanser) works well. Experience with many older (pre-1960) fixtures shows, however, that those not previously kept clean may have degraded to the point that even multiple cleanings make little difference. Areas populated with heavy smokers are especially difficult to clean due to tar buildup. In one location, the discoloration and dirt buildup was so severe that the only way to overcome it (short of replacing the fixture) was to cover the surface with a new specular reflector, which nearly doubled light output.[4]

Cleaning specular reflectors may require extra caution. While polished aluminum can stand fairly rough treatment, silver film reflectors are best cleaned with a soft moistened cloth. Both types should be allowed to air-dry to avoid building up static electricity by dry-wiping. Reflectors that can be easily removed (versus those screwed into place) are easier to handle when cleaning.

12.4.4 Room surface conditions

Dirt buildup on walls and ceilings can also influence maintained light levels. Depending on the fixture's light distribution, a drop of 5 to 20 percent in room light level (*not* fixture output) is possible over a three-year period, assuming lamps were installed in a newly painted space.[5]

THE TRUTH ABOUT DIRT

Fixtures located in dirty conditions (**Figure 12-5**) or where air is poorly filtered, can, over time, lose more than half their initial light output. Industrial processes that create moisture and dust can coat lamps with enough dirt to reduce total light output even further, long before they reach their rated lives.

Figure 12-5

Luminaire dirt depreciation factors for enclosed fixtures in environments of varying cleanliness

Even in a very clean environment, a fixture loses about 15% of its light output after three years to dirt build-up.

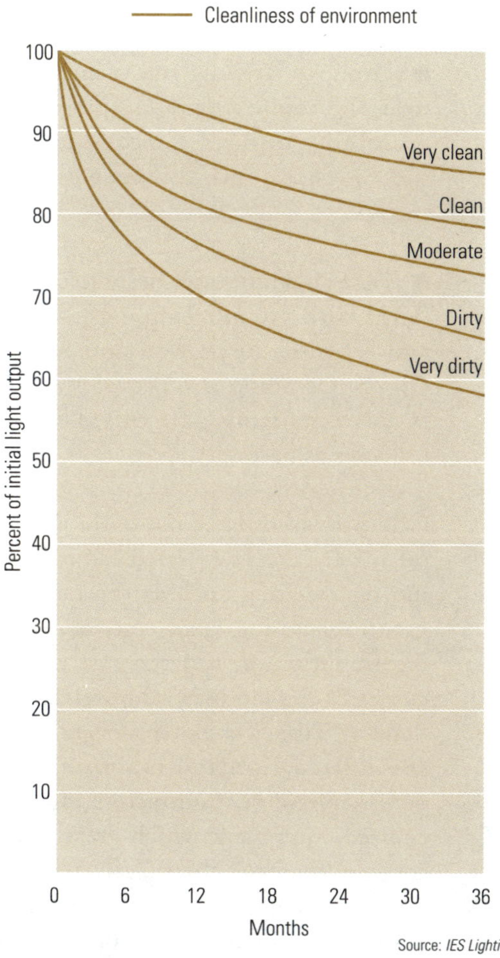

Source: *IES Lighting Handbook* [6]

Dirt is made up of several components:

- *inert* dirt lands by gravity on nonvertical or textured surfaces;
- *attracted* dirt adheres by electrostatic attraction;
- *adhesive* dirt clings to surfaces by its stickiness.

Methods for removing one type of dirt can influence how another type will build up later. For example, wiping a specular reflector with a dry cloth may remove adhesive dirt, but create static electricity, increasing adherence by attracted dirt. Cleaning methods are available (such as ultrasonic baths for removable reflectors, and antistatic cleaning solutions) that attend to all three types of dirt.

Examination of many older fixtures shows that several factors unrelated to lighting or industrial processes may affect dirt buildup, including timely maintenance/replacement of filters in ventilation and air conditioning systems, smoking policies, and humidity level control.

The data shown in Figure 12-5 is from studies performed in the 1950s. This old data, however, may overstate the modern effects of dirt depreciation because of the wider use of air filtration, reduced indoor smoking, and various indoor air quality requirements. Lower dirt depreciation factors may in turn allow lower installed lighting power densities because of the reduced need to overdesign to accommodate dirt buildup. A three-year project sponsored by the EPA and conducted by NALMCO is currently underway to measure actual dirt depreciation in a wide variety of facilities and create new design factors. Results of this study are expected near the end of 1998.[7]

Lack of proper filtration allows all three types of dirt to enter a space, smoking and high humidity increase the level of adhesive dirt, and very dry conditions foster static buildup. Efforts to improve indoor air quality may therefore also reduce light degradation due to dirt. Recent ordinances restricting smoking will go a long way to reducing the vaporized tar that clings relentlessly to the inside of fixtures. One lighting maintenance expert found that removal of smoking tar sometimes required using cleaning solutions with concentrations sufficient to remove a fixture reflector's white paint.[8]

Group relamping has higher lamp costs but lower labor costs than spot relamping.

The most severe impact occurs with indirect fixtures, which depend upon the reflectance of ceiling and upper wall surfaces to distribute light. Room shape will also influence light lost in this manner, with corridors and small rooms (that is, those with high room cavity ratios) being most affected. Light-colored porous surfaces, such as cinder block, demonstrate the most extensive depreciation as dust builds up in indentations. Merely repainting such corridor walls (with the same color) can raise light levels over 15 percent.[9]

12.5 SPOT VS GROUP RELAMPING

There are a variety of reasons to practice *group relamping* rather than *spot relamping,* in which lamps are only replaced when they burn out. Most of these reasons apply to fluorescent and HID lamps rather than incandescents, which have much shorter lifetimes and less lumen depreciation.

- On a per-lamp basis, group relamping requires much less labor than spot relamping. To replace a spot burn-out can take a worker as much as a half-hour to fetch the new lamp, set up, and install it. Having all the materials on hand and moving systematically from one luminaire to the next, reduces the required time to about three minutes per lamp,[10] and reduces disruption of normal activities, since the group relamping is normally done outside working hours.

- Group relamping is an easy task to schedule and delegate to outside contractors, who have special equipment and training. This reduces administrative overhead by enabling a smaller in-house maintenance staff (although some spot relampings will still be required).

- Group relamping provides brighter and more uniform lighting, because it gets rid of lamps before they are at the end of their lumen depreciation curve. This can pay direct energy benefits if the design uses a smaller safety factor (see **3.1**).

- Group relamping offers increased control over the replacement lamps, reducing the probability of mixing incompatible lamps—such as those with different color temperatures.

- Other maintenance activities can be combined with group relampings, such as ballast and reflector inspection and lens cleaning. Group relamping also provides an opportunity for retrofitting reflectors, lamps, ballasts, or lenses as necessary.

Group relamping is normally done at about 60 to 80 percent of rated lamp life, depending on variables such as the ratio of spot to group relamping labor cost, requirements for fixture cleaning, and the cost and mortality curves of the lamps. Group relamping can be done by a user's own maintenance staff or by a specialized lighting-maintenance firm or other contractor. Contractors often have specialized equipment which greatly reduces the cost and increases the effectiveness of lighting maintenance. This includes not only access equipment, such as telescoping scaffolding, but also ultrasonic lens- and louver-cleaners.

Economic comparisons typically show that group relamping has higher lamp costs but lower labor costs than spot relamping. One such comparison is shown in **Table 12-3**, which shows a 31 percent overall savings from group relamping. This type of calcula-

Table 12-3

Economics of group vs spot relamping for 1,000 3-lamp T8 lensed troffers

Group relamping has higher lamp costs but much lower labor costs, in this case providing a 31% overall savings. Group relamping also provides additional benefits in lighting quality and easier facility management.

	Relamp cycle (hours)	Average relamps per year	Average material cost per year	Average labor cost per year	Total average cost per year
Spot relamping on burnout [a]	20,000	525	$1,391	$3,150	$4,541
Group relamping at 70% of rated life) [b]	14,000	750	$1,988	$1,125	$3,113
Difference		225	$597	-$2,025	-$1,428 31% savings

Notes:
a: Assumes labor costs of $6.00/lamp for relamping and cleaning, material cost of $2.65/lamp, and 3,500 hours/yr operation.
b: Assumes labor costs of $1.50/lamp for relamping and cleaning, same material costs and operating hours as for spot relamping.

Source: EPA

tion is heavily dependent on the difference in labor costs between group and spot relamping. For example, if the group relamping cost of $1.50 per lamp jumps to $3.50, the balance tips in favor of spot relamping. Remember, however, the noneconomic benefits of group relamping discussed above before deciding between the two methods.

Lamps that burn out may not be replaced for weeks, providing a windfall energy savings that can be included in analyses for greater accuracy. In the Table 12-3 example, this savings would amount to a $212 credit for the spot relamping program, assuming two weeks of non-burn time, 30 watts of avoided power per lamp, and 10¢ per kWh total energy costs. Longer delays could conceivably shift the economics in favor of spot relamping, but slow response is hardly a useful energy savings (or maintenance) practice.

12.6 Purchasing and Inventory Control

Controlling choices of replacement equipment is crucial for maintaining lighting efficiency. If the right ballast or lamp is not immediately available when an occupant complains about a burnout, the need to provide a quick replacement will almost ensure that the wrong type is instead installed. Maintaining proper purchasing specs and inventory eliminates the option to use the wrong equipment. There are four key steps to controlling inventory:

- Convert the specifications used for purchasing the products in the retrofit or construction job into a list of acceptable substitutes for each lamp, ballast, or screw-in retrofit kit that was part of the initial job. This list then becomes "the rules" when ordering replacement parts. Otherwise, purchasing personnel tend to trust vendors, who may erroneously claim that their products are equivalent. It is not uncommon, for example, for a vendor to suggest a lower-cost 50-watt HPS lamp to replace the White SON or Iwasaki high CRI HPS lamps. The result may be base/socket incompatibility, or a major reduction in lighting quality and color.

- Get the inventory started on the right foot by ordering (as part of the upgrade) an extra one percent of lamps, ballasts, and screw-in retrofit kits (but no fewer than two of any item), and turn them over to the inventory department, along with a list of all components installed as part of the upgrade, plus sources for purchasing them.

Get the inventory started on the right foot by ordering an extra one percent of lamps, ballasts, and screw-in retrofit kits for inventory.

- Organize information from the upgrade into a database showing the lighting equipment used at each building, floor, or tenant. Totals for each type of component can then be used to find the quantity of spares to be stocked. Such a database may also expedite relamping (whether spot or group) by eliminating the need to visit a space to determine the types of lamps installed.

- Educate purchasing personnel regarding lamp and ballast pricing. "Sticker shock" is not uncommon when the first replacements are bought, creating an incentive to find cheaper alternatives. Personnel need to be shown that lamps that cost five times as much—but last five times as long—cost no more in the long run. Inventory budgets may need to be adjusted to account for increases in lamp costs that beget large decreases in energy costs.

Similar education and budget modification may be needed if group relamping is initiated. Some of the resulting savings in overall relamping labor should be reallocated to pay for the greater rate of lamp changeouts involved in such a program. Spreading out group relampings over time (instead of relamping an entire facility at one time) will also level out the cost of such work. Otherwise, an inventory budget could be depleted before a group relamping can be finished, disrupting this process in midstream.

1 Columbia University, 1991–1994, at the Uris Business School.

2 Data from measurements in the McVickar School of Social Work, Columbia University, 1993.

3 Parabolix 100, available from Merix Chemical Company, 2234 East 75th Street, Chicago, IL tel 312-221-8242.

4 Columbia University, 1992, at the Uris Business School.

5 Mark S. Rea, ed., *Lighting Handbook: Reference & Application,* 8th Edition (New York: Illuminating Engineering Society of North America, 1993), p. 399.

6 Mark S. Rea [5], p. 398.

7 Jim Burke, personal communication (October 13, 1997), Principal, Burke & Associates Lighting Consultants, 3703 Soaring Eagle Lane, Castle Rock, CO 80104, tel 303-660-8337, fax 303-660-8337, e-mail jblighting@aol.com. For results of the study, contact Rita Herald, Technical Director, Illuminating Engineering Society, 120 Wall Street, 17th Floor, New York, NY 10005-4001, tel 212-248-5000 x115.

8 Tim Culkin, personal communication (late 1991), Vice-President, Unalite Electric and Lighting Corp., 12-17 38th Avenue, Long Island City, NY 11101, tel 718-784-0630, fax 718-392-6398.

9 Columbia University, 1992, in the East Campus building corridors.

10 The Lighting Bureau, *Industrial Lighting Handbook*, p. 17. Further information on this and related topics is available from the ~100-member International Association of Lighting Management Companies (NALMCO), tel 215-569-3650, fax 609-799-7032.

RESOURCES

A Lighting Software

This appendix profiles sixteen software packages that address energy efficiency in lighting specification, design, and application. (Other software also exists to aid with lighting application and luminaire design.) A list of typical software characteristics follows; the programs are listed in the table on pages 286 and 287.

User interface. There are still one or two programs which require the use of FORTRAN compilers for data entry. Most programs, however, are now equipped with pull-down menus. A few also are DOS-driven, and use keyboard and function keys, rather than a mouse and menus. Most mouse-driven systems can still be controlled by the keyboard if desired. One of the newer programs uses a pen-tablet interface.

Economic analyses. Many of the programs include some type of economic analysis.

Simple payback. A simple payback compares the extra cost of installing a more efficient system to its projected annual energy savings. This provides a rough idea of when the savings in energy will have paid for the upgrade.

Lifecycle costs. A lifecycle cost analysis will show the financial impacts of a system over its life. It will incorporate initial costs, future payments, interest rates, maintenance, depreciation, and inflation.

Incremental rate of return on investment. This is the interest rate generated over the life of the system which offsets the initial investment. For example, if the return on investment were calculated to be 20 percent, it would be profitable to borrow money at as much as 19 percent to pay for the system.

Cost comparisons. A couple of programs will show side-by-side cost comparisons of different lighting systems.

Lighting calculations. Two types of illuminance calculations are available. This is true for electric lighting as well as daylighting.

Average illuminance calculations. The majority of programs will calculate an average illuminance for a work surface. This generates a good target value, but does not show how much levels may vary. An average illuminance assumes that the lighting system is arranged in a regular pattern, and that the walls, ceilings, and floors have no variations in finish.

Point-by-point calculations. A few programs will calculate illuminance values at many different points in a room. Unlike the average illuminance method, a point-by-point calculation predicts the varying levels of light across a surface. Thus, it is a more accurate tool. It will indicate whether the illuminance is higher below a fixture, and lower in the corner, and by how much. Point-by-point mechanisms can also account for changes in finishes, such as a dark wood door in a white wall.

General features.

Watts / ft². Some programs calculate lighting power density in watts per square foot for the designed lighting system. This is for connected load only, and does not assume any controls for daylighting or occupancy.

Controls consideration. Several programs will account for the impact of lighting controls such as occupancy sensors and daylight dimming controls.

Automatic design features. A few of the software packages provide an automatic design feature. The user enters a target illuminance value and a chosen luminaire, and the program will create a layout of those fixtures which meets the target illuminance. This feature will only consider average illuminances, and will create a

For more information, The Illuminating Engineering Society of North America conducts an annual survey of lighting software tools and publishes it in the July issue of its magazine, LD+A (Lighting Design and Application).

IESNA
120 Wall Street,
17th Floor
New York, NY 10005-4001
tel 212-248-5000
fax 212-248-5017

uniform spacing of luminaires in the room. In addition to electric lighting, some programs will automatically locate windows for daylighting.

Whole-building model. Several of the software packages will perform comprehensive studies on the building as a system. This includes analyzing HVAC or other systems besides lighting. The sophistication of the building models ranges from a single "snapshot" in time to hourly calculations of an entire year.

Rendering capabilities. A few programs generate shaded or contour renderings of lighting levels. Instead of interpreting a sea of numbers, the user can generate a picture which models the lighting variations in a space.

Expert system. This is a screening and direction device. It will prompt the user for information, and determine if the project is viable, or if more information is required. It will also suggest when a lighting consultant is needed. These programs operate according to a predetermined hierarchy of rules.

Hardware required. Unless otherwise listed, the programs will operate on an IBM compatible 286 with hard drive, at least one external disk drive, 640k RAM, and a monochrome monitor.

Software required. A few programs are run within other programs such as Windows 3.1. If software other than DOS 3.1 is required, it is listed.

Compatibility. Many of these programs are designed to work with each other. Any compatible software is listed.

Target audience. Some software packages are designed for trained energy or lighting consultants. Others are designed for a more general audience which would include facility managers and utility personnel.

Cost. All of the EPRI programs are available free to EPRI members, but they are also available to the general public for the costs indicated in the tables. Costs of all other packages are as shown.

How lighting software is used

Facility managers and lighting consultants can use software to assist in energy efficient design. While energy conservation is a measurable item, esthetics and quality are subjective, and harder to quantify. These programs, therefore, should be considered as valuable tools, but not as replacements for trained consultants when lighting quality is a concern.

Light only where it is needed. One of the first, but least obvious, steps in reducing connected load is to provide lighting only where it is needed. This is also known as a task-ambient lighting approach. A designer can take advantage of point-by-point calculation features to determine where fixtures can be eliminated from an otherwise uniform lighting solution. Lumen-Micro and LitePro, for example, can fulfill this function.[1]

Daylight and controls. Daylighting can save energy only if the electric lighting is deactivated. The designer needs a tool for assessing the impacts of daylighting controls on energy consumption. A few of the newer software products are able to predict daylighting levels and related energy savings. BEEM, for example, will predict illumination levels due to daylight. This information can then be fed into LightCAD which will predict impacts due to controls.[2]

Occupancy patterns and controls. The designer can also use software to study occupancy control impacts. Some programs analyze controls automatically, while others do so indirectly. For example, the Green Lights GL-DSS program will ask if occupancy controls are used, and then automatically reduce energy consumption. With MicroAxcess, however, the user must manually alter the workday hours to simulate occupancy and other control devices.[3]

Saving money as well as energy. Perhaps the most significant tool these programs offer is the ability to predict economic impacts due to energy conservation methods. Software will analyze individual rooms, or entire buildings. At times, the user is only interested in comparing the costs of different energy efficient equipment or systems. Lumen-Specifier, for example, can give an initial cost comparison of two lighting systems.[4] Some designers, however, need more detailed information about building systems. Since reducing the lighting load also means there will be less generated heat, the owner wishes to downsize the cooling system. By using a program such as DOE-2, the consultant can study the impact of the lighting system on the cooling system. With this "whole-building" approach, the efficiency of the entire building can be fine-tuned.

Table of software products

Company	Program	User interface	Economic analyses	Lighting calculations	General features	Hardware required	Software required	Compat-ibility	Target audience	Cost ($)	Contact
EPA	GREENLIGHTS GL-DSS 2.0[5]	Menu-driven	Incremental rate of return	Electric only, average illuminance	Watts/ft^2, controls consideration, automatic design	10M hard drive	Will run on DOS 3.X, 5.X or 6.X, but not DOS 4.0	None	Facility and utility personnel, consultants	Free[a]	Don Anderson ICF Information Technology, Inc. 9300 Lee Highway Fairfax, VA 22031-1207 tel 703-934-3900 fax 703-691-3353
Electric Power Research Institute	BEEM[6]	Menu-driven	Through LightCAD	Daylight only, average illuminance	Exports information to LightCAD			LightCAD	Consultants	$50	R.A. Rundquist & Associates 56 Ward Avenue Northampton, MA 01060 tel 413-586-7743
	Comtech[7]	Menu-driven	Simple payback, lifecycle cost, incremental rate of return, cost comparisons	Electric only, average illuminance	Watts/ft^2, controls considered, whole-building model with 48-day approach	5M hard drive		LightCAD and LightPAD	Facility and utility personnel, consultants	$950	Regional Economic Research, Inc. 12520 High Bluff Drive, Ste. 220 San Diego, CA 92130-2062 tel 619-481-0081 fax 619-481-7550
	LightCAD[8]	Run through autoCAD, menu/mouse driven	Utility rebates, cost comparisons	Electric only, average illuminance	Watts/ft^2, daylight levels imported from BEEM, controls considered, automatic design	386+ coprocessor, 4–8M RAM, VGA monitor	AutoCAD 11 or 12	Comtech and BEEM	Consultants	$300	The Weidt Group 5800 Baker Road, Suite 100 Minnetonka, MN 55345 tel 612-938-1588 fax 612-938-1480
	LightPAD[9]	Version 1.0 menu-driven, version 2.0 pen-based	Simple payback, cost comparisons	Electric only, average illuminance	Watts/ft^2, controls considered	Version 2.0 requires a pen-tablet system	Version 2.0 requires Windows for pen	Comtech	Facility and utility personnel	$295	The Weidt Group 5800 Baker Road, Suite 100 Minnetonka, MN 55345 tel 612-938-1588 fax 612-938-1480
	MicroAxcess[10]	Menu-driven	Monthly utility bills	None	Controls considered, whole-building model with hourly profiles up to 365 days/year			None	Consultants	$1,000	Edison Electric Institute tel 202-508-5000
	WinDOE[11]	Menu-driven	Simple payback, lifecycle cost	None	Watts/ft^2 for HVAC only controls considered, automatic design, whole-building model with hourly profiles up to 365 days/year, expert system	386, 8M–16M RAM, 50M hard drive, mouse, VGA monitor	Windows 3.1	None	Facility and utility personnel, consultants	N/A[b]	Hirsch and Associates tel 805-532-1045
Genlyte	Genesys[12]	Menu-driven	Simple payback, lifecycle cost, incremental rate of return, cost comparisons	Point-by-point	Watts/ft^2, controls considered, automatic design, whole-building model, rendering	2M RAM, mouse, VGA monitor, 6M hard drive	DOS 3.1 or MAC with soft PC DOS emulator	None	Consultants	$450	Lightolier 100 Lighting Way Secaucus, NJ 07096 tel 201-392-3902 fax 201-864-1018
Lawrence Berkeley Laboratory	Controlite[13]	Menu-driven	Lifecycle cost, incremental rate of return	Daylight only, point-by-point	Watts/ft^2, controls considered	Math coprocessor		None	Consultants	Free	Francis Rubenstein Lawrence Berkeley Lab. 1 Cyclotron Road Building 90-3111 Berkeley, CA 94720 tel 510-486-4000 fax 510-486-4089

Notes: a: Requires attendance at a two-day seminar.
b: Scheduled for availability in December 1994.

Continued on next page

Table of software products (continued from previous page)

Company/program		User interface	Economic analyses	Lighting calculations	General features	Hardware required	Software required	Compat -ibility	Target audience	Cost ($)	Contact
Lawrence Berkeley Laboratory *(continued from previous page)*	DOE-2[14]	DOS-driven	Simple payback, lifecycle cost, cost comparisons	Daylight only, average illuminance	Watts/ft^2 for HVAC only, controls considered, whole-building model with hourly profiles up to 365 days/year			None	Consultants	\$100–\$1,000	Fred Winkelman Lawrence Berkeley Lab. 1 Cyclotron Road Building 90-3147 Berkeley, CA 94720 tel 510-486-4925 fax 510-486-4089
	Radiance[15]	Multiple text files	None	Electric and daylight, point-by-point	Rendering	UNIX with X11 or Apple MAC II/Quadra running A/UC or 386 running LINUX or SCO UNIX	UNIX, A/UC, LINUX, or SCO UNIX	None	Consultants	Free	Building Technologies Prgm. Lawrence Berkeley Lab. 1 Cyclotron Road Building 90-3111 Berkeley, CA 94720 tel 510-486-4000 fax 510-486-4089
	Superlite[16]	Text file in FORTRAN format	None	Daylight only, point-by-point	Will model daylight in complex spaces, graphics package is optional	Math coprocessor	Microsoft FORTRAN 3.2 compiler, optional graphics package available	None	Consultants	\$35, \$95 \$895 for optional graphics package	Jong-Jin Kim College of Architecture & Urban Planning University of Michigan 2000 Bonisteel Boulevard Ann Arbor, MI 48109 tel 313-763-3518
Lighting Technologies	Lumen-Micro 6.0[17]	Menu-driven	None	Electric and daylight, point-by-point	Watts/ft^2, controls considered, automatic design, rendering	386 with math coprocessor 16MHz, 2M RAM, 6M–10M hard drive, VGA monitor		None	Consultants	\$595	Lighting Technologies Inc. 1540 Frontier, Suite 107 Boulder, CO 80301 tel 303-449-5791 fax 303-449-5864
	Lumen Specifier[18]	Menu-driven	Lifecycle cost, cost comparisons	Electric only, average illuminance	Watts/ft^2, controls considered, automatic design, rendering			None	Facility and utility personnel, consultants	\$295	Lighting Technologies Inc. 1540 Frontier, Suite 107 Boulder, CO 80301 tel 303-449-5791 fax 303-449-5864
National Renewable Energy Laboratory	FLEX[19]	Menu-driven	Simple payback, lifecycle cost, cost comparisons	Electric average illuminance, daylight average illuminance will be added in fiscal 1994	Watts/ft^2, controls considered, automatic design, whole-building model using 24 hour profiles and 3 day-types, expert system	386 2M RAM, Super VGA monitor		None	Facility and utility personnel, consultants	Free to federal sector, others contact NREL for prices	National Renewable Energy Laboratory 1617 Cole Boulevard Golden, CO 80401-3393 tel 303-275-6021 fax 303-275-6055
USI Lighting, Inc.	LitePro[20]	Menu-driven	Simple payback, cost comparisons	Electric only, point-by-point	Watts/ft^2, automatic design, rendering	10M hard drive		None	Consultants	\$150	Columbia Lighting North 3808 Sullivan Road P.O. Box 2787 Spokane, WA 99220 tel 509-924-7000 fax 509-924-1134

1 Patrick Burns and Michael Smith, personal communication (November 24, 1993), Lighting Technologies, Inc., 2540 Frontier, Suite 107, Boulder, CO 80301, tel 303-449-5791; and Kendra Paschall and Barbara Barker, personal communication (November 24, 1993), Engineers, Columbia Lighting, North 3808 Sullivan Road, P.O. Box 2987, Spokane, WA, tel 509-924-7000.

2 Larry Ayers, personal communication (November 30, 1993), EPRI, tel 800-525-8555; and Tom McDougall, personal communication (December 1, 1993), Energy Department, The Weidt Group, 5800 Baker Road, Suite 100, Minnetonka, MN 55345, tel 612-938-1588.

3 Gary Schroeder, personal communication (November 30, 1993), Consultant, E-Cube, 1900 Folsom, Suite 112, Boulder, CO 80302, tel 303-443-2610.

4 Patrick Burns [1], personal communication (November 24, 1993).

5 Randy Cook and Don Anderson, personal communication (December 1, 1993), Contractors for EPA, tel 703-934-3150.

6 Larry Ayers [2].

7 Scott Criswell, personal communication (December 2, 1993), Senior Software Engineer, Regional Economic Research, 12520 High Bluff Drive, Suite 220, San Diego, CA 92130-2041, tel 619-481-0081.

8 Tom McDougall [2].

9 Tom McDougall [2].

10 Gary Schroeder [3].

11 Jeff Hirsch, personal communication (December 1, 1993), Hirsch and Associates, 12185 Presilla Road, Camarillo, CA 93012-9243, tel 805-532-1045.

12 David Keith and Steve Stannard, personal communication (January 1994), Vice President and President, Illumination Arts, Inc., 493 Reed Ranch Road, Boulder, CO, 80302, tel 303-541-0120.

13 Francis Rubenstein, personal communication (November 1993), Lawrence Berkeley Laboratory, MS 90-3111, Berkeley, CA 94720, tel 415-486-4096.

14 Fred Winkelmann, personal communication (November 30, 1993), Lawrence Berkeley Laboratory, University of California, Building 90-3147, Berkeley, CA 94720, tel 510-486-4925.

15 Francis Rubenstein [13].

16 Francis Rubenstein [13].

17 Patrick Burns and Michael Smith [1].

18 Patrick Burns and Michael Smith [1].

19 Bob Westby, personal communication (November 1993), Senior Project Coordinator, National Renewable Energy Laboratory, 1617 Cole Boulevard, Golden, CO 80401-3393, tel 303-275-6021; and Mark Fishbaugher, personal communication (November 1993), Fishbaugher and Associates, P.O. Box 301, Cheshire, OR 97419, tel 503-998-2978.

20 Kendra Paschall and Barbara Barker [1].

B U.S. EPA Energy Star Specification for Residential Light Fixtures

To assist in the development and penetration of efficient residential lighting equipment, the United States EPA has created a program to apply the Energy Star label to qualifying residential light fixtures. This appendix lists the technical specifications that fixtures must meet to be certified as Energy Star fixtures, and is drawn from a larger memorandum of understanding (MOU) that defines other aspects of the Energy Star program, such as testing procedures, labeling guidelines, and use of the Energy Star logo (**Figure 1**). **Table B-1** shows the performance specifications for indoor fixtures, **Table B-2** shows the light source requirements for outdoor fixtures, and **Table B-3** shows the operating time requirements for outdoor fixtures.

Figure B-1

Energy Star® logo for residential light fixtures
Residential light fixtures bearing this logo meet some fairly rigorous standards for energy efficiency, power quality, and optical performance.

Source: U.S. EPA

For more information on joining the Energy Star Residential Light Fixture Program, contact:

Lena Nirk
Program Manager
U.S. EPA (MC: 6206J)
401 M Street SW
Washington, DC 20460

Or call the Energy Star hotline at:

888-STAR-YES
(888-782-7937)

Table B-1

Indoor fixtures
Fixtures intended for indoor use must meet the following performance specifications.

Energy efficiency and power characteristics	Performance specifications
System efficacy (lumens per watt)	
up to 20 watts input power	50 lpw
20 to 29 watts input power	55 lpw
30 watts input power and lamp length < 24 inches	60 lpw
30 watts input power and lamp length >24 inches	70 lpw
Total harmonic distortion (THD)	< 33%
Power factor	0.9
Current crest factor	1.7
Interference	Comply with FCC Class B Part 18.305 and 18.307 Sub-part for electromagnetic interference (EMI) and radio frequency interference (RFI).
Operating characteristics	
Convenience	Lamp must continuously illuminate within one second of being switched on.
Dimming	Portable light fixtures with input power of 22 watts or more shall be continuously dimmable from 100% down to 50% or less of maximum light output, or be switchable to two levels of brightness (in addition to off) with included switch.[a]
Light quality	Color rendering index (CRI) of at least 80 for compact fluorescent lamps (lamp diameter smaller than 3/4-inch)[b], and a CRI of at least 70 for all other lamps.
Noise	Ballasts if included must have sound rating of Class A or quieter.
Reliability/durability	
Warranty for defects in materials and manufacturing	Repair or replacement of defective parts of the fixture housing or electronics (except lamp) for two years from the date of purchase.
Longevity	Lenses or diffusers shall be constructed of glass or UV-stable optical grade plastics and be heat stable. Reflectors, if included, shall be UV resistant and heat stable. Comply with ANSI/IEEE C62.41 specifications for withstanding electrical transients.
Safety	
	Fixtures must be tested and listed by UL, ETL, CSA or other independent agency acceptable or compliance with NFPA 70, National Electric Code. Ballasts shall be Class P thermally protected.

a: Dimming ballasts need only comply at full brightness.
b: If the lamp is not included in the fixture package, see section IV.D.1 of the MOU.

Table B-2

Light source requirements for outdoor fixtures

Energy efficiency characteristics	Performance specifications
Maximum input power	150 watts
System efficacy (lumens per watt)	
up to 70 watts	40 lpw
71 to 150 watts	50 lpw[a]
Mechanical	Lamp holder will operate only lamps that perform to the input power range of the fixture.
Operating characteristics	
Reset	Resets automatically to automatic mode within 24 hours of a manual override or testing operation.
Shut-off	Automatic shut-off during daylight hours
Reliability	
Warranty for defects in materials and manufacturing	Repair or replacement of defective parts of the fixture housing or electronics (except lamp) for two years from the date of purchase.
Warranty for consumer satisfaction with fixture performance	One year
Safety	
	Fixtures must be tested and listed by UL, ETL, CSA or other independent agency acceptable for compliance with NFPA 70, National Electric Code including listing for damp or wet locations (Articles 410-4a and Article 100).

a: and all light must be distributed below a horizontal plane located at mid-luminaire

Table B-3

Operating time requirements for outdoor fixtures

Energy efficiency characteristics	Performance specifications
Maximum input power	250 watts
Shut-off	Automatic shut-off during daylight hours; and automatic shut-off within 20 minutes of either a manual on signal, or no motion in the fixture's field of view.
Operating characteristics	
Reset	Resets automatically to automatic mode within 24 hours of a manual override or testing operation.
Lamp start	Lamp must restart within three seconds of receiving the start stimulus.
Reliability	
Warranty for defects in materials and manufacturing	Repair or replacement of defective parts of the fixture housing or electronics (except lamp) for two years from the date of purchase.
Warranty for consumer satisfaction with fixture operation	One year
Safety	
	Fixtures must be tested and listed by UL, ETL, CSA or other independent agency acceptable for compliance with NFPA 70, National Electric Code including listing for damp or wet locations (Articles 410-4a and Article 100).

C SUMMARY OF LIGHTING PROVISIONS OF ASHRAE 90.1 STANDARD

Under the National Energy Policy Act of 1992, states are required to adopt ASHRAE 90.1-1989 as a minimum standard for new construction. This adoption process is still underway as of 1997. Section 6 of the standard, which deals with lighting, applies only to new construction and is only applicable if codified in a local, state, or federal statute, or in a facility's specification. The standard itself is a 147-page document that is undergoing substantial revision as of 1997; the revised standard should be officially approved by 1998 or 1999. The key implications for lighting are summarized here.

■ Many types of specialty lighting are exempted.

■ There are few equipment component requirements in ASHRAE 90.1. Only fluorescent ballasts are mentioned, and they must meet these criteria: minimum power factor of 90%, and minimum ballast efficacy factor (BEF) is specified as follows:

> Fluorescent lamp ballasts which have all of the following characteristics shall meet or exceed the minimum ballast efficacy factor (BEF) shown in **Table C-1.**
>
> (a) operate at nominal input voltages of 120 or 277 volts
>
> (b) input frequency of 60 Hz
>
> (c) maximum lamp operating current less than 1,000 milliamperes
>
> (d) used to operate one of the following lamp types:
>
> 1. one or two 4-foot, nominal 40-watt, rapid start lamps
> 2. two 8-foot, nominal 75-watt, slimline lamps
> 3. two 8-foot, nominal 110-watt, high-output rapid start lamps
>
> (e) not specifically designed for starting at temperatures below 40° F
>
> (f) not specifically designed for use with dimming controls

Table C-1

Fluorescent ballast efficacy factory (BEF)

Ballast type	Minimum ballast efficacy factor
One-lamp, 4 foot, nominal 40 W, rapid start	1.805
Two lamp, 4-foot, nominal 40 W, rapid start 120 V	1.060
Two lamp, 4-foot, nominal 40 W, rapid start 227 V	1.050
Two lamp, 8-foot, nominal 75 W, slimline	0.570
Two lamp, 8-foot, nominal 100 W, high output, rapid start	0.390

Ballasts that do not have all of the characteristics listed above are not required to meet the BEF in Table C-1.

■ The ASHRAE 90.1 methods for determining maximum allowable lighting wattage include the wattage of all task lighting, whether built in or plugged in.

■ Lighting must be controlled by some *type* of control (except in special cases). The *degree* of control provided by each *type* of control is reflected by the number of "points" assigned to that type (for example, a standard wall switch is assigned 1 point; a dimmer, up to 3 points; a 3-way switch, 1 point).

■ Don't confuse lighting control (type) *points* with lighting *controls* (*device types*). There must be at least one *control* (device type) for each 1,500 watts of connected lighting power (CLP). (Note: a 3-way switch pair is still considered *one* control.)

■ One lighting control *point* must be provided for every enclosed space and another lighting control *point* for each task location within the space (unless the space is 450 ft² or less, in

ASHRAE/IES 90.1-1989 "Energy Efficient Design of New Buildings Except Low-Rise Residential Buildings" and a several-hundred-page user's manual for making sense of the standard are available from:

American Society of Heating, Refrigerating, and Air-Conditioning Engineers (ASHRAE)
1791 Tullie Circle NE
Atlanta, GA 30329
tel 404-636-8400
fax 404-321-5478
web www.ashrae.org

which case all tasks in that space are considered as one task).

■ Once a design has been roughed out, its CLP may be adjusted downward (for purposes of compliance) by multiplying it by a power adjustment factor (PAF) related to the type of automatic controls installed (if any). The result is a lighting power control credit (LPCC), which is then subtracted from the CLP to yield an adjusted lighting power (ALP). It is the ALP that must not exceed the ASHRAE power limits.

■ Two alternative procedures exist for determining if a given design's adjusted lighting power (ALP) complies with the interior lighting power allowance (ILPA).

The *prescriptive method* (ASHRAE Section 6.5) is calculated for the interior of an *entire building* and is based on its gross lighted area (GLA). Find the appropriate maximum W/ft^2 in ASHRAE *Table 6-5,* called the unit lighting power allowance (ULPA) and multiply it by the GLA. The result is the maximum building interior lighting power allowance (ILPA). Compare the ILPA to the ALP to verify compliance.

The *system performance method* (also called the lighting power budget [LPB] method) is calculated for *individual spaces*. This method uses an area factor (AF) to adjust for a space's area and ceiling height (see the graphs in ASHRAE *Figure 6-1*). Individual LPBs are calculated by multiplying unit power densities (UPD) from ASHRAE *Tables 6-6 (a-c)* by the room area and the appropriate area factor. The results are summed to find the building's ILPA. That total is compared to the ALP.

■ Outdoor lighting has an exterior lighting power allowance (ELPA) calculated using ASHRAE *Table 6-1* and the appropriate area of length being illuminated.

A Guide To The Alphabet Soup Of ASHRAE 90.1- 1989

AF	area factor
ALP	adjusted lighting power [= CLP - (CLF x PAF)]
CLP	connected lighting power
ELPA	exterior lighting power allowance
GLA	gross lighted area
ILPA	interior lighting power allowance
LPB	lighting power budget (= A x UPD x AF)
LPCC	lighting power control credit (= CLP x PAF)
LS	listed space (i.e., space where tasks are known)
PAF	power adjustment factor (see ASHRAE *Table 6-3)*
ULPA	unit lighting power allowance (from ASHRAE *Table* 6-5)
UPD	unit power density (from ASHRAE *Tables* 6-6 *a-c)*

Remember: ILPA and ELPA are *limits*, not design goals. If one uses good components (for example, electronic ballasts), these limits can be easily met without any automatic lighting controls. They are independent of each other and no wattage exchanges are permitted. To make life easier, ASHRAE provides a PC-based program to help in applying the standard.

In the future, ASHRAE 90.1 may be updated to required some form of automatic lighting controls, may limit the use of standard incandescent lamps, and will likely become more stringent.

This specification was developed by E SOURCE research consultant Lindsay Audin for his work as Energy Conservation Manager at Columbia University. While specific to Columbia's situation and to the applicable codes of New York City and New York State, most of these guidelines are relevant to commercial and institutional lighting systems anywhere.

Interior Lighting Technical Specifications

1.0 General

1. *Compliance With Applicable Codes And Laws.* When not covered in the specifications below, ASHRAE/IES Standard 90.1-1989 shall apply, unless the subject is also covered by the State Energy Conservation Construction Code, which shall then take precedence. All designs and installations shall comply with State Building Code, National Electric Code, Americans with Disabilities Act, and the codes of the National Fire Protection Association, unless City codes covering the same matters pertain.

2. *Supporting Data.* Data supporting compliance with the above codes and the technical requirements of this specification shall be available upon its request.

3. *Operating Parameters.* All components shall be designed to operate properly between 50° F and 115° F without increased failure rate, in addition to any stated conditions listed below.

1.1 Components

1. *Lamps* by General Electric, Sylvania, Osram, Philips, Venture or equivalent, as indicated by test measurements, using standard IES procedures, within 5% of any of the above for the same type of lamp. All lamp ratings shall comply with the following testing standards in their most recently published versions:

(a) IES: LM-40-87, LM-9-88, LM-16-84, LM-62-85

(b) ANSI: C78.1, C78.2, C78.3

2. *Incandescent lamps* shall only be used for theatrical or decorative purposes, not for general or task illumination, and only when higher efficacy sources are unavailable to perform the same function. When available, halogen cycle lamps (e.g., Capsylite) or halogen-infrared lamps shall be used in such applications.

3. *Standard fluorescent lamps* must meet appropriate IES testing and measurement standards; straight tubular and U-shaped lamps shall be the rapid-start T-8 "Octron" type designed to operate at 265 milliamps (ma). The following types of fluorescent lamps shall *not* be used:

(a) any lamp with a color rendering index less than 75

(b) heater cutout lamps

(c) any straight tubular lamp 3, 5, 6 or 8 feet long

(d) any T-12 lamp.

Lamp color shall be 3,500 K, unless the task involves special requirements; in such unusual cases, prior approval for an alternative must be obtained from the Energy Conservation Manager.

4. *Compact fluorescent lamps* must meet all appropriate testing and measurement standards for standard fluorescent lamps, and shall conform to the following requirements:

(a) lowest available color temperature shall be used (preferably between 2,700 and 3,100 K)

(b) minimum color rendering index of 75

(c) must use separable ballast.

5. *High intensity discharge lamps* must meet appropriate IES testing and measurement standards; all shall conform to the following requirements when applied to interior lighting:

(a) low pressure sodium shall only be used for emergency and security lighting.

(b) high pressure sodium must be color corrected to provide at least 2400 K and color rendering index of at least 75.

(c) metal halide shall have color temperature greater than 3000 K but less than 4200 K and color rendering index of at least 75.

(d) mercury vapor lamps shall have separable ballasts, minimum efficacy of 50 lumens per watt (ballast losses included), and minimum CRI of 75.

6. *Ballasts* by Advance, Magnetek, Osram Sylvania, or equal, as indicated by test measurements, using standard IES procedures, within 5% of any of the above for the same type of ballast. All ballast ratings shall comply with the following testing standards:

(a) ANSI: C82.11-1993

(b) ANSI/IEEE: C62.41-1991

(c) IEC: 921 (regarding power factor calculation).

Ballasts with ballast factors (BF) above 0.93 (or below 0.88) may be used despite the above standards if higher (or lower) light levels are needed, but all other characteristics aside from BF must still comply with the above. BF shall not exceed 1.15 without prior approval of the Energy Conservation Manager.

7. *Power reducers and current limiters* shall not be used except in built-in task lighting where replacement of the existing ballast and lamp will not yield the desired results. Magnitude of power reduction to be approved by Energy Conservation Manager. Use of a power reducer shall:

(a) not void warranty of ballast to which it is connected.

(b) not delay lamp starting.

(c) not cause a perceptible increase in flicker.

8. *Fluorescent ballasts, except screw-in types* when available in the preferred type of fixture, shall be fully solid-state electronic (that is, includes a rectifier, high-frequency inverter, and power control and regulation circuitry) and meet the following requirements:

(a) UL listed, Class P (thermally protected).

(b) contains no PCBs, or other hazardous materials in quantities that require disposal as hazardous waste.

(c) operate Octron-type lamps in the instant-start mode unless no instant-start type is available for the desired configuration.

(d) when controlling multiple lamps, shall power them in parallel, not in series (so one burnout leaves others lit).

(e) meet FCC Rules, Part C (RF Lighting Devices), concerning radio frequency interference, or RFI; such interference shall not exceed that of the lowest RFI CBM-certified ballast usable for the same application.

(f) designed to withstand power transients as per IEEE 587, category A, and comply with ANSI 62.41-1984.

(g) minimum ballast factor of 0.88, unless exceptionally low or high light output is desired.

(h) hold an "A" sound rating, or better; noise level shall not exceed that of the quietest CBM-certified ballast usable for the same application.

(i) minimum power factor of .95.

(j) current crest factor shall not exceed 1.7.

(k) minimum ballast efficacy factor (BEF) shall be:

2.4 when operating one 32-watt T-8 lamp

1.4 when operating two 32-watt T-8 lamps

0.95 when operating three 32-watt T-8 lamps

0.75 when operating four 32-watt T-8 lamps

as defined in the National Appliance Energy Conservation Amendments of 1988 (Public Law 100-357)

(l) maintains light output within ±10% when voltage varies within ±10%.

(m) flicker factor shall be 5% or less (or flicker index below .01), with any lamp suitable for the ballast, as determined by analysis of light output variations over 1/60 of one second.

(n) has been in production for at least two years, unless approval obtained from the Energy Conservation Manager.

(o) total harmonic distortion (THD) shall not exceed 20% when powering the number of lamps proposed by the design.

(p) comply with State and Federal Ballast Laws, and requirements for utility rebates.

(q) ballast casing shape and dimensions shall be same as for standard magnetic ballasts, unless otherwise approved by the Energy Conservation Manager.

(r) shall be warranted for a minimum of 3 years, with a minimum $10 labor allowance for replacement of failed ballasts.

(s) dimming ballasts shall be furnished with complete wiring diagrams and a list of compatible dimming controls listing manufacturer's name and address and model numbers.

9. *Compact fluorescent ballasts (all types)* shall meet all fluorescent ballast requirements if electronic, or the following requirements if magnetic:

(a) have an "A" sound rating, or better.

(b) minimum power factor of 0.6.

(c) be warranted for at least 2 years.

(d) be theft-resistant, if to be installed less than 12 feet above the floor.

(e) maximum THD of 35%.

(f) minimum ballast factor of 0.80.

10. *High intensity discharge ballasts* shall meet the following requirements:

(a) have automatic starter cutoff circuitry, or similar circuitry that protects the ballast/starter against damage when the lamp fails.

(b) have an "A" sound rating, or better.

(c) minimum power factor of 0.65.

(d) maximum THD of 35%.

(e) minimum ballast factor of 0.85.

(f) if available, be solid-state electronic, as previously defined.

(g) be warrantied for at least three years.

11. *Lamp shielding* shall be defined as any material that alters or redirects light output from a fixture, and shall comply with the following requirements:

(a) shall be made of slow-burning plastic; e.g., acrylic or polycarbonate, or metal.

(b) creates appropriate beam spread and visual comfort probability (VCP) for task and space types, as defined by the IES "Recommended Practice" appropriate for the tasks involved, or by the administrator of the space, whichever requirement is less stringent.

(c) has a standard shape and dimensions, replaceable by more than one manufacturer at the time of installation.

(d) parabolic-wedge louvers shall have no dimension less than 1 inch between parallel louver blades, measured center-to-center, or as height of blade.

(e) "egg-crate" style louvers shall not be used.

(f) diffusers (defined as shielding that does not also alter light through reflection, refraction, or polarization) shall not be used, except in compact fluorescent fixtures.

(g) lenses (defined as shielding that refracts and/or polarizes) shall be at least 0.125" thick (unless mounted in track lights, where 0.05" is acceptable) and have minimum visible transmittance of 90%.

(h) shielding shall be supplied strain-free, uniform in appearance, destaticized, and show no visible evidence of sagging in the installed position; edges shall be evenly cut and free of cracks.

(i) shall be warrantied against cracking or discoloration for at least five years.

12. *Lighting fixtures,* other than exit signs. All fixture ratings shall comply with the following test standards:

(a) IES: LM-44-90, LM-41-85

(b) ANSI: C78, C82

(c) ASTM: D3359, G53, D2247, B117

The specifications that follow do not apply to fixtures designed for use in hazardous locations.

(e) all horizontally-mounted fixtures using linear fluorescent lamps shall have minimum Coefficient of Utilization (CU) as indicated below, when utilized under these conditions:

- room cavity ratio = 1.0, ceiling cavity reflectance = 0.8, wall reflectance = 0.5, floor cavity reflectance = 0.2

- minimum CU values:

general room lighting, corridors, stairwells, machine rooms, dining areas, conference rooms, clinics, auditoriums - CU $\geq$ 0.65

bathrooms, lounges, executive suites, traditional spaces, galleries - CU $\geq$ 0.55

(f) fixtures shall have minimum Visual Comfort Probability (VCP) value of 0.70 when used in rooms containing more than two video display terminals in continuous use during occupancy, based on room dimensions of 20 feet by 20 feet with fixtures mounted 10 feet above the floor.

(g) ballasts and lamps shall be easily removable for replacement.

(h) fixtures in elevators shall be vandal-resistant and have lamp shielding that remains connected to the fixture during relamping.

(i) wall-mounted fixtures in gymnasiums and dormitories shall be vandal-resistant or covered by wire guards.

13. *Compact fluorescent and HID fixtures* shall be exempt from VCP requirements listed above (unless utilized in VDT areas), and shall meet the following other requirements:

(a) decorative fixtures (e.g., wall sconces, chandeliers, globes) shall be exempt from CU requirements, but shall have a minimum total fixture efficiency of 50%.

(b) downlights in which the lamp is oriented vertically shall have a minimum CU value of 0.5 under the same conditions as listed above for other fixtures.

(c) downlights in which the lamp is oriented horizontally shall have a minimum CU value of 0.35 under the same conditions as listed above for other fixtures, but shall only be used when space restrictions preclude fixtures with vertically mounted lamps, or when no such fixture exists to provide the necessary beamspread and illumination level.

(d) when used in conjunction with other light sources in the same room, these fixtures need not meet any minimum criteria for coefficient of utilization, but must have a total fixture efficiency of at least 50%.

14. *Exit signs* are exempt from lighting fixture requirements, but shall meet the following requirements:

(a) use only light emitting diodes (LED) as light sources.

(b) must be warranted for at least 20 years.

(c) be UL listed.

(d) comply with all appropriate City codes and have all appropriate City approvals, especially with regard to minimum luminance requirements.

(e) have left and right arrows available on all signs.

(f) single-sided 6-inch signs shall use no more than 2 watts.

(g) double-sided 6-inch signs shall use no more than 4 watts.

(h) single-sided 8-inch signs shall use no more than 5 watts.

(i) double-sided 8-inch signs shall use no more than 8 watts.

(j) THD shall not exceed 10%.

(k) minimum power factor of 0.9.

(l) shall be wired to accept 120 or 277 volts.

(m) shall have continuous stroke lettering.

(n) shall be edge-lit or use other indirect method to distribute light within the fixture, unless unavailable in the required configuration or size.

(o) shall fit into existing locations, or else shall be supplied with any hardware, including black-painted back plates to cover existing recesses, needed to mount fixture.

(p) maintain minimum luminance to meet codes in effect at time of installation for at least 10 years.

15. *Compact fluorescent retrofit kit* shall be defined as any device that, when installed in an incandescent fixture, allows use of a compact fluorescent lamp in place of the incandes-

cent lamp, and complies with the following requirements:

(a) shall meet the appropriate ballast standards previously delineated.

(b) shall include all hardware and wiring to allow complete installation.

(c) if a reflector is part of the kit, it shall be constructed of metal or plastic (not glass) and be designed to maximize light output and concentration from its light source, within the diameter constraints of the existing fixture for which it is intended.

(d) shall be theft-resistant, if to be installed less than 12 feet above the floor.

(e) shall be warranted for at least two years.

(f) shall be supplied with appropriate lamp meeting above lamp specs for CRI and color temperature.

(g) shall include full instructions for installation and maintenance.

(h) shall have comparable or better photometrics than the existing fixture in which it will be installed (defined as candela measurements within an average of 20% of those for the existing fixture within 45 degrees from nadir, except for maximum beam candlepower, which may be up to 35% less if the remaining measurements comply with the 20% average; all measurements to be taken after 10 minutes warmup time for the compact fluorescent retrofit kit).

(i) when mounted in cylindrical fixtures, shall be recessed at least 1/2-inch inside the fixture.

16. *Specular reflector retrofit kit* shall be defined as any device designed for installation in an existing fluorescent fixture and which uses a specular reflector to increase or alter distribution of light output, and shall comply with the following requirements:

(a) shall have UL listing.

(b) shall not violate the UL listing of the existing fixture.

(c) provides new lampholders/sockets and socket bars that relocate lamps to maximize light output with the reflector supplied.

(d) shall provide appropriate beam spread for the existing location and tasks (as defined by spacing-to-mounting height ratios, or spacing criterion, that provides illumination levels that do not vary more than 30% as measured between and below fixtures).

(e) complies with all local electric and fire codes, especially with regard to thickness and flame spread.

(f) reflector is held by socket bars, or else is slotted so that it may be removed by merely loosening (not fully removing) screws.

(g) aluminum reflectors shall be used in open industrial-style fixtures, or in any very dirty environment; silver film shall be used in any fixture enclosed by a lens, louver, or diffuser in a clean environment.

(h) aluminum reflectors shall have a minimum reflectance of 88%; silver film reflectors shall have a minimum reflectance of 92% (reflectances to be determined as an average of 10 randomly selected samples).

(i) reflector shall be warranted for at least 10 years against bubbling, delaminating or yellowing.

(j) silver film reflectors shall not have bends tighter than a 90 degree acute angle.

(k) shall be accompanied by clear instructions, including a line sketch in which all parts are identified by replacement model number and quantity.

17. *Task lighting* shall be defined as any fixture, built-in or plugged-in, that provides localized illumination from a location below the ceiling plane and is supplemented by general lighting above the task light, and shall comply with the following requirements:

(a) shall use only fluorescent lamps (except for laboratory equipment or in a device that uses the light as an integral part of its operation, such as a copy machine).

(b) if altered to receive a compact fluorescent lamp, the ballast shall not raise the fixture's center of gravity sufficiently to create a tipping hazard

(c) if altered to receive a compact fluorescent lamp, the ballast (or retrofit kit) shall be theft-resistant or out of the visual range of those not using the fixture.

(d) if utilized in quiet locations (such as libraries, clinics, or reading rooms), shall be equipped with an electronic ballast with an "A" sound rating, or better; sample shall be tested in such location, and accepted by the Energy Conservation Manager prior to installing others.

18. *Automatic lighting controls* shall conform to requirements of the latest version of the State Energy Conservation Construction Code, including requirement for two-level lighting in new buildings.

(a) General items common to all units:

1. shall be mounted using theft-resistant screws requiring a special tool for removal
2. sensitivity and time settings shall be factory-set in accordance with user's written desires
3. contains no triac (i.e., reversing thyristor) circuitry affecting power wave form to the lighting fixture
4. RFI resistant
5. no visible means to manually override without special key
6. warranted for at least 3 years
7. fully compatible with all electronic ballasts; i.e., will not suffer reduced lifetime, or cause reduced lifetime of ballast, when loaded at full rated wattage of electronic ballasts.

(b) The following options are all acceptable:

1. motion sensors using infrared or ultrasonic technology, either individually or combined, and infrared and audible sound when combined
2. photocell controls
3. manually-cycled timers
4. "sweep" systems connected to an energy management system.

(c) The following types of systems are *not* acceptable:

1. astronomical clocks (unless equipped with 24-hour rechargeable battery backup warranted for at least 10 years)
2. those using only microwave or audible sound sensors.

(d) Ultrasonic sensors shall comply with the following requirements:

1. an Initial Report has been filed with the Bureau of Radiological Health, of the Federal Food and Drug Administration, under section 21 of the Code of Federal Regulations, section 1002.10 (1990)

2. does not emit ultrasonic sound in excess of the following decibel (dB) levels, as measured no more than 5 feet from the source, on axis:

Midfrequency of sound pressure third-octave band in (kHz)	Maximum dB level within third-octave band (in dB ref. 20 micropascals)
less than 20	80
20 to less than 25	105
25 to less than 31.5	110
31.5 or more	115

3. shall not emit sound below 32 kHz if used in areas commonly occupied by the hearing-impaired

(e) Minimum sensor requirements (regardless of type):

1. shall be able to detect at 20 feet or more the motion of a forearm measuring at least 12 inches from elbow to wrist joint which is rotating through 90 degrees of arc in a plane perpendicular to the line-of-sight between the arm and the sensor at an angular rotation rate (measured at the elbow) of no more than 90 degrees per second (i.e., 0.78 radians per second); line-of-sight shall be perpendicular to the surface upon which the sensor is mounted

2. wall-mounted sensors shall have an angular range when facing into an occupied space of at least 160 degrees measured in the horizontal plane and 40 degrees in the vertical plane, unless narrower beam sensitivities are required to screen out other motion

3. ceiling-mounted sensors shall have an angular range of 360 degrees around the axis pointing vertically down from the sensor, unless narrower beam sensitivities are required to screen out other motion

4. variable controls for sensitivity and on-off cycle

5. cycle time adjustable up to at least 20 minutes

6. LED "on" cycle lamp indicates when motion is sensed

7. instant-on after power restoration following interruption

8. failure of sensor circuitry or relay shall leave lights on, not off

9. teflon-coated low-voltage wiring, or acceptable equivalent, when being used in a return air plenum

10. include timer bypass key, or abil ity to set time cycle for less than 20 seconds for troubleshooting.

(f) minimum timer requirements:

1. fully electronic timing system (no springs or gears)

2. on-cycle adjustable from 10 minutes (or less) up to 12 hours

3. have no visible means to override or adjust.

19. *Replacements.* All components which fail within 90 days of installation, including lamps, shall be replaced by vendor at no charge to client, unless a longer warranty is specified herein.

1.2 Lighting designs

Lighting designs shall conform to lighting power limits of the latest version of the State Energy Conservation Construction Code. All proposed designs exceeding 1.5 watts/square foot (exclusive of task lighting) shall be provided with calculation utilizing the lumen method (or other method approved by the Energy Conservation Manager) to demonstrate that the average maintained illumination level at a 30-inch task height does not exceed values given below.

1. *Illumination Levels:*

(a) maintained lighting design illumination levels for general lighting shall be as follows:

	Footcandles
Waiting areas	20–30
Corridors and lobbies	10–20
Conference and classrooms	30–50
Offices, labs, libraries	40–60 (no task lights) 20–25 (with task lights)

(b) maintained level shall allow for a total reduction of 25% due to lamp lumen loss and dirt depreciation prior to relamping and cleaning

(c) verification of illumination levels of installed designs shall take place after lamps have operated for at least 100 hours, and shall utilize the IES method for "Determination of Average Illuminance on a Horizontal Plane" as delineated in the 1993 edition of the IES Reference Handbook.

2. *Design Criteria:* All designs shall comply with the following requirements:

(a) follow the appropriate IES "Recommended Practice" for the predominant tasks involved

(b) shall be checked and signed by a certified lighting designer, defined as any of the following individuals:

- AEE Certified Lighting Efficiency Professional
- NALMCO Certified Lighting Maintenance Contractor
- holder of an IES Certificate of Technical Knowledge
- full member of IES, IALD, or AIA

(c) shall provide IES-recommended maintained illumination levels, unless indicated otherwise in this specification

(d) shall comply with the State Energy Conservation Construction Code

(e) shall maximize the stated desires of the owner (e.g., energy savings, internal rate of return, product sales, occupant satisfaction) as agreed in writing prior to initiation of design, but still compliant with the remainder of this specification

(f) shall allow the owner or its representative one review prior to final version, during which changes will be considered by vendor

(g) whenever a question arises during design review, shall utilize the lumen method (or other calculation method previously accepted by the owner or its representative) to determine if acceptable illumination will be provided

(h) shall be appropriate to the space, and not merely a one-for-one replacement based solely on existing fixture photometrics

(i) shall not exceed spacing-to-mounting height ratios or spacing criterion of proposed equipment by more than 20%

(j) shall consider visual adaptation of those entering or leaving the designed spaces

(k) shall consider the option to alter reflectances of ceilings and walls, and the shading coefficient of the space's glazing

(l) shall consider appropriate automatic lighting controls, including daylighting, that will meet the facility's stated goals

(m) shall utilize tandem-wired ballasts whenever possible.

1.3 Lighting Audits

Lighting audits shall be defined as documents providing information on existing lighting systems and proposing changes to them, and complying with the requirements of this specification.

1. *General:* separate forms shall be used for incandescent, HID, and fluorescent lighting fixtures and controls except as otherwise indicated below; audits will cover interior lighting only, but will otherwise be filled out completely, except as indicated below.

2. *Coverage:* all rooms are to be audited by the fluorescent auditor, who shall indicate which rooms contain incandescent or HID fixtures; only those containing incandescents shall be visited during the incandescent audit; locked or inaccessible rooms shall be so noted under "remarks."

3. *Measurement Methods:* the following methods shall be used when performing an audit:

(a) light level readings will be taken with a meter acceptable to the owner and checked against a second meter; variations between them, when they are subjected to the same conditions, shall not exceed 10%

(b) light level readings shall be taken at task height (or 30 inches above the floor if no task height is apparent) both under fixtures and between fixtures; both shall be reported on fluorescent audits, but an average of the two may be reported on incandescent audits

(c) light level readings in spaces with windows shall be taken both with lights on and with lights off; only the difference between these two measurements shall be reported on the audit; in areas (such as corridors) where lights cannot be easily turned off, auditor shall block any direct outdoor light with a clipboard during measurements

(d) if a dimmer exists, levels will be taken with the dimmer turned up to full power

(e) where track lighting is the sole source of illumination, auditor shall measure peak beam level and indicate (under "remarks") if surface where measurement was taken was vertical or horizontal; supply estimate of distance between fixture and the point at which the measurement was taken; manufacturer of track fixture or track must be listed under "remarks"

(f) illumination levels to be taken in at least 25% of all rooms and in 100% of any rooms later covered in the design's "problem letter" (see below)

(g) determination of ceiling heights shall be performed by counting cinder blocks (or other visible standard construction units), or through use of a tape measure, folding ruler or ultrasonic distance measuring device, and not through an unverified estimate

(h) fixture height will be determined by subtracting the distance between the ceiling and the bottom of the fixture from the ceiling height; this distance may be estimated by eye when the ceiling height is less than 16 feet, but shall be measured as indicated above for ceiling height when the ceiling is higher

(i) reflectances may be estimated by eye, but (when estimating floor reflectance) shall indicate a weighted average of all reflective surfaces that make up the work plane or below it, and (when estimating wall reflectance) an average of all surfaces (e.g., bookcases, tapestries) covering walls; unless measured in the field, bookcases shall be assumed to have an average reflectance of 0.30

(j) only those reflectances varying from the standard 80% for ceiling, 50% for walls and 20% for floors need be noted on the audit.

4. *Information Not Listed On Audit Form Column Headings:* all the following should be noted under "remarks" (or other appropriate location on the form) when seen:

(a) code violations

(b) occupant lighting complaints and/or preferences

(c) indications of recent construction

(d) errors in plans

(e) if existing lighting levels are very uneven

(f) poor fixture condition

(g) difficult access to fixture

(h) if track lighting, manufacturer of track head or track

(i) presence of dimmers, and if they work properly

(j) faulty light switches, or the absence of room light switches

(k) additional data for incandescent audits: the orientation of the lamp (if not base-up), the lamp type (if not standard "A" type), the length of any pendants holding fixtures or tracks, recommendations for addition or deletion of socket extenders, and diameter of any fitters and/or globes in which retrofit kits could be installed.

5. *Items Pertaining To Both Fluorescent And Incandescent Audits:*

(a) "quantity" refers to the number of each type of fixture found in a single room; a separate line should be started for each fixture type found in a room

(b) "room number" refers to the number on the plans, not what may be found in the field

6. *Items Related Only To Incandescent Audits:* the following items relate only to incandescent auditing:

(a) "socket depth/extend" is the depth in inches from opening of the female lamp socket to the opening of the fixture, and the presence of any socket extenders; indicate as in this example:

• "3+3X" means 3 inches from socket to opening, and three contiguous extenders were seen in the fixture

(b) "opening diameter" refers to the diameter in inches through which unobstructed light escapes the fixture, or the opening in a globe-style fixture, through which a retrofit kit must fit

(c) "existing lamp (watts)" is the verified wattage of the typical lamp found in a group of identical fixtures in a single room; if more than one lamp is installed in a fixture, show watts and quantity as in this example:

- "3X100" means three 100 watt lamps per fixture

(d) proposed designs should give preference to full fixture replacement whenever possible

(e) emergency lighting is not to be audited.

7. *Presentation Of Results:* a copy (not an original) of each audit shall be completed and filled out legibly in black ink, without any crossing out, marginal notes or unclear entries, and supplied to the owner or its representative prior to submission to any other parties involved in construction. If more than 5 sheets, it shall be bound, with a cover delineating the name of the building, date of the audit/design, name (or initials) of the auditor unless the owner requests no binding. All first submissions shall be stamped "preliminary audit/design" and any that follow shall be stamped "revision no. __" and dated appropriately.

8. *Additional Documentation:* the following additional documentation should be supplied and may be gathered together in a "problem letter":

(a) a list of locked or inaccessible rooms

(b) a list of areas where light levels need to be increased or decreased to comply with appropriate IES standards for the tasks involved

(c) a list of areas where a retrofit may generate complaints

(d) location of possible demonstration rooms

(e) location and number of any fixtures to be removed for closer examination

(f) for fluorescent audit/designs, a preliminary ballast count indicating quantity, voltage, and type.

DISCLAIMER

This technical specification guide should not be used as the sole source of advice for an electrical lighting specification. Local and state codes often vary significantly, and these differences need to be incorporated in any final document controlling design and installation of a lighting system.

This guide does not address: coordination of trades; work quality assurance (beyond design); substitutions; submittals; shop drawings; material delivery, storage, handling, or disposal; cleanup; commissioning; or manuals. Such items vary with the needs of the client and facility, and should be appended to this guide by its user.

Neither E SOURCE nor the authors of its publications are responsible for any errors of omission or commission in this presentation, nor are they responsible for any electrical, architectural, or lighting designs resulting from the use or misuse of this guide. No warranty regarding accuracy, completeness, or timeliness is expressed or implied regarding the data or statements offered in this guide.

E DIRECTORY OF LIGHTING-RELATED ORGANIZATIONS

Listed in this appendix are lamp manufacturers, ballast manufacturers (full-size fluorescent and compact fluorescent), lamp recycling firms, ballast recycling firms, PCB incineration facilities, hazardous waste facilities, independent photometric laboratories, and government agencies, trade associations, and other lighting-related organizations.

LAMP MANUFACTURERS

Durotest Lighting Corp.
9 Law Drive
Fairfield, NJ 07004
tel 907-344-3876
fax 907-344-4751
web www.concentric.net/~durotest

GE Lighting
1975 Noble Road
East Cleveland, OH 44112
tel 800-626-2004
fax 800-327-0663
web www.ge.com.lighting/business

Lights of America
611 Reyes Drive
Walnut, CA 91789
tel 909-594-7883
fax 909-594-6758
web www.lightsofamerica.com

Osram Sylvania
100 Endicott Street
Danvers, MA 01923
tel 800-544-4828
fax 508-750-2152
e-mail corpcom@osi.sylvania.com
web www.sylvania.com

Panasonic Lighting
1 Panasonic Way, 4-A4
Secaucus, NJ 07094
tel 201-348-5381
fax 800-553-0384

Philips Lighting
200 Franklin Square Drive
Somerset, NJ 08875
tel 908-563-3000
fax 908-563-3641
web www.lighting.philips.com

Venture Lighting International, Inc.
32000 Aurora Road
Solon, OH 44139
tel 800-451-2606
web www.adlt.com/venture

FULL-SIZE FLUORESCENT BALLAST MANUFACTURERS

Advance Transformer Company
10275 W. Higgins Road
Rosemont, IL 60018-5603
tel 800-322-2086
fax 847-390-5109
web www.advancetransformer.com

CRA Enterprises
27515 Enterprise Circle West
Temecula, CA 92590
tel 909-676-7810
fax 909-695-4491
web www.craenterprises.com

Kingtec, Inc.
85 Challenger Road
Ridgefield Park, NJ 07660-2144
tel 800-448-5483
fax 201-229-4680

Magnetek
26 Century Boulevard, Suite 600
Nashville, TN 37214
tel 800-624-6383
fax 615-316-5181
e-mail productinfo@magnetek.com
web www.magnetek.com

Motorola
887 Deerfield Parkway
Buffalo Grove, IL 60089
tel 800-654-0089 or 847-576-5000
fax 847-215-6311
web www.mot.com/AECS/MLI

Power Lighting Products, Inc.
1770 Commerce Park Drive
El Paso, TX 79912
tel 915-877-4380
fax 915-877-1499
web www.powerlighting.com

Robertson Transformer Company
13611 Thornton Road
Blue Island, IL 60406
tel 708-388-2315

Compact Fluorescent Ballast Manufacturers

Advance Transformer
10275 West Higgins Road
Rosemont, IL 60018
tel 800-372-3331
or 847-390-5000
fax 847-390-5109
web www.advancetransformer.com

Energy Savings
2266 North Palmer Drive
Schaumburg, IL 60173
tel 847-925-8400
fax 847-925-8490

ETTA Industries
5680 Central Avenue
Boulder, CO 80304
tel 303-444-2244
fax 303-444-1222

Hatch Transformer
5403 West Gray Street
Tampa, FL 33609
tel 813-288-8006
fax 813-288-8105
e-mail hatch@interaccess.net
web www.hatchtrans.com/hti

Lightolier
100 Lighting Way
Secaucus, NJ 07096
tel 508-679-8131
fax 508-674-4710
web www.lightolier.com

Lights of America
611 Reyes Drive
Walnut, CA 91789
tel 909-594-7883 or 800-321-8100
fax 909-598-5931
web www.lightsofamerica.com

Lutron Electronics Company
7200 Suter Road
Coopersburg, PA 18036-1299
tel 610-282-3800
fax 610-282-0298
web www.lutron.com

Magnetek
26 Century Boulevard, Suite 600
Nashville, TN 37214
tel 615-316-5340
fax 615-316-5186
web www.magnetek.com

Mitor Industries
418 South 2nd Street, 3rd Floor
Mankato, MN 56001
tel 507-387-1599
fax 507-387-7491
web www.light-link.com/mitor

Motorola Lighting
887 Deerfield Parkway
Buffalo Grove, IL 60089
tel 708-215-6300
fax 708-215-6311
web www.motorola.com/AECG/MLI

Osram Sylvania
100 Endicott Street
Danvers, MA 01923
tel 508-777-1900
fax 508-750-2992
web www.sylvania.com

Power Lighting Products
(formerly Valmont Electric)
1770 Commerce Park Drive
El Paso, TX 79912
tel 915-877-4380
fax 915-877-1499

Prescolite Controls
1251 Doolittle Drive
San Leandro, CA 94577
tel 510-562-3500
fax 510-577-5022

Robertson Transformer
13611 Thornton Road
Blue Island, IL 60406
tel 708-388-2315
fax 708-388-2420

FLUORESCENT LAMP RECYCLING FIRMS

Advanced Environmental Recycling Corp.
2591 Mitchell Avenue
Allentown, PA 18103
tel 800-554-2372
fax 610-797-7696
web www.aerc-mti.com

Bethlehem Apparatus Company, Inc.
890 Front Street
Hellertown, PA 18055
tel 610-838-7034
fax 610-838-6333
e-mail bacohg@aol.com
web www.thomasregister.com/bethco

DYNEX Industries, Inc.
4751 Mustang Circle
Moundsville, MN 55112
tel 612-784-4040
fax 612-717-3161

Full Circle Recyclers
509 Manida Street
Bronx, NY 10474
tel 800-775-1516
fax 718-328-4462
e-mail fullcircle@evtc.com
web www.evtc.com

Global Recycling Technologies
P.O. Box 651
Randolph, MA 02368
tel 617-341-6080
fax 617-341-6088

Lamp Recyclers of Louisiana, Inc.
P.O. Box 2962
Hammond, LA 70404-2962
tel 504-345-4147
fax 504-345-4775
web www.i-55.com/lamprecyclers

Light Cycle, Inc.
1222 University Avenue
St. Paul, MN 55104
tel 612-641-1309
fax 612-641-1422

Lighting Resources, Inc.
1522 East Victory Street, Suite 2
Phoenix, AZ 85040
tel 800-57-CYCLE
fax 602-376-5432

Mercury Recovery Services
2021 South Myrtle
Monrovia, CA 91016
tel 818-303-2053
fax 818-358-2703

Mercury Technologies of MN Inc.
Pine City Industrial Park
P.O. Box 13
Pine City, MN 55063-0013
tel 800-864-3821
fax 320-629-7799

Mercury Technology International
30677 Huntwood Avenue
Hayward, CA 94544
tel 510-429-1129
fax 510-429-1498

Midwest Recycling & Mercury Recovery
860 White Street
Dubuque, IA 52001
tel 319-556-8037
fax 319-582-3072

Northeast Lamp Recycling, Inc.
250 Main Street, East Windsor, CT 06088
tel 860-292-1992 fax 860-292-1114
e-mail nlr@eci.com

NSSI, Inc.
5711 Etheridge Street
Houston, TX 77087
tel 713-641-0391
fax 713-641-6153

P-3, Inc.
4515 Oak Fair Boulevard, Suite 103
Tampa, FL 33610
tel 813-628-8888
fax 813-628-8889
e-mail onesource@p-3.com

Recyclights, Inc.
401 West 86th Street
Bloomington, MN 55420
tel 800-831-2852 or 612-948-0627
fax 612-948-0627

Salesco U.S.A.
5736 West Jefferson
Phoenix, AZ 85043
tel 602-233-2955
fax 603-278-0608
web www.ssusa.com

Superior Lamp Recycling, Inc.
P.O. Box 1323
Fond du Lac, WI 54936-1323
tel 800-556-5267
fax 414-923-9010

USA Lights
2007 County Road C-2
Roseville, MN 55113
tel 612-628-9370 fax 612-628-9371

BALLAST RECYCLING FIRMS

For an update to this list call the Toxic Substances Control Act Hotline at 202-554-1404 or the Green Lights Hotline at 202-775-6650.

Advanced Environmental Recycling Corp.
2591 Mitchell Avenue
Allentown, PA 18103
tel 800-554-2372
fax 610-797-7696
web www.aerc-mti.com

Bethlehem Apparatus Company, Inc.
890 Front Street
Hellertown, PA 18055
tel 610-838-7034
fax 610-838-6333
e-mail bacohg@aol.com
web www.thomasregister.com/bethco

DYNEX Industries, Inc.
4751 Mustang Circle
Moundsville, MN 55112
tel 612-784-4040
fax 612-717-3161

Full Circle Recyclers
509 Manida Street
Bronx, NY 10474
tel 800-775-1516
fax 718-328-4462
e-mail fullcircle@evtc.com
web www.evtc.com

Global Recycling Technologies
P.O. Box 651
Randolph, MA 02368
tel 617-341-6080
fax 617-341-6088

HID Recycling Technologies
32000 Aurora Road
Cleveland, OH 44139
tel 800-200-9716
fax 800-200-9718
e-mail chris_cole@adlt.com
web www.adlt.com/venture

Lamp Recyclers of Louisiana, Inc.
P.O. Box 2962
Hammond, LA 70404-2962
tel 504-345-4147
fax 504-345-4775
web www.i-55.com/lamprecyclers

Light Cycle, Inc.
1222 University Avenue
St. Paul, MN 55104
tel 612-641-1309
fax 612-641-1422

Lighting Resources, Inc.
1522 East Victory Street, Suite 2
Phoenix, AZ 85040
tel 800-57-CYCLE
fax 602-376-5432

Mercury Recovery Services
2021 South Myrtle
Monrovia, CA 91016
tel 818-303-2053
fax 818-358-2703

Mercury Technologies of MN, Inc.
Pine City Industrial Park
P.O. Box 13
Pine City, MN 55063-0013
tel 800-864-3821
fax 320-629-7799
e-mail ecmis.mtm@norsol.com

Mercury Technology International
30677 Huntwood Avenue
Hayward, CA 94544
tel 510-429-1129
fax 510-429-1498

Midwest Recycling & Mercury Recovery
860 White Street
Dubuque, IA 52001
tel 319-556-8037
fax 319-582-3072

Northeast Lamp Recycling, Inc.
250 Main Street
East Windsor, CT 06088
tel 860-292-1992
fax 860-292-1114
e-mail nlr@eci.com

NSSI, Inc.
5711 Etheridge Street
Houston, TX 77087
tel 713-641-0391
fax 713-641-6153

P-3, Inc.
4515 Oak Fair Boulevard, Suite 103
Tampa, FL 33610
tel 813-628-8888
fax 813-628-8889
e-mail onesource@p-3.com

Recyclights, Inc.
401 West 86th Street
Bloomington, MN 55420
tel 800-831-2852 or 612-948-0627
fax 612-948-0627

S.D. Myers
180 South Avenue
Tallmadge, OH 44278
tel 330-633-2666
fax 330-633-0756

Salesco U.S.A.
5736 West Jefferson
Phoenix, AZ 85043
tel 602-233-2955
fax 602-278-0608
web www.ssusa.com

Superior Lamp Recycling, Inc.
P.O. Box 1323
Fond du Lac, WI 54936-1323
tel 800-556-5267
fax 414-923-9010

Transformer Service, Inc.
74 Regional Drive
Concord, NH 03301
tel 603-224-4006
fax 603-228-2430
e-mail tsisales@tsi1.mv.com

Hazardous Waste Landfills

For an update to this list, call the Toxic Substances Control Act Hotline at 202-554-1404 or the Green Lights Hotline at 202-775-6650.

Chemical Waste Management
Call 1-800-843-3604 for information on CWM disposal facilities nationwide
e-mail cwmi@ix.netcom.com

CWM Chemical Services Control, Inc.
1550 Balmer Road
Model City, NY 14107
tel 716-754-8231
fax 716-754-0211
web www.wmx.com

Envirosafe Services Inc. of Idaho
P.O. Box 400
Grandview, ID 83624
tel 800-274-1516
fax 208-834-2997

U.S. Ecology, Inc.
Box 578
Beatty, NV 89003
tel 702-553-2203
fax 702-553-2742
web www.americanecology.com

PCB INCINERATION FACILITIES

For an update to this list, call the Toxic Substances Control Act Hotline at 202-554-1404 or the Green Lights Hotline at 202-775-6650.

Aptus, Inc.
P.O. Box 1328
Coffeyville, KS 67337
tel 316-251-6380
fax 316-252-1483

Chemical Waste Management
P.O. Box 2563
Port Arthur, TX 77643
tel 409-736-2821
fax 409-736-1636

Rollins Co.
P.O. Box 609
Deer Park, TX 77536
tel 713-930-2300
fax 713-930-2316

Roy F. Weston, Inc.
One Weston Way
West Chester, PA 19380
tel 610-701-3000
fax 610-701-3124
web www.rfweston.com

INDEPENDENT PHOTOMETRIC LABORATORIES

Building Acoustics & Lighting Laboratories, Inc.
1618 Headland Drive
Fenton, MO 63026
tel 314-343-6006
fax 314-343-6051

Calcoast Industrial Testing Laboratories
4072 Watts Street
Emeryville, CA 94608
tel 510-652-2979
fax 510-652-3085

Independent Testing Labs, Inc.
3386 Longhorn Road
Boulder, CO 80302
tel 607-753-6711
fax 607-756-9891

ITS Testing Services, Inc.
Industrial Park
Cortland, NY 13405
tel 607-753-6711
fax 607-756-9891
web itsglobal.com

Lighting Research Laboratory
P.O. Box 6193
Orange, CA 92667
tel 714-771-1312
fax 714-639-9226

Lighting Sciences, Inc.
7830 East Evans Road
Scottsdale, AZ 85260
tel 602-991-9260
fax 602-991-0375

Luminaire Testing Lab
905 Harrison Street
Allentown, PA 18103
tel 610-770-1044
fax 610-770-8912
e-mail ltl100@enter.net

Professional Service Industries
850 Poplar Street
Pittsburgh, PA 15220
tel 412-922-4000
fax 412-922-4014

Government Agencies, Trade Associations, and Other Lighting-Related Organizations

American Society of Heating, Refrigerating, and Air-Conditioning Engineers (ASHRAE)
1791 Tullie Circle NE
Atlanta, GA 30329
tel 404-636-8400
fax 404-321-5478
web www.ashrae.org

Certified Ballast Manufacturers
355 Lexington Avenue, 17th Floor
New York, NY 10017-6603
tel 212-661-4261
fax 212-370-9047
e-mail certbal.org.com
web www.certbal.org

Electric Power Research Institute (EPRI) Lighting Office
Bevilacqua Knight, Inc.
501 Fourteenth Street, Suite 200
Oakland, CA 94612
tel 510-444-8707
fax 510-444-2072
e-mail eprilio@bki.com
web www.epri.com

Green Lights Program
U.S. EPA Atmospheric Pollution Prevention Division
401 M Street SW (6202J)
Washington, DC 20460
tel 888-STARYES
fax 202-233-9569
web www.epa.gov/greenlights.html

Illuminating Engineering Society of North America (IESNA)
120 Wall Street, 17th Floor
New York, NY 10005-4001
tel 212-248-5000
fax 212-248-5017
e-mail iesna@iesna.org
web www.iesna.org

International Association of Lighting Designers (IALD)
1133 Broadway, Suite 520
New York, NY 10010-7903
tel 212-206-1281
fax 212-206-1327
web www.aecnet.com/iald/iald.html

International Dark Sky Society
3545 N. Stewart Avenue
Tucson, AZ 85716-1241
tel 520-293-3198
fax 520-293-3192
e-mail SaveOurSky@aol.com
web www.darksky.org

Lighting Research Center
Rensselaer Polytechnic Institute
110 8th Street
Watervliet Facility
Troy, NY 12180
tel 518-276-8716
fax 518-276-2999
e-mail lrc@rpi.edu
web www.lrc.rpi.edu

National Council on the Qualification of Lighting Professionals (NCQLP)
4491 East-West Highway, Suite 305
Bethesda, MD 20814
tel 301-654-1212
fax 301-654-4273

NIST Lighting and HVAC Interaction Test Facility
National Institute of Standards and Technology, Building Environment Division
Gaithersburg, MD 20877
tel 301-975-6444

Underwriter's Laboratories
333 Pfingsten Road
Northbrook, IL 60062
tel 847-272-8800
fax 847-272-8129
web www.ul.com

absorptance The ratio of the flux absorbed by a medium to the incident flux.

absorption A general term for the process by which incident flux is converted to another form of energy, usually to heat.

accent lighting Directional lighting to emphasize or draw attention to an object or area.

ambient lighting Lighting throughout an area that produces general illumination.

ANSI American National Standards Institute

ballast An electrical device designed to convert line current into the proper voltage, amperage, and waveform to operate a fluorescent or high-intensity discharge lamp.

ballast efficacy factor Percentage of light output relative to standard/watts input.

ballast factor The flux of a fluorescent lamp(s) operated on a ballast as a fraction of the flux when operated on the standard (reference) ballasting specified for rating lamp lumens.

base The end of a lamp containing electrical contacts.

beam angle An angle created by two beams of light, one on either side of a beam axis, each equal to 50 percent of the maximum beam candlepower.

beamspread The angle created by two points of equal light intensity on either side of the beam's axis and the point where the axis and lamp surface intersect; the intensity at the two side points is typically 50 percent of the maximum beam candlepower.

BEF Ballast efficacy factor.

BF Ballast factor (decimal light output relative to standard ballast).

BIAX Biaxial (long twin-tube compact fluorescent lamp).

blackbody A temperature radiator of uniform temperature whose radiant exitance in all parts of the spectrum is the maximum obtainable from any temperature radiator at the same temperature. Such a radiator is called a blackbody because it will absorb all the radiant energy that falls upon it. All other temperature radiators are called nonblackbodies. They radiate less in some or all wavelength intervals than a blackbody of the same size and the same temperature.

bulb The glass envelope of a lamp.

candela (cd) The SI unit of luminous intensity, equal to one lumen per steradian (lm/sr). Formerly "candle."

candlepower Luminous intensity expressed in candelas.

CIE Commissione Internationale d'Eclairage

clerestory That part of a building rising clear of the roofs or other parts and whose walls contain windows for lighting the interior.

coefficient of utilization (CU) The ratio of the luminous flux (lumens) from a luminaire calculated as received on the work plane to the luminous flux emitted by the luminaire's lamps alone.

coffer A recessed panel or dome in a ceiling.

color rendering A general expression for the effect of a light source on the color appearance of objects in conscious or subconscious comparison with their color appearance under a reference light source.

color rendering index (of a light source) (CRI) A measure of the degree of color shift objects undergo when illuminated by the light source as compared with those same objects when illuminated by a reference source of comparable color temperature.

color temperature (of a light source) The absolute temperature of a blackbody radiator having a chromaticity equal to that of the light source.

cones Retinal receptors in the eye that dominate the retinal response when the luminance level is high and provide the basis for the perception of color.

cove A horizontal ledge or recess, connected to a wall, that directs light across a ceiling and upper wall.

CW Cool White.

diethylhexyl phthalate (DEHP): solvent used in dielectric fluids in ballasts having wet capacitors, which were common between 1979 and 1984. Like PCB ballasts, DEHP ballasts cannot be discarded in sanitary landfills.

diffuser A device that redirects or scatters light, primarily by diffusing transmission of the light.

dimmer A device used to control the intensity of light emitted by a luminaire by controlling the voltage or current available to it.

direct component That portion of the light from a luminaire which arrives at the work plane without being reflected by any room surfaces.

discomfort glare Glare producing discomfort. It does not necessarily interfere with visual performance or visibility.

downlight A small direct lighting unit which directs the light downward and can be recessed, surface mounted, or suspended.

drum A fixture shaped like a drum or thick pancake; typically mounted on ceilings or upper walls.

efficacy The total luminous flux emitted by a lamp divided by the total lamp power input, expressed in lumens per watt.

eggcrate A type of louver having square cells oriented to allow light to pass through the louver; cells may be opaque or translucent, and cell walls are typically parallel or perpendicular to each other and not curved.

electric-discharge lamp A lamp in which light (or radiant energy near the visible spectrum) is produced by the passage of an electric current through a vapor or gas.

electrodes Coated heating elements inside a lamp, used to start and operate the lamp.

electroluminescence The emission of light from a phosphor excited by an electromagnetic field.

electronic ballast Device using solid-state components to provide current to a lamp at high frequency (typically 25,000 to 60,000 cycles per second) in order to produce more light using fewer watts than magnetic ballasts.

EMS Energy management system.

equivalent sphere illumination (ESI) The level of sphere illumination which would produce task visibility equivalent to that produced by a specific lighting environment.

field angle Same as beam angle, except that the two beams each equal only 10 percent of the maximum beam candlepower.

FL Flood; footlambert.

flicker fusion frequency (fff) The frequency of intermittent stimulation of the eye at which flicker disappears.

flicker index A measure of the cyclic variation in output of a light source, taking into account the waveform of the light output. It is the ratio of the area under the light output curve that is above the average light output level to the total area under the light output curve for a single cycle.

flood Type of lamp where beam angle is usually 30 percent or more.

floodlight A projector designed for lighting a scene or object to a luminance considerably greater than its surroundings. It usually is capable of being pointed in any direction and is of weatherproof construction. The beamspread of floodlights may range from narrow field angles (10°) to wide ones (more than 100°).

fluorescent lamp A low-pressure mercury electric-discharge lamp in which a fluorescing coating (phosphor) transforms some of the UV energy generated by the discharge into light.

flutes Elements of a lens designed to reduce glare by breaking up a lamp's image, but not altering the fixture's beamspread.

footcandle (fc) A unit of illuminance, equal to 1 lumen/ft^2 or 10.76 lux.

glare The sensation produced by luminance within the visual field that is sufficiently greater than the luminance to which the eyes are adapted to cause annoyance, discomfort, or loss in visual performance and visibility.

globe A fixture shaped like a sphere; typically pendant-mounted.

HAL Halogen.

harp Assembly used to attach a table lamp shade to the body of that fixture.

HID High-intensity discharge.

high hat A cylindrical fixture recessed into a ceiling.

high output Ballasts and lamps designed to operate with 800-milliamp current in order to provide greater light output.

high-bay lighting Interior lighting where the roof truss or ceiling height is greater than approximately 7.6 meters (25 feet) above the floor.

high-intensity discharge (HID) lamp An electric-discharge lamp in which the light-producing arc is stabilized by wall temperature, and the arc tube has a bulb wall loading in excess of 3 W/cm^2. HID lamps include groups of lamps known as mercury, metal halide, and high-pressure sodium.

high-pressure sodium (HPS) lamp A high-intensity discharge (HID) lamp in which light is produced by radiation from sodium vapor operating at a partial pressure of about 1.33×10^4 Pa (100 Torr). Includes clear and diffuse-coated lamps.

HIR Halogen infrared.

HO High-output (800-milliamp fluorescent lamp).

HPS High-pressure sodium.

hybrid ballast A magnetic ballast that shuts off heat to a lamp's electrodes after the lamp has started.

IES Illuminating Engineering Society.

incandescence The self-emission of radiant energy in the visible spectrum due to the thermal excitation of atoms or molecules.

indirect lighting Lighting by luminaires distributing 90 to 100 percent of the emitted light upward.

instant start A lamp and ballast system designed to start a lamp without preheating electrode by providing a high-voltage spark.

instant-start fluorescent lamp A fluorescent lamp designed for starting by a high voltage without preheating of the electrodes. Also known as a cold-start lamp in some countries.

intensity A shortening of the terms *luminous intensity* and *radiant intensity*. Often misused for level of illumination or illuminance.

inverse square law The law stating that the illuminance E at a point on a surface varies directly with the intensity I of a point source, and inversely as the square of the distance d between the source and the point. If the surface at the point is normal to the direction of the incident light, the law is expressed $E = I/d^2$.

IS Instant start.

K Kelvin (color temperature).

kW Kilowatt.

kWh Kilowatt-hour.

lamp An artificial source of light consisting of a bulb, electrodes, phosphor coatings, electrical contacts, and a mixture of gases designed to convert electricity into visible light.

lamp contacts In fluorescent lamps, the conductive points at the ends of a lamp that connect the lamp to its socket.

lamp lumen depreciation factor (LLD) The fractional loss of lamp lumens at rated operating conditions that progressively occurs during lamp operation.

lamp sockets In fluorescent lamps, the tombstone-shaped components in which lamp contacts are inserted to make electrical contact with the ballast.

LDD Luminaire dirt depreciation factor.

LED Light-emitting diode.

lens A glass or plastic element designed to change the direction and control the distribution of light rays.

light Radiant energy that is capable of exciting the retina and producing a visual sensation. The visible portion of the electromagnetic spectrum extends from about 380 to 700 nm.

LLD Lamp lumen depreciation factor.

LLF Light loss factor (= LDD x LLD).

louver A series of baffles used to shield a light source from direct view, absorb unwanted light, or redirect light by reflection.

low-pressure mercury lamp A discharge lamp (with or without a phosphor coating) in which the partial pressure of the mercury vapor does not exceed 100 Pa during operation.

low-pressure sodium (LPS) lamp A discharge lamp in which light is produced by radiation from sodium vapor operating at a partial pressure of 0.1 to 1.5 Pa (approximately 10^{-3} to 10^{-2} Torr).

LPS Low pressure sodium.

lumen (lm) The SI unit of luminous flux. Radiometrically, it is determined from the radiant power. Photometrically, it is the luminous flux emitted with a unit solid angle (1 sr) by a point source having a uniform luminous intensity of 1 cd.

luminaire Generic term for a complete lighting unit, consisting of lamp(s), parts designed to distribute light from the lamps, components to connect the lamp to its power source, and an electrical device to provide that power.

luminaire dirt depreciation factor (LDD) The fractional loss of task illuminance due to luminaire dirt accumulation.

luminaire efficiency The ratio of luminous flux (lumens) emitted by a luminaire to that emitted by the lamp or lamps used therein.

luminance contrast The relationship between the luminances of an object and its immediate background.

lux (lx) The SI unit of illuminance. One lux is one lumen per square meter (lm/m^2).

mA Milliampere.

magnetic ballast Typical fluorescent ballast consisting of a magnetic coil and capacitor, designed to limit current and provide necessary starting voltage for fluorescent lamps.

mercury lamp A high-intensity discharge (HID) lamp in which the major portion of the light is produced by radiation from mercury operating at a partial pressure in excess of 10^5 Pa (approximately 1 atmosphere). Includes clear, phosphor-coated (mercury-fluorescent), and self-ballasted lamps.

metal halide lamp A high-intensity discharge (HID) lamp in which the major portion of the light is produced by radiation of metal halides and their products of dissociation—possibly in combination with metallic vapors such as mercury. Includes clear and phosphor-coated lamps.

MH Metal halide.

monopoint Another name for a spotlight.

nadir The point directly below a ceiling-mounted light source.

nit (nt) A unit of luminance equal to 1 candela per square meter. Note: the candela per square meter (cd/m^2) is the International Standard (SI) unit of luminance.

PAR Parabolic aluminum reflector.

paracube A series of baffles arranged like an eggcrate, but whose cross sections consist of segments of parabolas, designed to shield a light source while directing light through a narrowed angle, usually by specular reflection.

paralouver A series of parallel baffles whose cross sections consist of segments of parabolas, designed to shield a light source while directing light through a narrowed angle, usually by specular reflection.

PCB Polychlorinated biphenyl.

pendant luminaire A luminaire hung from a ceiling by supports.

pendant The structural element of a light fixture that suspends the fixture from a ceiling and which holds electrical wiring; typically a metal tube or conduit.

phosphors Fluorescing substances that transform ultraviolet light generated by an electric arc into visible light; there is no phosphorus in these phosphors.

photopic vision Vision mediated essentially or exclusively by the cones. It is generally associated with adaptation to a luminance of at least 3.4 cd/m^2. See *scotopic vision*.

plenum The empty space between a hung ceiling and a structural slab; usually contains ductwork, wiring, communications cables, and lighting systems; often used to return air back to the air-conditioning system.

polarizer Typically a lens designed to redirect transverse vibrations of light into parallel planes.

preheat start Ballast system designed to start a lamp using a circuit having a manual or automatic starting switch to preheat the lamp's electrodes in order to initiate an arc; usually seen on magnetically ballasted smaller lamps (under two feet long) and compact fluorescent lamps.

prismatic Adjective describing the shape of elements in a lens designed to redirect light over wider angles in all directions.

radio-frequency interference (RFI) Interference to the radio-frequency band caused by other high-frequency equipment or devices in the immediate area.

rapid start A ballast and lamp system designed to operate a lamp containing a low-voltage coated electrode that initiates an arc within the lamp without using high voltage.

RCR Room cavity ratio.

recessed can Another name for a cylindrical fixture recessed into a ceiling.

reference ballast A ballast specially constructed to have certain prescribed characteristics for use in testing electric-discharge lamps and other ballasts.

reflector A device used to direct light from a source through reflection, either specular or diffuse.

reflector lamp An incandescent filament or electric-discharge lamp in which the outer blown glass bulb is coated with a reflecting material so as to direct the light (such as R- or ER-type lamps). The light-transmitting region may be clear, frosted, patterned, or phosphor-coated.

rods Retinal receptors which respond at low levels of luminance, even below the threshold for cones. At these levels there is no basis for perceiving differences in hue and saturation.

room cavity The cavity formed by the plane of the luminaires, the work plane, and the wall surfaces between these two planes.

room cavity ratio (RCR) A number indicating room cavity proportions, calculated from the length, width, and height.

sconce Wall-mounted fixture providing indirect light from a wall and ceiling, and direct light through a lens.

scotopic vision Vision mediated essentially or exclusively by the rods. It is generally associated with adaptation to a luminance below about 0.034 cd/m^2. See *photopic vision*.

SHO Super high output. See *very high output*.

S/MH Spacing-to-mounting-height ratio.

socket bars Metal parts that hold lamps sockets in positions appropriate to distribution of light from a lamp and reflector.

spacing-to-mounting-height ratio (S/MH_{wp}) The ratio of the actual distance between luminaire centers to the mounting height above the work plane.

specular angle The angle between the perpendicular to a surface and the reflected ray.

specular reflectance Reflectance of light at the same angle and in the same plane as it arrived at the reflector.

spot Type of lamp where the beam angle is usually 12° or less.

super high output (SHO) See *very high output*.

T# As in T8, T12 and so on. T stands for tubular; the number describes lamp diameter in 1/8-inch increments. A T8 lamp is 8/8 inches in diameter; a T12 is 12/8 inches (or 1-1/4 inches) in diameter.

table lamp A portable luminaire with a short stand, suitable for standing on furniture.

tandem wiring A wiring option in which a ballast is shared by two or more luminaires. This reduces labor, materials, and energy costs. Also called "master-slave" wiring.

task-ambient lighting A combination of task lighting and ambient lighting within an area such that the general level of ambient lighting is lower than and complementary to the task lighting.

task light Typically a fixture, either mounted in office furniture or having its own base and electric cord, with limited beamspread and designed to illuminate a small area.

torchiere An indirect light fixture, typically floor-mounted.

track light A fixture mounted on a gimbal having electrical contacts that recess into a powered track that will hold the gimbal anywhere along its length.

troffer A recessed lighting unit, with its opening flush with the ceiling in which it is installed; this term is derived from "trough" and "coffer."

tungsten halogen A lamp using a tungsten filament and halogen gas cycle to restore vaporized tungsten to the filament.

Underwriters Laboratories (UL) An independent organization whose responsibilities include rigorous testing of electrical products. When products pass these tests, they can be labeled (and advertised) as "UL listed." UL tests for product safety only.

valance A shield mounted in front of a horizontal wall-mounted fixture, directing light up and down the wall.

VCP Visual comfort probability.

veiling reflection Regular reflections that are superimposed upon diffuse reflections from an object and that partially or totally obscure the details to be seen by reducing the contrast. This sometimes is called *reflected glare.*

very (or super) high output (VHO or SHO) Ballasts and lamps designed to operate using 1,500 milliamp current in order to provide greater light output.

visual comfort probability (VCP) A rating of a lighting system expressed as a percent of people who, occupying a specified location, and looking in a specified direction, will be expected to find it visually acceptable.

wall washer Any fixture installed sufficiently close to a wall to project most of its light onto that wall.

wavelength The distance between two successive points of a periodic wave, in the direction of propagation, at which the oscillation has the same phase.

work plane The level at which work is done and at which illuminance is specified and measured. For office applications, this is typically a horizontal plane 30 inches above the floor (desk height).

WW Warm white.

INDEX

INDEX

STAYING CURRENT!

CONGRATULATIONS! You are the owner of the most useful and comprehensive resource on energy efficiency available—the *Technology Atlas Series* from E SOURCE. The *Series* provides you with up-to-date and objective information, in a field that is rapidly changing and filled with conflicting and exaggerated performance claims.

The E SOURCE research team is continuously evaluating the latest energy efficient products and practices—discovering little-known techniques that can translate into big energy savings. On an ongoing basis, E SOURCE incorporates this new research into completely revised and updated editions to the *Series;* typically, two volumes per year.

You can ensure that your *Technology Atlas Series* stays up-to-date for the low price of $195 per year (or $250 for both print and CD-ROM formats). Send no money now, just complete and return the attached card, and we'll send you an invoice and the updated volumes when they are ready. If you are not 100% satisfied, simply return the new editions for a full refund.

MAKE SURE YOU HAVE THE LATEST AND BEST INFORMATION AVAILABLE—RESERVE YOUR UPDATES TODAY!

E SOURCE MEMBERSHIP

E SOURCE is a membership-based information services company providing organizations with unbiased, independent analysis of end-use technologies and application issues, innovative energy service offerings, and strategic developments. At the company's core is an exceptional team of research professionals whose technical and analytical capabilities, intellectual breadth, and assimilation of real-world experience have earned E SOURCE international acclaim. E SOURCE members include over 350 electric and gas utilities, other energy service providers, large corporate energy users, government agencies, and other organizations from two dozen countries worldwide.

These members receive an integrated package of benefits designed to keep them on top of critical issues related to the delivery and management of energy services, and surrounding developments in technology, at a fraction of the cost of customized research. Information is delivered through a series of publications, newsletters, special reports, conferences and seminars, CD-ROMS, and the Internet. A variety of membership packages is offered to meet the needs of our diverse clientele.

IF YOU ARE INTERESTED IN MORE INFORMATION ON E SOURCE MEMBERSHIP BENEFITS, JUST CHECK THE BOX AT THE BOTTOM OF THE ATTACHED CARD, AND WE'LL SEND YOU A FULL INFORMATION PACKAGE.

E SOURCE®

E SOURCE
1033 Walnut Street
Boulder, Colorado 80302
TEL 1.800.E SOURCE
(303.440.8500)
FAX 303.440.8502
E-MAIL esource@esource.com
WEB www.esource.com

YES!

I would like to keep my *Technology Atlas Series* current. Please reserve my copies of:

- ❑ **Both print and CD-ROM updates** **$250** (U.S.)
- ❑ **Print updates only** **$195**
- ❑ **CD-ROM updates only** **$195**

I understand that from now on I will receive all updates to the *Technology Atlas Series.* I also understand that I will be billed annually at the then-current subscription fee for this service. All material will be sent on condition that it may be returned in salable condition within thirty days for a full refund. I may cancel my subscription at any time and this agreement can be terminated by either party with thirty days notice. There will be an additional shipping charge for orders outside the United States. Only the terms and conditions on this form apply.

AUTHORIZED SIGNATURE ______ DATE ______

CONTACT

NAME ______

TITLE ______ DEPARTMENT ______

ORGANIZATION ______

ADDRESS ______

CITY ______ STATE ______ ZIP ______ COUNTRY ______

PHONE ______ FAX ______

E-MAIL ______

❑ PLEASE SEND ME INFORMATION ABOUT E SOURCE MEMBERSHIP.

E SOURCE, INC.
1033 WALNUT ST
BOULDER CO 80302-9836